AF790251

History of Mechanism and Machine Science

Volume 30

Series editor

Marco Ceccarelli, Cassino, Italy

Aims and Scope of the Series

This book series aims to establish a well defined forum for Monographs and Proceedings on the History of Mechanism and Machine Science (MMS). The series publishes works that give an overview of the historical developments, from the earliest times up to and including the recent past, of MMS in all its technical aspects.

This technical approach is an essential characteristic of the series. By discussing technical details and formulations and even reformulating those in terms of modern formalisms the possibility is created not only to track the historical technical developments but also to use past experiences in technical teaching and research today. In order to do so, the emphasis must be on technical aspects rather than a purely historical focus, although the latter has its place too.

Furthermore, the series will consider the republication of out-of-print older works with English translation and comments.

The book series is intended to collect technical views on historical developments of the broad field of MMS in a unique frame that can be seen in its totality as an Encyclopaedia of the History of MMS but with the additional purpose of archiving and teaching the History of MMS. Therefore the book series is intended not only for researchers of the History of Engineering but also for professionals and students who are interested in obtaining a clear perspective of the past for their future technical works. The books will be written in general by engineers but not only for engineers.

Prospective authors and editors can contact the series editor, Professor M. Ceccarelli, about future publications within the series at:

LARM: Laboratory of Robotics and Mechatronics
DiMSAT—University of Cassino
Via Di Biasio 43, 03043 Cassino (Fr)
Italy
email: ceccarelli@unicas.it

More information about this series at http://www.springer.com/series/7481

Michela Cigola
Editor

Distinguished Figures in Descriptive Geometry and Its Applications for Mechanism Science

From the Middle Ages to the 17th Century

Springer

Editor
Michela Cigola
Department of Civil and Mechanical Engineering
University of Cassino and South Latium
Cassino
Italy

ISSN 1875-3442 ISSN 1875-3426 (electronic)
History of Mechanism and Machine Science
ISBN 978-3-319-20196-2 ISBN 978-3-319-20197-9 (eBook)
DOI 10.1007/978-3-319-20197-9

Library of Congress Control Number: 2015944151

Springer Cham Heidelberg New York Dordrecht London
© Springer International Publishing Switzerland 2016
This work is subject to copyright. All rights are reserved by the Publisher, whether the whole or part of the material is concerned, specifically the rights of translation, reprinting, reuse of illustrations, recitation, broadcasting, reproduction on microfilms or in any other physical way, and transmission or information storage and retrieval, electronic adaptation, computer software, or by similar or dissimilar methodology now known or hereafter developed.
The use of general descriptive names, registered names, trademarks, service marks, etc. in this publication does not imply, even in the absence of a specific statement, that such names are exempt from the relevant protective laws and regulations and therefore free for general use.
The publisher, the authors and the editors are safe to assume that the advice and information in this book are believed to be true and accurate at the date of publication. Neither the publisher nor the authors or the editors give a warranty, express or implied, with respect to the material contained herein or for any errors or omissions that may have been made.

Printed on acid-free paper

Springer International Publishing AG Switzerland is part of Springer Science+Business Media
(www.springer.com)

Series Editor's Preface

I am very happy, for the following reason, to present this impressive book in our series. It is a first book of a series of stories about notables who have contributed to developments of Mechanisms and Machine Science (MMS) from the field of Descriptive Geometry. It is important to recognize the merits of these people and to give proper credit for their achievements that are still of modern interest and application. Thus, let us hope to have more of these contributions that are aimed at building an encyclopaedia of who-is-who in the wide areas of MMS, in combination with the other series of 'Distinguished Figures in MMS'. This book is a brilliant example of the multidisciplinary content and interest in MMS.

In addition, as one looks at the outstanding names that appear in this book, a reader will find already famous scientists presented with novel perspectives on their activities, even highlighting aspects that elsewhere might be considered of minor importance. But those contributions and efforts were significant for the evolution of MMS, both in theory and practice, with influential impact even in technological developments. Similarly, some of these notables are presented for the first time in MMS frames, bringing specific attention to outlining their achievements that still have possibilities for modern implementation.

I am sure readers will not only find satisfaction in reading this book but will receive inspiration and hope for more historical evaluations and technical evolutions.

Thus, I congratulate the editor and authors of this book for the very interesting results and I wish enjoyment to all its readers.

Cassino
March 2015

Marco Ceccarelli
Chief Editor of Series on History of MMS

Contents

Descriptive Geometry and Mechanism Science from Antiquity to the 17th Century: An Introduction

Michela Cigola

Abstract The focus of this brief introduction is the common birth and parallel destiny of Descriptive Geometry and Mechanism Science. This argument will compare some scientists from the chosen period who can be considered of common interest between the two disciplines, devoting a chapter to each of them. And especially in this introductory chapter we will discuss four major personalities, one for Antiquity (Vitruvius), one for the Middle Ages (Villard de Honnecourt), one for the Renaissance (Filippo Brunelleschi), and finally one for the Baroque period (Giovanni Branca).

Introduction

Descriptive Geometry and Applied Mechanics, and more particularly the Theory of Mechanisms, which are at first sight disciplines belonging to separate and disjoint fields, actually hide a common birth and parallel destiny.

Since ancient times, with Vitruvius and then in the Renaissance with Brunelleschi the two disciplines began to share a common direction which, over the centuries, took shape through less well-known figures until the more recent times in which Gaspard Monge worked.

Understood in its modern sense, the Theory of Machines and Mechanisms can be traced back to the founding of the École Polytechnique in Paris and particularly to Monge and Hachette, personalities who made a fundamental contribution to the development of Descriptive Geometry.

Over the years, a gap has been created between the two disciplines, which now appear to belong to different worlds. In reality, however, there is a very close relationship between Descriptive Geometry and Applied Mechanics, a link based on

M. Cigola (✉)
DART - Laboratory of Documentation, Analisys, Survey of Architecture & Territory, Department of Civil & Mechanical Engineering - University of Cassino & Southern Latium, via G. Di Biasio 43, 03043 Cassino, Italy
e-mail: cigola@unicas.it

© Springer International Publishing Switzerland 2016
M. Cigola (ed.), *Distinguished Figures in Descriptive Geometry and Its Applications for Mechanism Science*, History of Mechanism and Machine Science 30, DOI 10.1007/978-3-319-20197-9_1

extremely solid foundations; just think that without the theoretical foundations of geometry it would not be possible to draw and design mechanical parts such as gears, while in kinematics it would be less easy to design and predict the reciprocal movements of the parts in a complex mechanical assembly (Fig. 1).

Thereafter, a long list of scholars can be identified who spanned the two disciplines, studying and researching in Descriptive Geometry and Mechanics in alternating phases of their scientific paths, and who were authors of books in one or another subject matter.

In this book we will try to better delineate this matter, considering some scientists from the chosen period who can be considered of common interest between the two disciplines, devoting a chapter to each of them. And especially in this introductory chapter we will discuss four major personalities, one for Antiquity (Vitruvius), one for the Middle Ages (Villard de Honnecourt), one the Renaissance (Filippo Brunelleschi), and finally one for the Baroque period (Giovanni Branca).

In our discussion we will try not to lose sight of the development of the two disciplines, devoting to each personality a brief summary that puts their contribution to the state of the art of the time into proper context. Continuing our research, we will make use of the representations that our authors include in their works,

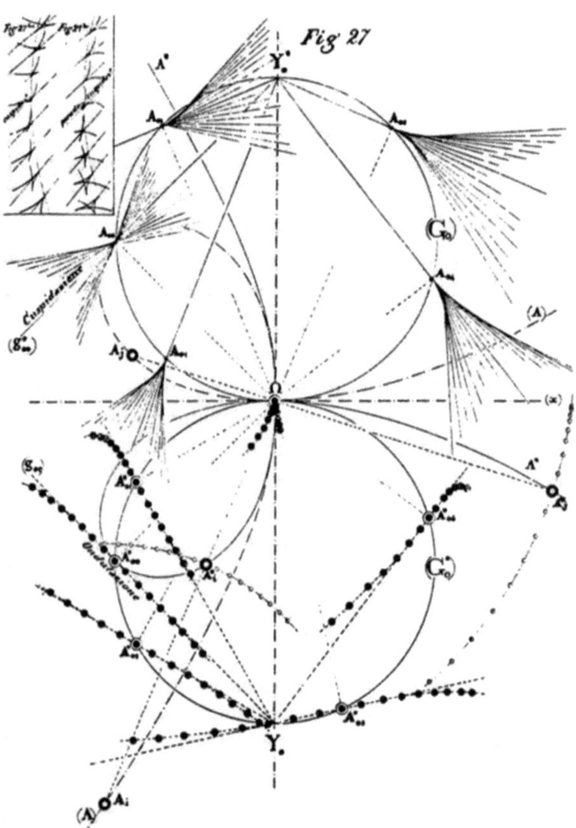

Fig. 1 Graphic representation of kinematic scheme of motion, in "Cinematica della biella piana", 1895 by L. Allievi, Fig. 27

analysing alternately drawings of machines and architectural drawings, or functional representations in geometric theories.

The analyses of these graphical representations (which are often design projects) will also lead us to talking about drawings, analysing wherein at times the two disciplines intertwine and sometimes confront each other, because they were always deeply rooted in the culture of their epochs.

In fact, theory, based on pure knowledge and technology, based on experiments, for centuries remained two opposite worlds. Since ancient times, in the wake of Aristotelian philosophy, theory was the domain of those who in the Middle Ages were called practitioners of the "liberal arts", while technology belonged to the "mechanical arts".

In later times, enlargement of the terrestrial and celestial worlds, happening between 400 and 700 due to geographical and astronomical discoveries, triggered a profound ideological change that accompanies the early developments of technology; a new philosophical interpretation of the mechanical arts makes its début—a way of knowing linked to active exploration of the nature that exceeds the Aristotelian concept of contemplative science.

Technological development has in every age greatly influenced evolution of the representation of design sketches, whether it was related to the representation of architecture or if it was an expression of a project related to mechanics. In particular, the drawings of mechanical elements such as mechanisms, from an initial, highly descriptive, representation, has evolved towards an ever greater clarity of representation that would facilitate a rapid and unequivocal interpretation.

In fact, from the Middle Ages to the Renaissance, technical drawing in general and especially that of machines and mechanisms, is not an autonomous encoded language but a figurative representation of an object that only rarely turns into a schematic simplification, with the use of some simple convention that is traditionally applied but not formally declared; in addition, there was still a deeply rooted belief that ornamentation enhances the value of the material, and the beauty of its use.

Technical drawing in today's sense comes only with the Industrial Revolution, when production is broken down into various structural levels and the use of the executive project rather than summary schematisation is introduced to clarify the features and perform the design calculations.

On the other hand, in terms of Descriptive Geometry, the situation is certainly more complex.

On the one hand, it can be considered as a discipline that collects in an organic whole the rules devised by painters and designers to delineate on a flat surface certain figures that are faithful images of objects in three dimensions. In this case, it is obvious that its origins date back to the origin of man. This understanding of geometry will lead to the development of perspective in painting, with the fundamental contribution of Brunelleschi and, for architecture, refinement of the rules for the representation of construction projects, whose development is still ongoing today.

Another important phase of the study is related to the geometric measurement of the land, as is well highlighted by the Greek root of this word referring to γή = "land" e μετρία, = "measure". We have evidence of such use in Egypt, and then gradually in all periods of human history, through the important phase of Roman "Agrimensores" and also not forgetting the long period associated with "Practical Geometry" that falls in the Middle Ages, the Renaissance and beyond.

On the other hand, we can trace a purely scientific route of Geometry that sees this discipline investigated since ancient time for purely theoretical reasons, closely related to mathematical research; to arrive then, step by step, to the time of Gaspard Monge when, thanks to the contributions of many personalities, it is rationalised to the form in which it comes down to in our days.

Antiquity and Vitruvius

The origin of ancient mechanics is traditionally associated with the name of Archytas of Tarentum (second half of the IV century BC) who was the first to deal methodically with the issues of mechanics. Archytas is also remembered for being the inventor of the first automaton, a flying dove. Aristotle has also made an important contribution: his work shows that as early as the fourth century BC mechanics and other physical-mathematical disciplines were so deeply rooted in Greek culture as to be the subject of regular philosophical reflection.

Among the few scholars of antiquity who deal with machines and mechanics in the ancient world, the memory of whom has reached our days, the most remembered one would be Ctesibus (active in Alexandria in the first half of III century BC), the author of a Treatise on Pneumatics and also for having invented the hydraulic pump and the water organ; Archimedes (born in Syracuse and lived between 287 and 212 BC) considered to be the founder of theoretical mechanics and also to have given important contributions to geometry; and Philo of Byzantium (end of III century BC) who wrote a treatise on the systematization of the machines.

Without a doubt, the best known scholar of Antiquity is Heron of Alexandria (I c. AD). In addition to works in mechanics, he wrote of automata, and pneumatic and war machinery. His study of an instrument used to measure the land, the diopter, testifies that he also worked with practical geometry.

The Roman era left us treaties and examples of war machines and machines for construction (cranes, winches, etc.), quite normal for a civilization founded on territorial expansion and, above all, a desire for practical application rather than for scientific speculation for science's sake.

Geometry probably originated in Mesopotamia and Egypt, and was linked to land surveying purposes. It was, therefore, empirical geometry, consisting of a number of practical rules for measurements. This is evidenced by some discovered depictions that show a gridded base used to enlarge, and perhaps also to modulate, the representation.

Ancient Greek civilization saw the rise of geometry as a rational science, detached from every application need and every method of empiricism. Euclid (active around 300 BC), collected all the wealth of knowledge amassed by scholars who preceded him (Pythagoras, Thales, Eudoxus of Cnidus, etc.), offers in his “Elements”, a work in 13 volumes, the first example of what today we would call a scientific treatise. In his “Optics”, Euclid was the first to attempt to formalise the problems of representation and to insert the elements, which we will later find in the formulation of geometric perspective.

In addition to this purely speculative branch, from Egyptian and Mesopotamian civilizations the Greeks also inherited some substantial knowledge of practical geometry and agricultural surveying that used to be applied regularly, up to arrival of the period of the Alexandrian school, which is the continuity and development of Greek culture.

In Roman times we come to a moment of interruption of that progression, linked to the personalities who worked with geometric applications. In Egypt and Mesopotamia these were the priests, and in Greece—philosophers and men of culture, closely connected with priestly power, who dealt with geometry. In Roman times it evolved to become a profession in itself, recognised by the state and endowed with its own prerogatives: the “Agrimensores” or “Gromatici”, from the name of the instrument used. They were divided into two categories: the ones who dealt with the scientific side, and the others, who worked with practical problems and the executive side. They left us some texts and testimonies of another type, such as “Forma Urbis Romae”, dated to the III century AD. This document is of fundamental importance for the high knowledge of practical geometry that it demonstrates, but also for the expertise of graphical representation, which can be considered, to all effects, orthogonal projections in plans, with little or no difference from those used today—and this testifies to the strong progress also in the field of drawing.

A persona apart is Vitruvius, who stood on a firm foundation of the legacy of ancient knowledge in the fields of mechanics and geometry.

Marcus Vitruvius Pollio, Fig. 2, was a Roman architect and engineer, who was active in the second half of the I century B.C. He is considered the most significant theoretician of Architecture of the time because of his treatise “De Architectura” in ten books (today named as chapters).

Very little is known of his life. It is not even sure that his “cognomen” (surname) was Pollio. As birth place several cities around Italy are cited like Rome, Fondi, Fano, Verona, or Formia.

Most of his life information is indicated in the introductory notes in the ten chapters of “De Architectura”. From those notes it can be understood that he was an officer responsible for war machines under Julius Caesar. After the death of Caesar, Vitruvius continued design collaboration with Emperor Augustus, to whom Vitruvius dedicated his treatise “De Architectura”. The treatise can be dated among 27 and 13 B.C., when Vitruvius was old, as he mentions in the treatise notes.

Fig. 2 Engraved portrait of Vitruvius by Iacopo Berardi in 1830 ca

The treatise "De Architectura" is composed of ten chapters (named as books in manuscript in Latin and Italian editions). Each of them is introduced with a preface (named as "proemio"). The contents of the ten chapters are distributed by subjects, namely:

Chapter I on definition of Architecture and required expertise of architects; Chapter II on materials, wall constructions and building techniques; Chapters III and IV on religious buildings and orders of architecture (Doric, Ionic, and Corinthian); Chapter V on public buildings; Chapter VI and VII on domestic buildings; Chapter VIII water supplies and aqueducts; Chapter IX on elements of Astronomy for solar and water clocks; Chapter X on machines.

If Vitruvus' work was known since the time of Charlemagne, but with the Renaissance, you reach a real "rediscovery" and his assiduous study.

The first rediscovery of Vitruvius' treatise is considered as due to Poggio Bracciolini (1380–1459), who declared to have found a manuscript with a copy of Vitruvius' treatise in the library of Montecassino Abbey in 1414. Unfortunately, today no Vitruvius manuscript exists in the Montecassino library. At that time the fame of the discovery attracted great interest in the treatise that was considered as an important reference source even for new Architecture studies in the early XVth century. Thus Lorenzo Ghiberti (1378–1455) used the treatise for his work "Commentarii", Leon Battista Alberti (1404–1472) cited it explicitly for the first time in his work "De re aedificatoria", and Francesco di Giorgio Martini (1439–1502) worked on the first translation in Italian. Many other Renaissance artists and architects mentioned that they studied Vitruvius' treatise, such as Raphael (1483–1520), Luca Pacioli (1447–1517), Antonio di Pietro Averlino known as Filarete (1400–1469), and Mariano di Jacopo named il Taccola (1382–1453). This motivated publication of printed editions, the first being the one by Giovanni Sulpicio da Veroli (second half of XV century) and Pomponio Leto (1425–1498) in 1486–1490 with no drawings.

During the 16th century editions multiplied, so that there are four in Latin and nine in Italian.

We note that the Latin manuscript was rediscovered with only text whose translation was also a subject of disputes both for wording and design interpretations. Drawings were added for the first time in the Frà Giocondo edition in 1511.

Fra' Giovanni Giocondo (1434–1515) was a Franciscan friar and a humanist with expertise as an architect with great interest in machines. He published the first edition of "De Architectura" with drawings in 1511 in Venice. This edition was reprinted in 1513 and again in 1522 and 1523 with 136 illustrations made by xylography.

The first edition in Italian was published in 1521 by Cesare Cesariano (1483–1543), who was an Italian architect and painter. This edition is provided with illustrations of improved quality and complexity as compared with those by Frà Giocondo.

In the first Chapter of "De Architectura", part II, Vitruvius exposed the meaning of Architecture, describing its "dispositio" (from Latin: arrangement) and developing what might be a statement of the systems used in his day for graphic description of architecture, or as we might say today, the systems of representation and drawing.

Dispositio autem est rerum apta conlocatio elegansque compositionibus effectus operis cum qualitate. Species dispositionis, quae graece dicuntur ideae, sunt hae: ichnographia, orthographia, scaenographia. Ichnographia est circini regulaeque modice continens usus, e qua capiuntur formarum in solis arearum descriptiones. Orthographia autem est erecta frontis imago modiceque picta rationibus operis futuri figura. Item scaenographia est frontis et laterum abscedentium adumbratio ad circinique centrum omnium linearum responsus. Hae nascuntur ex cogitatione et inventione. Cogitatio est cura studii plena et industriae vigilantiaeque effectus propositi cum voluptate. Inventio autem est quaestionum obscurarum explicatio ratioque novae rei vigore mobili reperta. Hae sunt terminationes dispositionum.

The English translation made by Morris Hicky Morgan in 1914 reads: "Arrangement includes the putting of things in their proper places and the elegance of effect which is due to adjustments appropriate to the character of the work. Its forms of expression (Greek ἰδέαι) are these: groundplan, elevation, and perspective. A groundplan is made by the proper successive use of compasses and rule, through which we get outlines for the plane surfaces of buildings. An elevation is a picture of the front of a building, set upright and properly drawn in the proportions of the contemplated work. Perspective is the method of sketching a front with the sides withdrawing into the background, the lines all meeting in the centre of a circle. All three come of reflexion and invention. Reflexion is careful and laborious thought, and watchful attention directed to the agreeable effect of one's plan. Invention, on the other hand, is the solving of intricate problems and the discovery of new principles by means of brilliancy and versatility. These are the departments belonging under Arrangement."

Since the work of Vitruvius was rediscovered in the early fifteenth century, the authors of its many editions have proposed for this step many interpretations that, exploiting the ambiguity inherent in the works that require translation, reflected the culture of the time to which each particular edition belonged.

In fact, the three "species" that make up the "dispositio", namely "ichnographia", "ortographia" and "scaenographia" in almost all the editions of Vitruvius are treated as conventional graphics systems. The "icnographia" would then be assimilated to a plan, the "ortographia" to the raise and elevation, that is, to what we now call the façade drawing, and finally the "scaenographia" to what we today call a perspective drawing.

There are several examples that demonstrate how the ancient Romans knew the representation methods described above; the representation of a Roman city in "Forma Urbis Romae" (III century AD) is clearly planimetric; and one can find examples of pseudo-perspective depictions in many frescoes of antiquity.

So for most of the authors who worked on the various editions of Vitruvius' works, the "species" of "dispositio" match graphic conventions very similar to our own, plan and elevation as projections on the horizontal planes, and a perspective sketch of the whole. Aligned to this interpretation are also the first illustrations accompanying the Latin text of Vitruvius, famously handed down without illustrations, i.e. those contained in the 1511 edition of Frà Giocondo (1434–1515), the first to add images to this text (Fig. 3).

Another interesting interpretation, and not without its charm, is that which, returning to a faithful translation of the Latin text, puts forth "ichnographia" as an actual tracing made directly on the ground at the construction site. This operation was certainly preceded by a phase of graphic work, but reduced to a pure and simple scheme to support the tracing on site.

This would seem to be the general line of the edition of Vitruvius from 1521 by Cesare Cesariano (1483–1543), developed in that Lombard atmosphere that, albeit already fully in the Renaissance era, was still much influenced by the Gothic ways (Fig. 4).

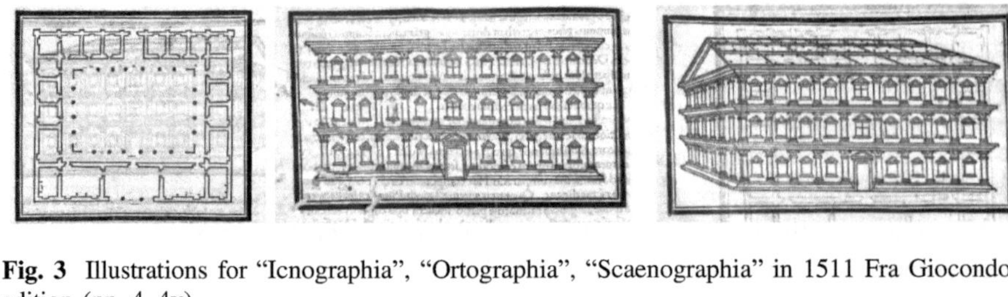

Fig. 3 Illustrations for "Icnographia", "Ortographia", "Scaenographia" in 1511 Fra Giocondo's edition (pp. 4–4v)

Fig. 4 Illustrations for "Icnographia", "Ortographia", "Scaenographia" in 1521 Cesariano's edition (pp. 14, 15, 15v)

Therefore, for Cesariano the "ichnographia" is the excavation on the ground, an imprint of the building on the terrain, which to be demarcated requires only a simple preliminary sketch. Similarly, "ortographia" is the main prospect traced according to a proportional scheme based on "ichnographia", where the decorative motifs are studied and embodied. Finally, the "scaenographia" is always the prospect, where the geometrical outlines that govern it are brought out.

Today, we can not really define what Vitruvius truly meant, as illustrations of the original text have been lost. Yet, certainly, the underlying assumption in both is a remarkable knowledge of the graphical representation, the basis of both interpretations of the triad of "species" which forms the "dispositio".

To analyze how Vitruvius was also a great expert in mechanics, we will examine the X chapter of his work, namely some lifting machines, described in Sections 2-3-4. We will use illustrations of the 1511 Edition of Fra Giocondo, one of the most beautiful Renaissance editions of "De Architectura" by Vitruvius and illustration from the first edition in Italian, published in 1521 by Cesare Cesariano.

Chapter X is specifically dedicated to machines both for war and peace times that an architect needs to design and operates not only for their specific use but mainly as means for works in Architecture.

Machine analysis starts with a concise definition of "machine" given as "a combination of materials (as components) that performs a large action for motion of weights". This definition is important since it clarifies the structure of machines and their mechanical generation as finalized to weight manipulation. It can be considered the first machine definition with engineering understanding. Relevant is the description of the machines with technical terms that are useful both for operation understanding and mechanical design.

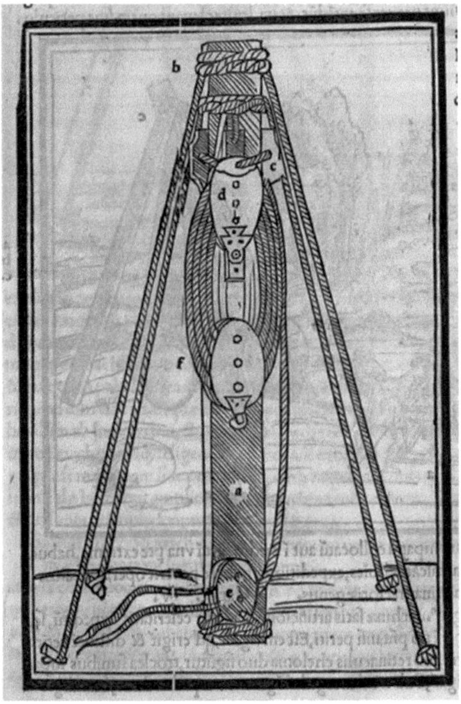

Fig. 5 Lifting system in 1511 Fra' Giocondo's edition (Pag. 96v)

The illustration of Frà Giocondo (Fig. 5) describes a lifting system with movable and fixed elements; the device is represented in a natural but very concisely outlined environment, traced with a few lines that indicate the horizon that is interrupted, to make it less monotonous, by the inclusion of mild hills. The machine is placed in the foreground, and it is interesting to note that the illustration inset is not completely understood, because it leaves out some parts of the representation, as if to suggest a continuation of the fictitious reality beyond the caesura represented by a prominent black frame.

It is important to remember that these woodcuts are the first illustration of the work of Vitruvius. The machines are drawn in a third dimension that is not completely certain and in some cases it is not completely clear how the parts join together (see the top of the device where the ropes meet).

In the illustration of Cesariano (Fig. 6a), from only a decade later than that of Frà Giocondo, it is detached completely, because the image is much more mature both in the three-dimensional handling and in the composition of the figure. In fact, the figure that at the top bears the titration (in Latin) in capital letters, inside a slender frame Cesariano inserts, almost crowding the space, many devices for raising weights, in a representative selection completely opposite to the broad and rarefied one of Frà Giocondo. The machines are placed one behind the other in order to punctuate and highlight the depth of the space represented, as if to take particular advantage of the master's ability to depict the third dimension and use shadows.

(a) (b)

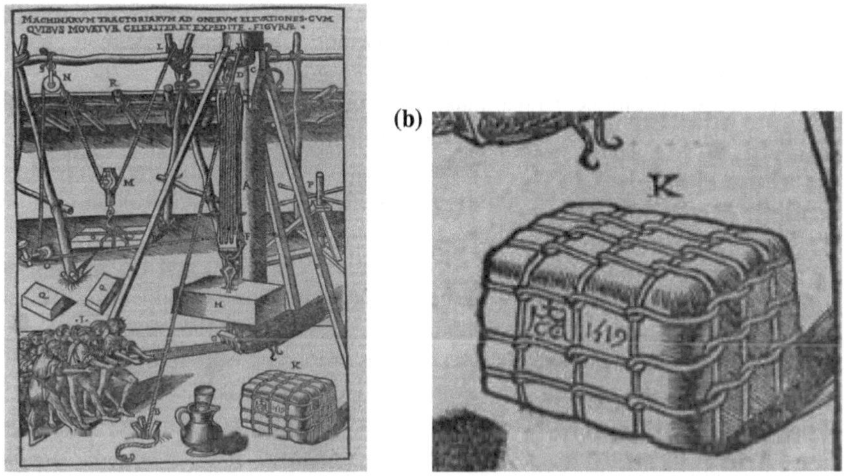

Fig. 6 **a** Lifting devices in 1521 Cesariano's edition (Pag. CLXVL). **b** Detail of the table

At the bottom right one can notice the presence of a group of workers that allow one to correctly estimate the dimensional scope of the machines. Also in the bottom left appears a load, just laid on the ground (or waiting to be raised). In one of its the rectangular spaces defined by the ropes that meticulously bind weight, we can see Cesariano's monogram (Fig. 6b), with the date 1519 that is evidently referring to the composition of the illustration and not to the date of the edition itself.

Middle Ages and Villard de Honnecourt

One must immediately eliminate the concept, still widespread today, of a cultural regression from the classical world in the early medieval or medieval period, or even that the idea that scientific activities were no longer practised is incorrect.

One of the principal factors for retention of ancient knowledge in the medieval world was the great spread of monastic orders, in the first place the Benedictines, who set up numerous monasteries in Italy, France and England that soon became not only centres of religion and trade but also important cultural centres where the knowledge of antiquity was collected, stored and transmitted.

Another factor of fundamental importance for the formation of pre-medieval culture was the contribution of Islam. With the expansion and consolidation of the Islamic world in the Mediterranean basin, the Arabs took on a decisive role in development of the sciences. Their knowledge of Greek culture was profound, and it is only thanks to their work that some Greek treatises have come down to us. Contacts with the Arab world were direct and frequent from the start of the Moslem expansion that rapidly occupied a large part of the Mediterranean basin and provided many opportunities for meeting and cohabitation between Christian and

Islamic peoples; not least through the prolonged presence of Moslem dynasties in Spain, intensified exchanges between Islamic and Christian culture.

On the other hand, no model of the early medieval society, as far as being in crisis, is plausible unless it postulates a substantial continuity of certain practices essential to economic and social life, such as arithmetic, measurement or division of land, regulation of the course of the waters, or construction of buildings: all operations which require a minimum of level of mechanical or geometric knowledge.

There is therefore no break between the classical world and the Middle Ages; rather, there is a continuity between the two periods. In the Middle Ages, therefore, persists a great scientific and technical activity, thanks to the experiences, having matured in the classical age and had been preserved and transmitted from the Byzantine and Islamic world; there are prerequisites to achieving significant progress, but there is no methodological consciousness that would make them possible.

In the field of mechanics, the writings of medieval technicians are "manuals", sets of rules and precepts. Their authors "merely demonstrate what things can be done and how they should be done. They make no attempt to explain why they need to be done in that particular way, do not attempt to provide the reader with a series of general concepts on the basis of which he may face problems not foreseen by the author." (Panofsky 1953). These kinds of works are "workshop texts", (bottega texts), limited texts, in which the technical side does not follow a rational design, a heterogeneous and fragmented material formed on the basis of personal notes. The designs of machines and mechanisms that are found in this type of texts are not projects in the modern sense, because they have no connotation of feasibility, they are given only for illustrative purposes and not by way of explanation of the written text. The methods of representation are personal, the proportions and dimensions are approximate, the placement of the various mechanisms is subject to the overall view of the machine. Sometimes these early approaches to mechanical illustration contain schematizations of material objects, with some conventional representations.

After the end of the classical world, the science of representation has evolved slowly, and without any systematic development. Pictorial and/or architectural representations normally do not address the problem of spatial rendering. One has to look further, to the pre-Renaissance painters, to find the first systematic attempts at spatial rendering. The result appears as something of a hybrid between axonometry and perspective. However, this is precisely the extended period that laid the foundations on which will merge the theory and practice of the mode of representation of the following centuries.

Great travels, translation and study of classical texts, desire for direct knowledge, large construction projects—these are, on the whole, the essential factors that support the development and deepening of subsequent studies in the field of representation, which evolves rapidly in the following centuries. The reasons for this are many and can be considered of a cultural nature (renewed interest in mathematics, translation and study of classical works, university studies), as well as rooted in practical reasons (geometric and construction problems, construction of large architectures, need to know the existing order to better evaluate the characteristics, ageing of the practice of construction site, etc.).

With the passing of the centuries, the separation between the conceptual and executive duties at construction sites becomes more and more marked. This explains why it is becoming increasingly important to execute preliminary project graphics, hitherto non-existent. Thanks to this new structure of the construction site, there is a significant advance in the use of the graphic medium. The drawings begin to show evident use of a system of orthogonal projections, always sensed, but properly coded only much later. This presupposed the acquisition by the medieval architects of the abstract concept of projection from infinite distance.

The function of the medieval drawing as the main vehicle of the figurative tradition is represented in typical fashion by the "notebooks", where artists gathered, as a case-study for themselves, for their workshop, and for scholars, drawings depicting their own works of art of those of others: a collection of notes and observations of various kinds, a kind of manual, a handbook with which to solve many of the problems encountered in their daily practice.

Among them, the most interesting document for the general knowledge of the background and interests of a Gothic architect is the notebook or "Livre de Portraiture" of Villard de Honnecourt, a designer with a broad spectrum of interests.

Villard de Honnecourt (Honnecourt-sur-Escaut, 13th century) was born in Picardy, in a village beside the abbey of St. Pierre, where he may have studied in the workshops and made use of the library which had some ancient manuscripts on technical and scientific subjects.

His work is a collection of 33 sheets of bound parchment, each of which contains illustrations on various topics, together with brief annotations or captions. The fame of Villard is mainly due to this written and drawn body of work. It also contains an illustration that could be his self-portrait in military garments (Fig. 7). Currently, this notebook is kept in the National Library of France, after, in a stroke of luck, being rediscovered at the library of St. Germain des Pres in Paris in 1849

In the 13th century work was done on parchment: expensive material and limited in size. That's why the drawings in his Notebook, dated as 1225–35, are usually arranged four or five to a page, in a balanced composition of text and drawings in a summary framework. Some of the objects are drawn in plan, others in prospect or in section, in a very expressive graphic language.

This notebook's simplicity, at the same time combined with assuredness and strength of expression of the images, the properties of its graphic language show that we are not facing a first example, one of its kind, but merely one piece that has survived to our days out of many that possibly existed and had to be a result of widespread and intense production in the first half of the thirteenth century.

The drawings are traced as "wire frames", without shadows, without any concession to aesthetics, but in a simple and understandable manner. Villard differentiates the type of handwriting conventions and the way of representing depending on whether it's sculpture, a geometrical problem or measurement of a machine or an architectural piece. In the drawings of saints and animals the motion is continuous and thread-like, but reinforces the points of greatest movement or tension of the

Fig. 7 Presumed self-portrait of Villard de Honnecourt in military garments in "Livre de Portraiture" by Villard de Honnecourt (Fol. 2)

figure, making it clear that this graphics research was aimed at giving the idea of plasticity, the comprehensive summary of the sculptures for which this graphics, perhaps, had to serve.

In geometric drawings we see greater simplicity, a more uncertain motion, a schematic that is rather impelled. As to the representations of machines, these are often carried out with a view of pseudo-perspective, somewhat naive, and accompanied by explanatory texts.

On sheet 22v of his notebook, completely devoted to mechanics, it is the most beautiful in both layout and in its graphic components (Fig. 8a), we find the mechanism of a rotating angel.

The clock mechanism that caused the movement of the angel "which constantly indicates the sun" is sketched simply but exhaustively in elevation, showing the system of weights that allowed the angel to move, while the angel itself is not shown in order to focus the attention of the reader on the mechanism as opposed to the figure of the automaton. This is because the Notebook was addressed to technical readers who were not to be astonished by the marvellous appearance of the angel but needed to understand how the miracle took place.

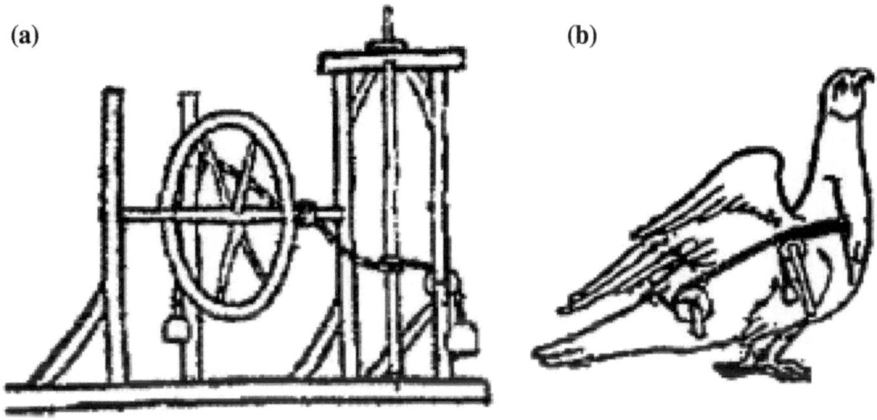

Fig. 8 **a** Angel pointing to the sun, **b** Eagle automata "Livre de Portraiture" by Villard de Honnecourt (Fol. 22v)

Presumably the angel was constructed to move about the pole that appears on the extreme left of the illustration and its rotation was permitted by a wheel driven by ties and counterweights.

The representation of the mechanism of the angel is in elevation, but the wheel is turned over to show the circular shape and to better summarise the different aspects of the object shown in a single drawing. There are also some of the very first attempts at graphics standardisation, due to a sketching and abstraction of the contingent aspects of the material object. For example you can recognize a use of a sinusoidal line to represent the rope, even though it is certainly in tension.

On the same page we also find the automaton of the eagle (Fig. 8b), which is presented in Villard's text as "must turn its head towards the deacon on the reading of the gospel". It is probably the oldest image of a western automaton that has come down to us. This representation, which is another proof of Villard's great ability as an artist, is an able combination of perspective and section, with pictorial connotations in the treatment of the feathers on the wing. The device for transmission and transformation of movement is clearly delineated, based on a mechanism of pulleys and counterweights that allow the bird's neck, fixed on a vertical axis, to rotate thanks to a cord and series of pulleys.

These two automata, operated by counterweights and rope transmission with pulleys, certainly connect to techniques going back to Heron's automata, and even if their application in Villard enters the spiritual field and becomes a religious function, it cannot but be linked to precedents in the tradition of the *deus ex machina* of the classical Greek theatre.

The architectural drawings, the best known of this album, are expressed through pseudo-perspective views, sometimes not completely finished and controlled, and through plans and elevations, where the orthogonal projections, used empirically, are plotted with confidence and skill.

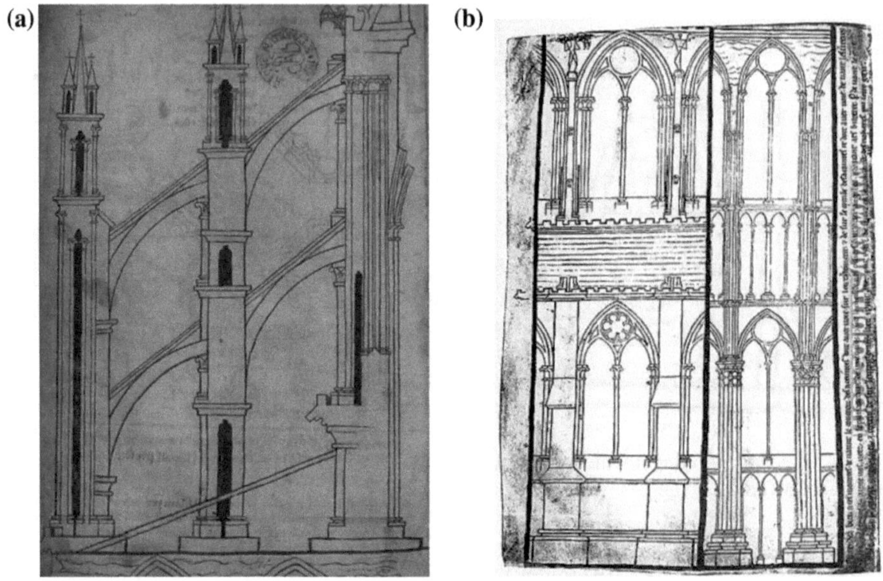

Fig. 9 Reims Cathedral **a** Section of the wall and arch of the absidial chapels, **b** Exterior and Interior Elevation of the Lateral Walls in "Livre de Portraiture" by Villard de Honnecourt (Fol. 52v-51v)

The example of drawings for the cathedral of Reims (Fig. 9a, b) shows us how Villard combines in the same composition a representation of the interior and exterior elevation of the building; again for Reims is the prospectus section where Villard proposes, drawing them meticulously, the external buttresses and a part of the section of a chapel. Both these compositions give us proof of his will to control and formally represent the various parts of the cathedral, but also his precise and strong mastery of graphics.

Real project drawings are those prepared for construction of the cathedral of Laon (Fig. 10a, b), which offer us a testimony to Villard's mastery of graphic media and orthogonal projections, although still only guessed rather than measured. Such control of the graphic media can not have been achieved in a short span of years, but it must be the result of a prolonged and comprehensive experience, about which we can only make assumptions, due to the surviving evidence being few and far between.

Renaissance and Filippo Brunelleschi

In the Renaissance, we see a merger of the liberal and mechanical arts, thanks to profound social, economic and cultural changes: the spread of production facilities brings with it the institutionalisation and protection of the craft and its forms of transmission.

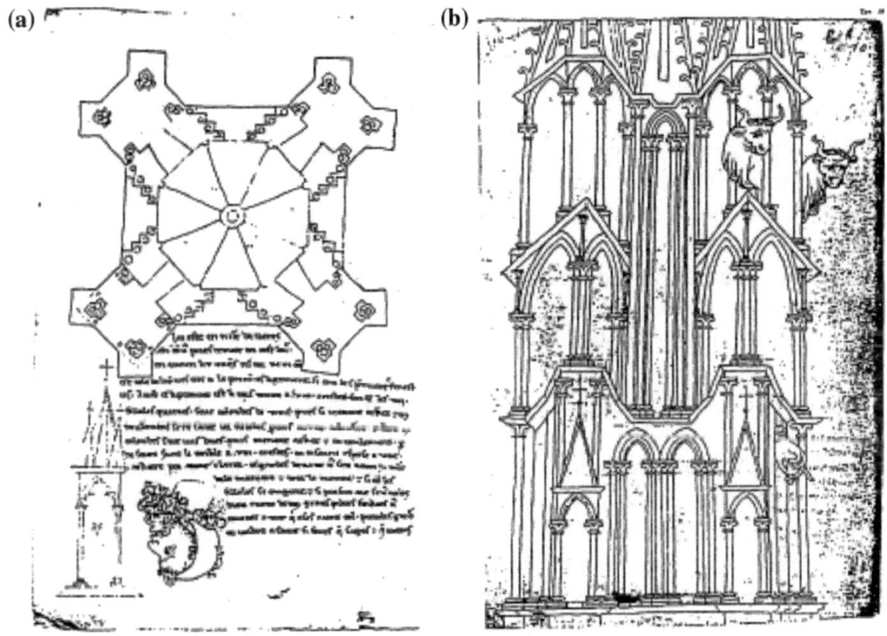

Fig. 10 Laon Cathedral **a** Plan of the tower, **b** Elevation of the tower in "Livre de Portraiture" by Villard de Honnecourt (Fol. 9v-10)

In the Renaissance, we witness a development of technical representation. In the drawings, visually sophisticated means are used for comprehensive and precise communications. Greater aesthetic attractiveness is an integral part of a new conception in the technical drawing, held out to highly technological content.

Technical drawings in general, and in particular the mechanical variety, see the first manifestations that are intentional, not plotted for decorative or illustrative occasional: it is the beginning of the evolution that will, thanks to the press, turn it into a literary genre with the "theatres of machines". During the first phase of this development, still connected to the medieval tradition and Aristotle, the illustrations are a mere support to the text; in notes or workshop texts, prepared for the exclusive use of the authors, drawings prevail over the text.

The same type of dualism manifests itself also in terms of authors of the drawings: artists and technicians. The first exhibit their skill in drawing, representing the machine in natural environments and using its own painting techniques; the second, technicians, schematize objects and impoverish the environment, improving the information on the object represented. Thus, both the number of components represented (also in rudimentary sections) and metric directions increase and expand.

The recurrent use of naturalistic representations is perhaps explained by the shared "workshop" (bottega) cultural training received by the authors of the drawings and the craftsmen. Although, it is still firmly the setting where the author is the only one who can effectively understand the drawing, due to special omissions, and realise the machine whose functioning as a whole, however, is clearly shown.

The development of research on the coded representation of the space is due to several factors. One was certainly the rediscovery of Vitruvius' work, which took place around the first quarter of the fifteenth century, and its subsequent wide diffusion. Other factors are social in nature, such as the phenomenon of extraordinary and unusual spread among large sections of the population of the understanding of beauty, which occurred around 1450/1550; as well as studies from life and practical training implemented in the schools-workshops of the first medieval artists.

Another interesting aspect to consider is the great architects of that time and their practice of observation and study of ancient architecture. This practice needs to be placed in the wider framework of the humanities. The graphic works that remained of this impressive study campaign conducted by the Renaissance architects are executed in a uniform, linear hand. It is still an empirical approach, strictly devoted to the author's personal training, but given the great personalities of the artists who live in this period, they are already distinguished by strong technical-analytical connotations. Between the fifteenth and sixteenth centuries, graphic artists enhance their works with depth and perfect the studies of perspective, concomitantly codifying a number of graphic conventions.

At the beginning of the sixteenth century, the drawing is already regulated, particularly in the systematic use of "orthogonal projections". Many artists use commonly represented plans, elevations, sections and details. In terms of architectural representations the problem of the non-visible is also taken into consideration, in the wake of the tendency to know everything about an old building.

In the Gothic period, at construction sites of the great cathedrals, there used to be a division between the figure of the Master Builder and that of the Designer, hitherto united in the same person of the medieval Master. The consequence of this division of roles was the isolation of the architect from the craftsmen, and the beginning of the phenomenon of Bottega ("workshop"), a real hotbed of wits.

So while, on the one hand, the importance of study from life was affirmed, on the other hand studies on perspective were developing, understood as the science of representation and intimately related to the search for the absolute values of mathematics and geometry.

A crucial opportunity for the development of scientific research in the field of geometry was the one connected to the study of perspective.

It has been found that the ancients had knowledge, although elementary, of perspective. Muslim and Christian scholars have ventured out on this route, after which the road to research related to our vision had been opened by Euclid (375–333 BC) in his "Optics". In the Middle Ages, research was developed by the Arabs, as well as by Roger Bacon (1214–1292/94), who was interested in the problems of vision, devoting to them the fifth part of his "Opus Maior".

The discovery and the first coding, however imperfect, of the rules of perspective date back to the early fifteenth century and is attributed to the combined action of some of the greatest Renaissance architects such as Brunelleschi, Leon Battista Alberti (1404–1472) and Piero della Francesca (1415–1492). These studies had great importance for all the architects of the Renaissance, both as reflections on their training, and in terms of profound impact on their operations.

The men of the fifteenth century engaged in research considered perspective as a method for the objective representation of space, that is—a tool. Yet also for them it was a basic component of the new linear conception of space, a consequence of a particular historical situation, the result of a reversal of the figurative taste. It was no accident that it was an architect, Filippo Brunelleschi, who arrived before others at the most important discovery of geometric construction of perspective representation, and which another architect, L.B. Alberti, began to codify scientifically. In fact, the research and the codification of the laws governing the construction of perspective exist first and foremost due to the need to control three-dimensional spaces through two-dimensional design drawings.

Filippo Brunelleschi (Florence 1377–1446)

Recognised already by his contemporaries as one of the founders of the Renaissance, for his architectural works, and his study of perspective and proportion, Brunelleschi (Fig. 11) was trained and first worked as a goldsmith and sculptor, entering the art of silk-making from 1398 and that of a master in 1404. The construction of the dome of the Florentine church of Santa Maria del Fiore occupied almost the entire career of Brunelleschi, who was present at the construction site from 1404 until 1446, the year of his death.

His fame as an architect is paralleled only by his renown as an inventor of machines. Certainly, his designs of construction machines used for the dome of St. Maria del Fiore introduced many innovations in the field of mechanical engineering, as well as in construction technology.

The church had been under construction since 1296 and was originally designed by Arnolfo di Cambio. The construction was interrupted at the height of the octagonal drum (architectural structure on which the dome is set) because of considerable technical and economic difficulties. In 1418 was announced a competition for the public presentation of the models or designs for the framework, bridges and construction machinery for the dome. Brunelleschi won it in 1420 and began construction of the dome, which was completed in 1434 and 2 years later blessed by Pope Eugene IV.

To complete the construction of the final part of the dome Brunelleschi conceives a “placer” for marble elements, a real mechanical arm to slide horizontally, vertically, and circularly, carrying loads. Of this, today we possess two depictions: one represents it inside a huge wooden scaffold standing on the locking ring of the dome; the other, more schematic, outside of the scaffold (Fig. 12). Both clearly

Fig. 11 Presumed portrait of Filippo Brunelleschi by Masaccio. 1423–1428, Cappella Brancacci in S. Maria del Carmine Church, Florence

show the importance of the whole for an effective explanation of the functioning of individual parts.

In fact, it can be observed that generally the mechanisms of force multiplication are represented in a clear way, namely once indicated, the sizes are not given in the pulley design and the guides are emphatically illustrated in bold lines and with no constructive details that can convey an understanding of the operation of each single mechanism inside the machine. However, what is highlighted with particular clarity are the design components of the mechanism, allowing, for example, the distinction between ropes and chains, and demonstrating the author's awareness of the full scope of the forces on the individual components.

Also for the dome, Brunelleschi designed a crane (Fig. 13). This is a design piece of great refinement: text and graphics, expertly balanced, contribute to the final effect, mainly due to the wisdom of the author in the use of perspective and contrast between light and dark. The mechanisms (gears, transmissions rope, linkages) are represented with a wealth of technical and functional demonstration inherent in a mature awareness of the possibility of an appropriate combination of elementary machines of different constitution.

We know these machines of Brunelleschi from the drawings of Bonaccorso Ghiberti (1451–1516), a nephew of the great sculptor Lorenzo Ghiberti (1378–1455), who in his manuscript called "Zibaldone" gathered copies of drawings of machines designed for the dome of Santa Maria del Fiore in Florence.

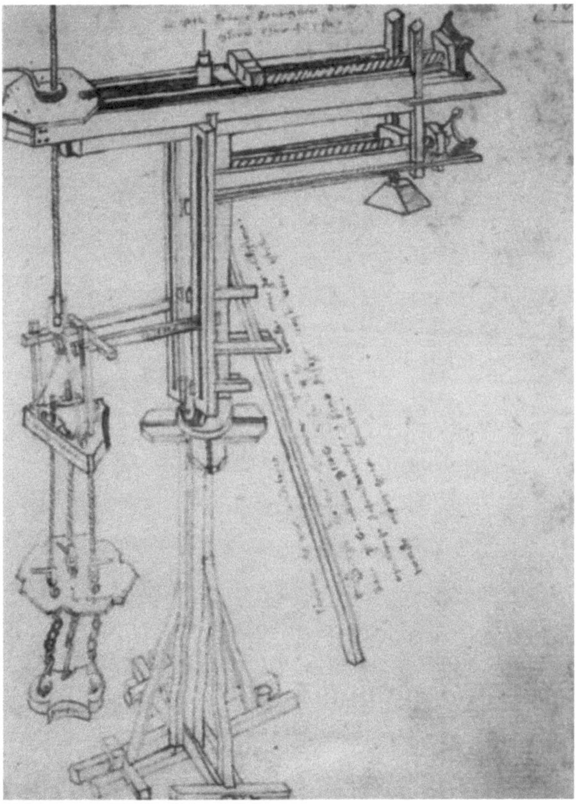

Fig. 12 Rotating crane by F. Brunelleschi. [Zibaldone di Buonaccorso Ghiberti, fol. 104r, 1420–21]

Yet Filippo Brunelleschi is also remembered as the discoverer, or better, as the encoder of perspective representation. From the earliest works of the artist, towards the turn of the 1300s, when he was a practising goldsmith, he began to work on issues of technical vision, which in those years had little to do with the practice of art.

He painted two tablets with views of Florence, one of the Baptistery, the other of Piazza della Signoria. These tablets were part of the Medici collection for a long time and served as an example for many of the 1400s artists who were also researching perspective. These tablets and the perspective experiment that were made possible with their use, reach our days as a description by Antonio Manetti (1423–1497), a mathematician of Brunelleschi's circle and his main biographer.

To repeat the perspective experiment with the first tablet (Fig. 14a), one fist needed to reach the same location of the point of view from which Brunelleschi had painted the work and, holding the tablet with one hand, look through a hole drilled in it, while in the other hand holding up a mirror in which you would see a reflection of the painted image. This vision would create a perception very much

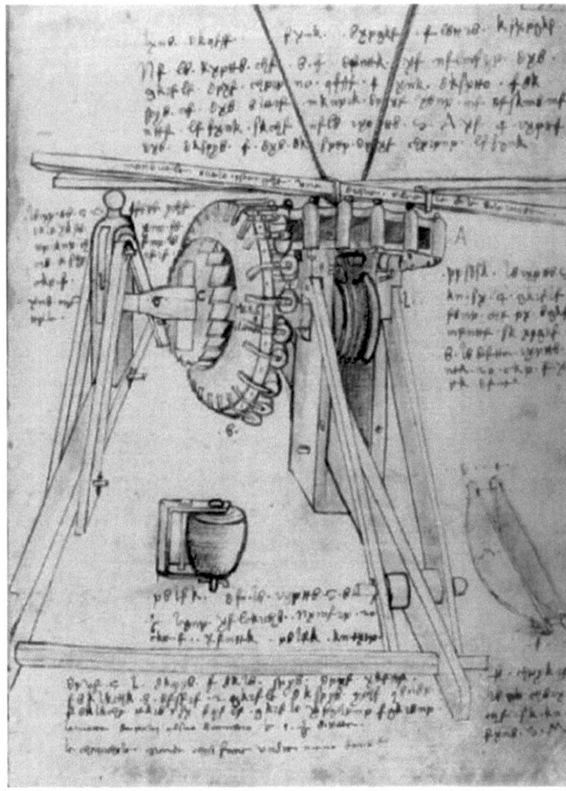

Fig. 13 Crane machine by F. Brunelleschi. [Zibaldone di Buonaccorso Ghiberti, fol. 106r, 1420–21]

similar to the real one. The illusion was reinforced by the ingenious replacement of the sky surface of the tablet with a piece coated in burnished silver, which would reflect clouds moving in the sky, making the vision all the more real.

The second tablet (Fig. 14b), as we learn from the description left to us by Manetti, depicted Piazza della Signoria. However, because, unlike the first one, it was not a central perspective but angled, Brunelleschi did not use the artifice of the hole and mirror. The composition appears as a skyline of today, that is, with the outlines of buildings, so that the background was that of the Florentine sky, to make a perfect fusion of rendered objects and real vision.

To-date we do not possess any texts to enable us to fully understand the geometric procedure underlying the perspective tablets of Brunelleschi, but certainly we recognise their great merit in bringing to life the new vision of space of the Renaissance, and tools to express it. New ideas and new spatial visions that Brunelleschi then expressed in the most revolutionary way with the design and construction of the dome of Santa Maria del Fiore.

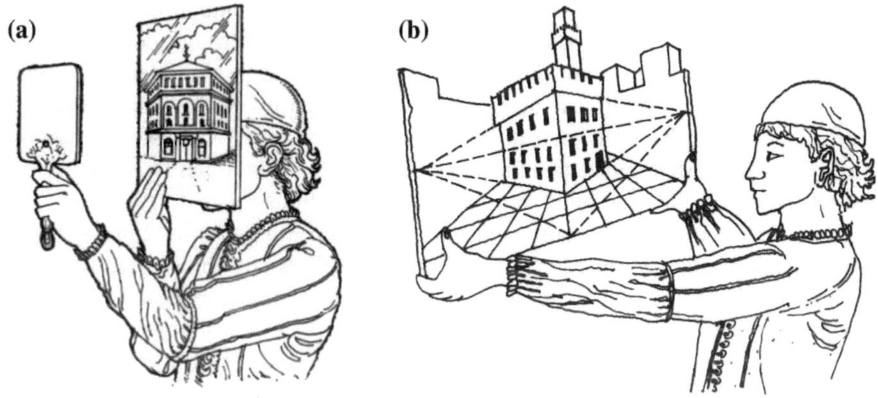

Fig. 14 **a** First perspective tablet **b** Second perspective tablet of Brunelleschi

Baroque Era and Giovanni Branca

In the last quarter of the sixteenth century the scientific world undergoes a deep conceptual change, which will be widely reported in the scientific and academic world, and particularly in the Mechanics and Drawing related to it. In addition to the work of Galileo Galilei (1564–1642), a newly appointed professor at Padua, who lays the rationalizing foundations of the applied mechanics, we must also remember Guidobaldo del Monte (1545–1607), who in his "Mechanicorum Liber", published in 1577, begins the process that from a figurative representation of a concrete object will lead to abstract geometric schematization.

It is remarkable that they knew each other at the Court of Tuscany and they were good friends as a documented correspondence shows also in term of discussions on their research interests in several fields.

However, Guidobaldo del Monte can be considered important in the History of Mechanics not only for his studies on mechanical design but also and mainly since he can be recognised as one of the first to use and standardise the perspective representation.

As we have seen, this technique seems to be used by the ancient Greeks and Romans in the form of central perspective. But a later flattening of figures and spaces during the late Roman Age and the Byzantine period produced an infrequent use of perspective until a complete loss during the Middle Age. The rediscovery and standardisation of perspective in the current form were started by Filippo Brunelleschi in the form of linear perspective. Later many others applied and enhanced the perspective technique: relevant are Albrecht Dürer (1471–1528), who disseminated the Italian methods of graphical representation in North Europe, and Guidobaldo del Monte who gave a rigorous formulation of the procedure.

Thus, although the perspective has been commonly used in the Renaissance, nevertheless only later it has been used continuously for the drawing of machines and mechanisms. This has been due to the dual nature of the Authors: artists or technicians. The first used the drawing to show their skill through perspective and shadows when they represent the machines with a figurative approach; the latter used sketches for mechanical design not only because of a technical synthetic approach but mainly since they do not have artistic skill. When the Author of a drawing is both artist and technician, then the representation can be recognised as perfect from both the graphical and technical viewpoints.

Between the end of the fourteenth and early fifteenth centuries European publishers discovered the topicality of the problem of the machine. The numerous volumes on this topic document an interest that slowly transforms and reveals this epoch's attitude towards the technology that will lead to that literary phenomenon called "theatres of machines", in which the technical and scientific knowledge are used to impress and amaze.

The volumes published in this period, whose authors are less and less "artists" and more and more "technical experts", are not treaties as such, but rather collections, largely including non-original images with captions and comments. The drawings are of the traditional type and usually depict both details and the machine as a whole. The only properly innovative aspect, present at times, is the inclusion of the reduction scale.

Up to the XVIth century there was no great difference between architect and engineer, and indeed usually they were the same person: a designer. The evolution of a specific knowledge in mechanics requires specialization in people preparation and technological language. This was due also to a need, being experienced more and more by people who constructed mechanisms but were not designers, for drawings from which all the components are well described in terms of their behaviour. Theoretical aspects were developed and a suitable graphical expression became required to help investigators to clearly express the basic characteristics of mechanisms. Therefore, specific drawings were drawn and they were so specific for the subject and the scheme that they can be considered as devoted to experts only.

Later, in the seventeenth century, the increasingly dominant idea is that of a possible collaboration between science and technology, as a culmination of a series of experiences that are rooted in the guiding function assumed by the bourgeoisie in the world of production, a collaboration that will lead to an unstoppable progress in every field.

Yet, before this big step, it is necessary to brush away prejudices and false problems facilitated by tradition. In fact, baroque science still feeds off the temptation to use new knowledge in the fields of ancient disciplines, and this attitude was responsible for the vein of "Tecnica curiosa" (The Curious Technology), placed at the service of wonder and fun.

To lead to practical cooperation between science and technology, everyone needed to start using the Galilean approach, that is the theoretical prediction of the mechanical behaviour for integral design of machines and mechanisms, as well as to set the technical and production problems in mechanical terms, laying the bases for their solution.

Giovanni Branca (S. Angelo in Lizzola 1571–Loreto 1645). Little is known of the biographical particularities of this person, and neither is there a surviving portrait. He studied mathematics and architecture in Rome, and in 1616 was appointed architect of the Holy House of Loreto, a post he held until his death. His known work is that of an architect in Assisi and in Rome, where in 1622 he was awarded Roman citizenship.

However, Branca is best remembered for his "Manuale di architettura" (Manual of Architecture), published for the first time at Ascoli in 1629, and the book "Le machine, volume nuovo e di molto artificio da fare effetti meravigliosi tanto spiritali quanto di animale operazione" (Fig. 15) (The Machines), published in Rome, again in 1629. Although the "Manual of Architecture" was a great success both in Italy and abroad for many years, it is mainly due to the last book that he still holds a prominent place among the writers of technical texts of the post-Renaissance and Baroque eras.

In the volume "The Machines", beyond the purpose announced in the title as "to describe the wondrous effects of mechanical contraptions…" that is to say, of machines on pneumatic energy certainly dating back to the experiences of Heron,

Fig. 15 Frontispiece of Giovanni Branca's Book "Le Machine"edited in 1629

that use animal energy, Branca summarises and sets out the main machines known and existing in his time.

In addition to devices that are not particularly novel, in Fig. 25 of his text (Fig. 16), Branca shows a completely new machine, most likely conceived as a refinement of *eolipile* (Heron engine) that can be considered the progenitor of modern steam turbines. It consisted of “a metal head with its chest, filled with water, placed over coals”, from whose mouth came a jet of steam that was used “to crush materials to make dust”. We have a few vague notes indicating that a model of this machine was built by Branca and that he conducted some experiments with it in Milan, but no physical traces of the device remain, nor any consistent documentation.

The illustrations that accompany his work have generally a concise representation, almost devoid of elements that can distract or remove weight from the importance of the machine and its mechanisms. The graphical technique is precise for both the choice of perspective and for chiaroscuro effects.

Fig. 16 Machine for crush materials to make dust in “Le Machine” by Branca 1629. Figure XXV

In this illustration, the surroundings in which the device is placed are absolutely schematic and are reduced to a few elements that show a rolling terrain and a slightly curved horizon. What is interesting is the inclusion of a human figure, working on a wooden table on trestles placed in the foreground. This detail serves to make the board more alive and eventful but especially to give a dimensional definition to the machine represented.

Despite this strong emphasis of conciseness and schematic representation, Branca had an undoubted graphic expertise, as shown by Fig. 9 in the same text (Fig. 17). In fact, unlike his usual style, here Branca displays his virtuoso ability in graphical representation, depicting water in the part where the device is immersed. The illustrations to his works mark a further development towards a technical approach intent on depicting only the machine: bare and essential are the traits, and the shadows are represented regardless of the actual size of the object, not to privilege the representation at the expense of the represented object.

In this figure a hydraulic plant is shown with a technical emphasis for gears and slider-crank mechanisms, although they are still plunged into a pictorial scenario.

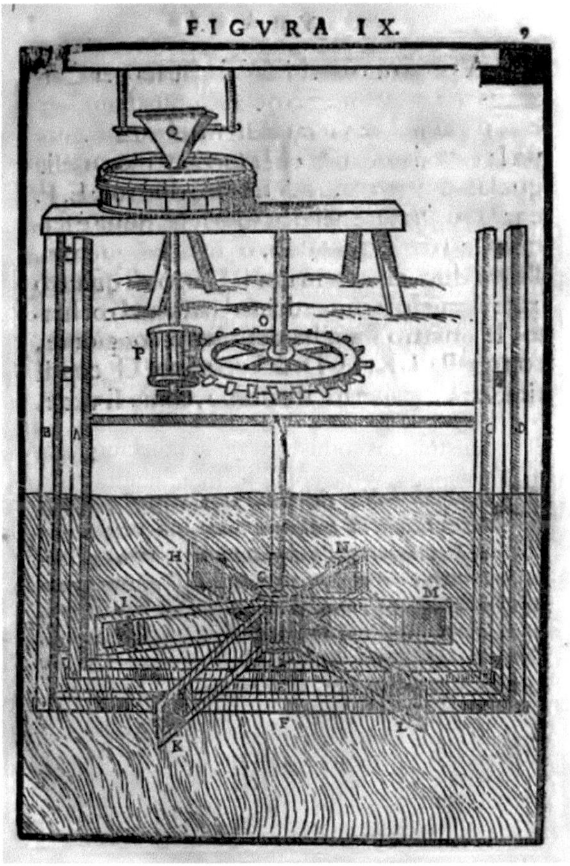

Fig. 17 Drawing of a mechanical assembly for a pumping plant in "Le Machine" by Branca 1629. Figure IX

(a) (b)

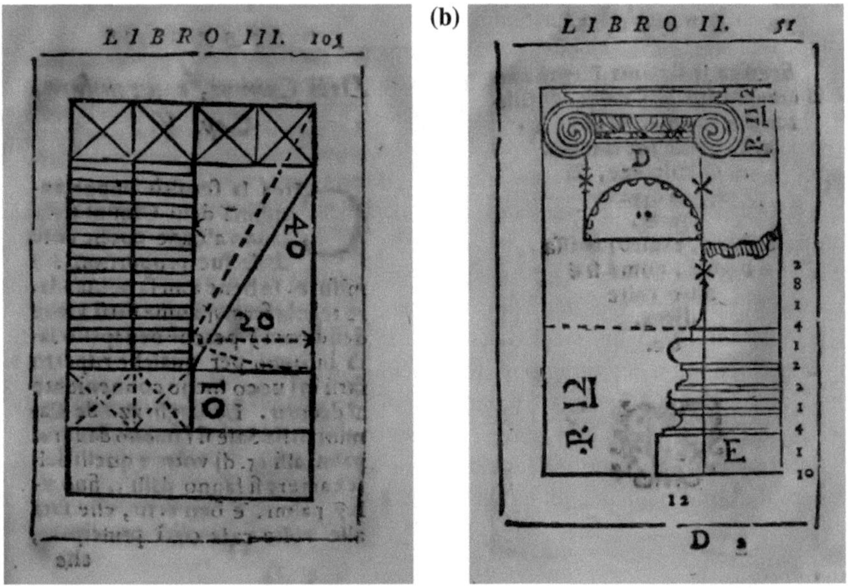

Fig. 18 **a** Stairs, **b** ionic order in "Manuale di Architettura" by Branca 1629. Figures 105 and 51

Branca's approach changes in his "Manual of Architecture" published in the same year as "The Machines" (1629).

Branca probably didn't write the two volumes in the same year. Most likely, they worked on and prepared over a long period, perhaps in parallel, or first one and then the other, depending on how his research progressed. Once both texts were ready, he published the two concomitantly with the highest point of his career, after being named in 1616 the Architect of the Holy House of Loreto, and gaining Roman citizenship in 1622, as well as completing architectural works as a designer in various parts of Italy.

The illustrations to his text on architecture are less concise than those in the text on the machines. What could be more natural, if we think that it was not possible to draw an architectural order depriving it of decorative pieces. Nevertheless, in his drawings Branca still manages to be far more dry and precise in comparison to many other texts on architecture of his contemporaries.

Figure 18 shows two examples. The first is (Fig. 18a), in the words of Branca, "a method of proportioning the stairs in every building". In this representation we find methods almost identical to the ones still used today, such as solid lines for the elements in view and dashed lines for concealed elements.

The second example (Fig. 18b) is an ionic order, represented as a rule with capital and base. In line with Branca's partiality to brevity, the overturning of the column that we see under the capital is represented by a semi-circumference, where, to distinguish from the drum that runs from the base and fits under the capital which is smooth, the column is drawn with grooves. The two parts are shifted relative to one another, probably to exploit the space on the page, and also to insert indications of the virtual (not visible) section of the shaft, drawn as a rupture of the material and shown by a dotted line. This rule, albeit in a less pictorial expression, is also used in the current graphic symbology.

Conclusions

Throughout Antiquity and the Middle Ages, the personalities of the greatest interest are certainly the ones that are diffusely described both in this and other chapters of our work.

The Renaissance proposes numerous personalities that have reached prominence in the fields of Geometry and Mechanics.

Luca Pacioli (Borgo San Sepolcro, Arezzo 1445–Rome 1517) was an italian franciscan friar. In Milan he wrote "De divina proportione"; in its first manuscript Leonardo da Vinci drew the illustrations of the regular solids. His work is fundamental for every later study that is based on geometry. In 1509 he published a latin edition of Euclid's Elements.

Albrech Dürer (Nuremberg 1471–1528) was a painter, printmaker and treatiser, generally regarded as the greatest German Renaissance artist, he introduced in north Europe perspective rules, that he had learned in Italy. Good connoisseur of precision mechanics since his father, Albrecht Dürer the Elder and his maternal grandfather, were goldsmiths. In 1525 he published "Unterweisung der Messung" and in 1527 "Unterrichtung zur Befestigung der Städte, Schlösser und Flecken".

Francesco Maurolico (Messina 1494–1572) was an Italian mathematician, architect and astronomer of Sicily. He is remembered for his contributions to the fields of geometry, optics, conics, mechanics, music, and astronomy. He edited the works of classical authors including Heron and Euclides. Maurolico published an edition of Aristotle's Mechanics and after his death, in 1654, was published "Emendatio et restitutio conicorum Apollonii Pergaei".

Francesco Luci Durantino (Casteldurante-Urbania XVend-XVI half) Francesco Luci or Luzi, said Durantino by his birth place, was an italian humanist, scholar and architect. He is remembered for his 1524 Edition of "De Architectura" by Vitruvius. In the X book Durantino illustrates many of the machines described in Vitruvius's text.

The post-Renaissance epoch is also rich in scholars who were interested in Geometry and Mechanics and reached important results in both disciplines.

Matteo Ricci (Macerata 1562–Beijing 1610) was a member of Jesuit order. He is known for opening the way for European technical culture in China. He compiled works in Chinese on science and technology based on European knowledge. In 1611 he published Euclid's Elements translated into Chinese.

Philippe de la Hire (Paris 1640–1712) was a pupil of Girard Desargues, who was a friend of his father, and who introduced him to projective geometry. De La Hire is remembered for his work on geometry and mechanics. In 1673 he published "Nouvelle Méthode en Géométrie pour les sections des superficies coniques et cylindriques". In 1695 he published "Traité de mecanique".

Guido Grandi (Cremona 1671–Pisa 1742) was a member of Camaldolese order. He was professor at Pisa University by carrying out an intense activity with specific interests on geometry, mechanics, astronomy and hydraulics. In 1740 he published "Elementi geometrici piani e solidi di Euclide" (Euclidean Geometry).

As a conclusion to this brief excursus on descriptive geometry and mechanism science from Antiquity to the 17th Century, we would like to quote the words written more than 750 years ago as the opening of a work that deals with these topics: "… Here you will find the technique of drawing and shapes as the science of geometry commands and teaches." "Villar de Honnecourt, "Livre de Portraiture", 1225/35.

Bibliography

Argan GC (1946) The architecture of Brunelleschi and the origins of the perspective theory in the fifteenth century. J Warburg Courtland Inst IX:90 ss

Argan GC (1978) "Brunelleschi" Oscar Saggi, Mondadori, Vicenza

Bartoli MT (1978) "Ichnographia, ortographia, scaenographia", in Studi e documenti di architettura VIII:197–208

Battisti E (1975) Brunelleschi, Electa

Bechmann R (1991) Villard de Honnecourt, Le pensée technique au XIIIe siècle et sa communications, Paris

Borsi F (1965) "Il Taccuino di Villard de Honnecourt", in "Cultura e Disegno", Firenze pp 29–49

Bossalino F (2002) a cura di Marco Vitruvio Pollione "De Architectura Libri X" traduzione in italiano di Franca Bossalino e Vilma Nazzi, Roma: Kappa

Branca G (1629) "Le machine: volume nuouo et di molto artificio da fare effetti marauigliosi tanto spiritali quanto di animale operatione arichito di bellissime figure conle dichiarationi a ciascuna di esse in lingua uolgare et latina", In Roma: Ad ista[n]za di Iacomo Martuci … per Iacomo Mascardi

Branca G (1629) "Manuale d'architettura: breue, e risoluta pratica", In Ascoli: Appresso Maffio Salvioni

Bruschi A, Carugo A, Fiore FP (eds) (1981) Vitruvius Pollio, De architectura, Milano: Il Polifilo

Camilli E (1971) "Giovanni Branca", Pesaro

Cesariano, Cesare (1521) Vitruvius, Di Lucio Vitruvio Pollione De architectura libri dece: traducti de latino in vulgare, affigurati, comentati, & con mirando ordine insigniti: per il quale facilmente potrai trovare la multitudine de li abstrusi & reconditi vocabuli a li soi loci & in epsa tabula con summo studio expositi & enucleati ad immensa utilitate de ciascuno studioso & benivolo di epsa opera, Como: Gotardo da Ponte

Ciapponi LA (1984) "Fra Giocondo da Verona and his edition of Vitruvius". J Warburg Courtauld Inst XLVII:72–90

Ceccarelli M (2008) Renaissance of machines in Italy: From Brunelleschi to Galilei through Francesco di Giorgio and Leonardo. Mech Mach Theory 1530–1542. doi:10.1016/j.mechmachtheory.2008.01.001

Ceccarelli M, Cigola M (1995) On the evolution of Mechanisms drawing. In: Proceedings of IXth IFToMM world congress, vol 4, pp 3191–3195, Milano

Ceccarelli M, Cigola M (2001) Trends in the drawing of mechanisms since the early middle ages. In: Journal of Mechanical Engineering Science, Proceedings of the institution of mechanical engineers Part C, vol 215, pp 269–289. Professional Engineering Publishing Limited, London UK. ISSN 0954-4062

Cigola M, Ceccarelli M (2014) Marcus Vitruvius Pollio: In: Ceccarelli M (ed) Distinguished Figures in Mechanism and Machine Science: Their Contributions and Legacies, Part 3, pp 307–344. Springer, Dordrecht. ISBN 978-94-017-8947-9, ISSN 1875-3442, doi:10.1007/978-94-017-8947-9

Cigola M, Ceccarelli M (2014) Machine designs and drawings in renaissance editions of De Architectura by Marcus Vitruvius Pollio. In: Proceedings of 2013 IFToMM PC workshop on history of MMS, pp 1–5. Napoli. ISBN 9788895430843

Cigola M (2012) In praise of parallel theories: descriptive geometry and applied mechanics. In: Carlevaris L and Filippa M (eds) In praise of theory. The fundamentals of the disciplines of representation and survey, pp 39–46. Roma Gangemi editore, ISBN 978-88492-2519-8

Damish H (1987) "L'origine de la perspective", Flammarion, Paris

Del Monte G (1577) "Mechanicorum liber", Pesaro

Del Monte G (1984) "I sei libri della prospettiva di Guidobaldo dei marchesi del Monte" Sinisgalli R, L'erma di Bretscneider, Roma

Docci M, Cigola M (1997) "Representación gráfica e instrumentos de medición entre la Edad Media y el Renacimiento". In "Anales de Ingeniería Gráfica", n. 2 Mayo-Deciembre 1995, pp 1–20 Madrid

Docci M, Maestri D (1993) "Storia del rilevamento architettonico e urbano", Laterza, 1° ed. Roma-Bari

Docci M, Migliari R (1992) Scienza della rappresentazione, fondamenti e applicazioni della geometria descrittiva, Roma, La Nuova Italia Scientifica

Erlande-Brandenburg A (1987) «Carnet de Villard de Honnecourt», Paris 1986 trad. it. Villard de Honnecourt, disegni, Milano

Edgerton S (1975) The renaissance discovery of linear perspective, New York

Koetsier T (1983) A contribution to the history of Kinematics—I. Mech Mach Theory 18(1):37–42

Federici Vescovini G (1965) "Sudi sulla prospettiva medievale", Torino

Frà Giocondo (1511) M. Vitruvius per Iocundum solito castigatior factus cun figuris et tabula et iam legi et intelligi possit, Venezia: Giovanni da Tridino

Gabucci G (1930) "La patria di Giovanni Branca", Fano, Tipografia Sonciniana

Galileo G (1964–1966) "Le meccaniche", 1600. In: Favaro A (ed) "Opere di Galileo", Firenze, 1964–66

Gioseffi D (1957) "Prospectiva artificialis. Per la storia della prospettiva. Spigolature e appunti", Trieste

Lassus JB (1858) "Album de Villard de Honnecopurt architecte du XIII siecle", Paris

Loria G (1921) Storia della Geometria Descrittiva dalle origini sino ai giorni nostri. Milano, Hoepli

Mancini P (1841) "Cenno biografico intorno Giovanni Branca", Pesaro

Marcolongo R (1919) "Lo sviluppo della Meccanica sino ai discepoli di Galileo". In Mem Acc Lincei, Cl. sc. fis. e mat., s. 5ª, XIII

Morgan HV (1914) Vitruvius. The ten books on architecture: Translated by Morris Hicky Morgan. Oxford University Press, London, Humphrey Milford

Oechslin W (1981) Geometry and line. Vitruvian science and architectural drawing. In Daidalos n. 1, Berlin Sept 1981

Panofsky E (1953) Galileo as critics of the arts. The Hague

Parronchi A (1964) "Le due tavole prospettiche del Brunelleschi. In: Paragone, IX(107):3–32 (1958); X(109):3–31 (1959) (ripubbl. in Studi su la dolce prospettiva, Milano)

Pellati F (1921) Vitruvio e la fortuna del suo trattato nel mondo antico, in Riv. di filologia, XLIX:305 ss

Portoghesi P (1965) "Infanzia delle macchine", Roma

Saalman H ed (1970) "The life of Brunelleschi by Antonio di Tuccio Manetti", University Park and London

Sanpaolesi P (1951) "Ipotesi sulle conoscenze matematiche, statiche e meccaniche del Brunelleschi". In: "Belle Arti" pp 25–54
Svanellini P (1911) "Giovanni Branca (1571–1645) precursore di Watt e di Parsons", Arona
Schöller W (1989) "Le dessin a'Architecture á l'époque gotique», in AA.VV. «Le bátisseire del cathédrale gotiques», Strasburg
Shelby LR, Barnes CF (1988) The Codicology of the Portfolio of Villard de Honnecourt, Scriptorium 42, pp 20–48
Sgosso A (2000) "La geometria nell'immagine. Storia dei metodi di rappresentazione", vol 1, Utet
Tobin R (1990) Ancient perspective and Euclid's Optics. J Warburg Courtland Inst, 53:14–41 (London)
Vagnetti L, Marcucci L (1978) "Per una coscienza vitruviana. Regesto cronologico e critico delle edizioni del De Architectura". Studi e documenti di architettura VIII:11–184
Vagnetti L (1979) "De naturalis et artificiali perspectiva", Libreria Editrice Fiorentina, Fiirenze
Wittkower R (1953) Proportion in perspective. J Warburg Courtauld Inst XVI:275–291

Gerbert of Aurillac (c. 940–1003)

Carlo Bianchini and Luca J. Senatore

Abstract Gerbert of Aurillac represents one of the most relevant personalities of the European medieval culture, being a prolific scholar as well as an acknowledged teacher especially as tutor of Emperors Otto II and Otto III. A disciple himself of Atto, during his long and successful career, first as a teacher in Reim's Cathedral School, then as Abbot of the monastery of Bobbio, Archbishop of Ravenna and finally as Pope Silvestre II (999–1003), Gerbert always encouraged and promoted the study of the *quadrivium* (arithmetic, geometry, music and astronomy) also through the reintroduction to western Europe of ancient Greek-Roman scientific culture, especially in the augmented Arab versions. Gerbert's influence on western scientific thought refers not only to theory (i.e. the arabs' decimal numeral system or some of Euclid's theorems) being instead always balanced with practical applications that involve instruments (abacus, armillary sphere, astrolabe, etc.) and that immediately affect the lives of common people.

Even though the present study has been developed together by both authors, different authorships can be recognized within the paper. In particular the *Biographical notes* have been written by Luca. J. Senatore while the section dedicated to *Review of Main Works of Gerbertus* has been developed by Carlo Bianchini. All other parts have been written in common.

C. Bianchini (✉) · L.J. Senatore
Department of History Drawing and Restoration of Architecture, Sapienza—University of Rome, Piazza Borghese 9, 00186 Rome, Italy
e-mail: carlo.bianchini@uniroma1.it

L.J. Senatore
e-mail: luca.senatore@uniroma1.it

© Springer International Publishing Switzerland 2016

M. Cigola (ed.), *Distinguished Figures in Descriptive Geometry and Its Applications for Mechanism Science*, History of Mechanism and Machine Science 30, DOI 10.1007/978-3-319-20197-9_2

Biographical Notes

Born in Auvergne (a mountainous region in central France) in a period spanning from 938 to 950, Gerbert's origins are likely to have been humble. Around 963 he entered the monastery of St. Gerald in Aurillac, a rather strict Benedictine settlement and above all an institution independent from any local control, being subject, like Cluny, only to the Pope.

Here Gerbert starts his training in the *trivium* subjects (grammar, logic, and rhetoric) learning Latin grammar as well. In 967 the first turning point in his life: Count Borrell of Barcelona visits the monastery and, under request of the abbot, decides to take Gerbert back to Spain with him in order to study mathematics there. He had proven to be a particularly brilliant pupil deserving the opportunity of improving his knowledge through the study of the *quadrivium* (arithmetic, geometry, music, and astronomy).

Gerbert moved thus from France to Catalunya where he started to attend the cathedral school of Vic. Catalunya at that time represented, both culturally and politically, the western frontier of Christianity; a stronghold somehow against Arabian penetration but at the same time a privileged point for exchange and communication. Christian Europe was in fact considerably under-developed in comparison with the Muslims of al-Andalus who settled in southern Spain and thus communications (especially in the south-north direction) were generally tolerated and often encouraged.

Books were a highly precious part of this trade: through this channel actually a large part of the ancient Greek/Roman knowledge came back to western Europe together with Persian, Indian and even Chinese achievements, filtered and improved by Arab culture: Muslim "scientists" were highly regarded, and perhaps nowhere in Islam as much as in al-Andalus. Muslim astronomy was the most advanced in the world, and Muslim astronomers proficient in using the astrolabe had done much to map the skies. Although the names of modern planets and constellations are Latin, the names of most major stars—Altair, Deneb, Rigel, Sirius, Fomalhaut, Aldeberan, Betelgeuse—are Arabic as are many of the other terms of astronomy, such as azimuth, almagest, almanac, and the Zodiac. The Arabs were even further advanced in the realm of arithmetic. They had adopted the concept of zero from the Indians and used a positional numeric system much like the modern system—in fact, our numerals are based on the Arabic notation. They had also borrowed the abacus from the Chinese and were proficient in its use. They had gone beyond arithmetic, had established algebra, and were investigating prime numbers and coordinate equations. Their study of proportions made it possible for them to approach music in a quite precise manner, distinguishing accurately between notes, developing theories of harmonies and discords, and constructing musical instruments with quite accurate tuning.

For its geographical position and thanks to the farsighted policy of their bishops, the Vic's Cathedral library became one of the largest and best equipped in Europe, positively influencing the cathedral school itself.

Gerbert took full advantage of these opportunities, quickly showing abilities and knowledge so far beyond average that his name soon became legendary, sometimes also in a negative way as a magician or worse as a devil's disciple.

In 969 Count Borrell and Atto, the bishop of Vic, traveled to Rome on a pilgrimage, taking Gerbert with them. Pope John XIII remained very impressed by the young monk and persuaded emperor Otto I to take Gerbert on as tutor for his young son, the future emperor Otto II. Gerbert attended to his duty for some years and finally the emperor let him move to the famous cathedral school of Reims to study advanced logic. Here he had the opportunity to show all his abilities: not only in the study of arithmetic (mastering arabic numerals he could use quick mind calculation to solve problems extremely difficult with roman notation) but also in applying his wide knowledge to practical issues and mechanical equipment.

A masterpiece from this standpoint is an organ with constant pressure supplied by water power he is said to have built in Reims: not only did Gerbert's machine provide an extended steady level of sound, but its pipes were also mathematically calibrated so that its harmonics were astonishingly superior to anything heard in Western Europe before. The abacus has certainly been one of the main instruments for his work (he actually built a giant one marking out the floor of the nave of the cathedral of Reims) and his book on the subject became a sort of standard in the new cathedral schools that were quickly populating the West. During a visit to Ravenna he met again Otto's son reviving their old acquaintance: a crucial episode for Gerbert, for in 983 his old pupil became Holy Roman Emperor with the name of Otto II. The emperor appointed his teacher as abbot of the famous Abbey of Bobbio (its library was one of the largest in Western Europe) and count of the correspondent district.

As a new abbot, Gerbert tried right from the start to invert the decay tendency of the monastery, too long lacking for strong guidance and under pressure by local nobles eager for its lands. Unfortunately, the death of Otto (983) led to new balances that forced him to move from Bobbio and return once more to Reims: this time, though, as master of the cathedral school and secretary to Adalberon, the archbishop. This new role represents for Gerbert an occasion to get involved in politics that, at that time, meant the contrast between the German Saxons and the Carolingians for the throne of France.

Counting on the weakness of the Empire due to the very young age of emperor Otto III (born in 980) and on the internal contrasts for the regency, Lothair made an attempt to conquer Lorraine (985): on that occasion Gerbert and the archbishop of Reims sided firmly with the count of Paris, Hugh Capet who, after the death of Lothair and of his son, became the first Capetian king of France.

At the death Adalberon (989), even if Gerbert seems to be a natural choice for succession, the king appoints Arnulf, an illegitimate son of Lothair. This episode represents the beginning of a very difficult period: the archbishopric of Reims, with the help of Arnulf, becomes the theatre of the conflict between Duke Charles (the Carolingians' heir) and Hugh Capet and Gerbert himself has to take refuge at the court of the king.

Although in 991 Arnulf's treason was declared and Gerbert appointed as archbishop, his nomination was harshly criticized and finally revoked by Pope Gregory V in 997. The deposed archbishop moves then to the court of Otto II, where he becomes the teacher and advisor of the young Otto III.

Gerbert reaches his new pupil in Ravenna, being appointed archbishop (998) and then, at the death of Gregory V, moves to Rome where the emperor imposes him as Pope Sylvester II in 999.

The Roman populace did not accept a foreign pope, and in 1001 both Otto and Gerbert were forced to return to Ravenna. He will come back only in 1003, the year of his death. Despite his brief reign, Gerbert supported the *renovatio imperii* that intended to regain for Rome the ancient classical values. He sensed the strategic importance of the eastern European countries (in particular Poland and Hungary) promoting their Christianization. Finally, he always defended the autonomy of the Pope from the Emperor's influence.

Gerbert's Work

Gerbert was undoubtedly one of the most relevant writers in the early Middle Ages. Even if is quite difficult to provide in this occasion a systematic classification of his works, we can divide his writings according to the specific character of each one:

- Mathematical writings
  - *Liber abaci*
  - *Libellus de numerorum divisione*
  - *Regula de abaco computi*
  - *De geometria*
  - *Libellus de rationali et ratione uti*
- Ecclesiastical writings
  - *Selecta e concil. Basol., Remens., Masom.,* etc.
  - *De corpore et sanguine Domini*
  - *Sermo de informatione episcoporum*
- Letters
  - *Epistolae ante summum pontificatum scriptae*
  - *218 letters, including letters to the emperor, the pope, and various bishops*
  - *15 letters to various bishops, including Arnulf, and abbots*
  - *Epistolae et decreta pontificia*
  - *5 short poems*
  - *one dubious letter to Otto III.*

- Other
  - *Acta concilii Remensis ad S. Basolum*
  - *Leonis legati epistola ad Hugonem et Robertum reges*

Review of Main Works of Gerbertus

Even if Gerbert's production includes many ecclesiastical writings, the majority of his works actually aim at teaching the *quadrivium* topics: arithmetic, music, geometry and astronomy.

And the transmission of knowledge represents for Gerbert an absolute priority: certainly a means for education but also, more deeply, the privileged tool to penetrate the secret structure of reality and finally to understand the magnificence of the Creation.

In this framework, arithmetic represents the first step in this actual ascent, the key, somehow, to the other three disciplines both from a speculative and practical standpoint. Music cannot be handled without knowing the theory of numbers; on the other hand, proportions and calculations are unavoidable tools both for the problems of the *geometria practica* and for astronomy.

In this period (and for many centuries to come) arithmetic will continue to coincide with the science of numbers in the form handed down by Boethius, while the *practica mathematica* would indicate calculation methods.

This ancient roman philosopher represents Gerbert's principal mathematical source: not only does he adopt the structure and content of Boethius' *corpus*, but also enhances its original goal of preserving the *quadrivium* through its promotion as a "key tool" for the study of Christian theology.

But, at the turn of the first millennium, the boethian *Institutio Arithmetica* hands down to the western medieval culture also the Pythagorean concept of "number" strictly related to the profound structure of the Universe as well as to the most esoteric aspects of Greek culture. The graft of Christian vision on this tradition actually transforms numbers in the ordering principles of the Universe, each one referring to a specific aspect of creation harmony. The disclosure of the properties of numbers actually meant to access the secret rules used by the Creator himself.

Omnia in numero et mensura et ponder (Sapienza Liber XI, 21) is stated in the Bible: an idea already accepted by the patrology (i.e. Augustine of Hippo) to which Gerbert himself refers directly in the *Prologus* of his *Geometria.* Nevertheless arithmetic is also much more than a symbolic representation of reality: it can help in fact in enlightening how nature itself works in physical terms. This level of knowledge, though, is a real challenge for the human mind: it cannot be reached without continuous training and refining. Thus *Arithmetica* becomes also a powerful brain training method for education.

Theory and practice are faces of the same medal for Gerbert: a more profound understanding of numbers cannot be achieved without improving calculation methods: refining procedures but also innovating the instruments used for calculations. From this last standpoint the Scholastic of Reims could count at the same time on two unique opportunities: access to the richest Christian libraries and first-hand information coming from the Arab scientific culture. And Gerbert succeeds in merging together *arithmetica* and the *practica mathematica* introducing and widely promoting the use of the *abacus*, the first calculator the western European culture would have ever counted on for centuries. Even if it cannot be considered a pure invention of the future Pope Sylvester II, nevertheless the instrument described in many writings by Gerbert is quite different from the traditional Latin one essentially because it uses the Arabic positional notation based on 10 Figs. (1, 2, 3, 4, 5, 6, 7, 8, 9 and 10). No Arabic source actually reports a description of such an instrument, reinforcing the hypothesis that a deep renovation both of the calculator and of its use appeared at the end of the first millennium in the Christian mathematicians communities, Gerbert being the creator of this tool or at least its most active "popularizer". (Reims Abbey itself has for long been considered as the center of abacus spreading and its users, the abacisti, at that time were simply called *gerbertisti*).

The use of this new tool not only changed the way calculations were made, but also the way *arithmetica* itself was taught: the abstract study of the *codices* progressively gives place to a more inter-personal relationship between teacher and pupils, allowing these last actual apprentices to learn the use of the abacus through imitation of their teacher's model.

For more than 100 years Gerbert's abacus will represent the most powerful calculation system (hardware and software we would say nowadays) available in Christian western Europe. It will be overcome in the XII century only by the spreading of the algorithm, a method based on the tracing on a table covered with sand or dust of all the calculation steps using all the figures, including zero.

As mentioned above, *arithmetica* provides all necessary fundamentals to the other disciplines of the *quadrivium*. This idea, quite clear in all Gerbert's writings, has actually also been the leitmotif of its teaching activity. Even a discipline apparently quite far from mathematics like Music does perfectly comply with this "quantitative" approach: within the *quadrivium* program, in fact, theory of music and theory of numbers are tightly connected, so that sounds correspond to numbers, and intervals to number ratios. These principles, very well-known in Ancient times since Pythagoras, originated also the metaphysical interpretation of music illustrated by Plato in his Timaeus where the movement of celestial spheres produces the perfect music harmony of the universe.

At the turn of the first millennium, though, this knowledge was essentially lost in the Christian western Europe where the Gregorian chant, a form of monophonic, unaccompanied sacred song, seemed to be the only expression of music. Gerbert radically changes the teaching of Music in his school, again trying to put together theory and practical demonstrations in the same way we have seen about arithmetic calculations.

This time, instead of the abacus, we find a new instrument, the *monochord*: a single string, stretched over a sound box, is fixed at both ends while one movable

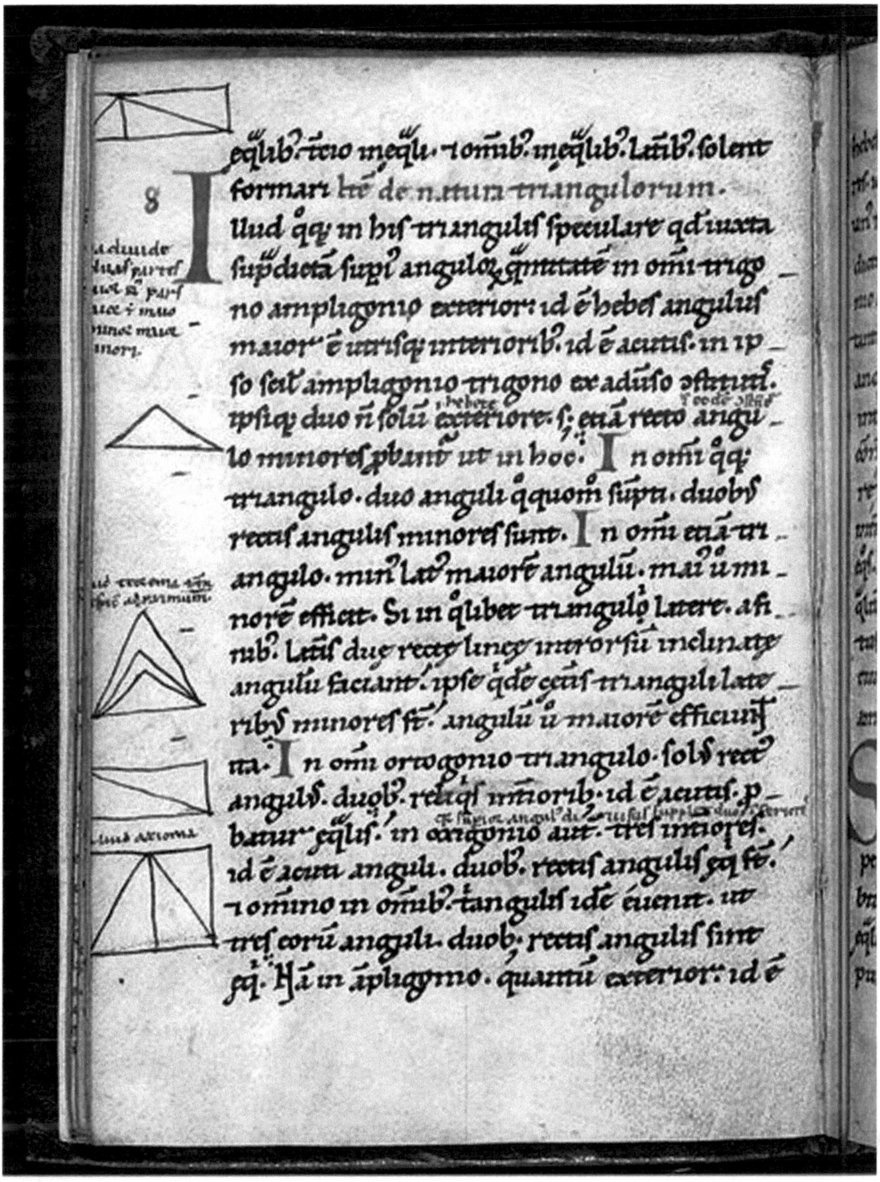

Fig. 1 Gerbert_d'Aurillac, De Geometria

bridge or weight is manipulated to demonstrate mathematical relationships between sounds. The chord can thus be divided according to simple ratios in order to precisely produce the corresponding note which clearly comes as a result of a theoretical reasoning but at the same time becomes immediately audible for any student. Where in a Gregorian chorus the master guided the chant arbitrarily

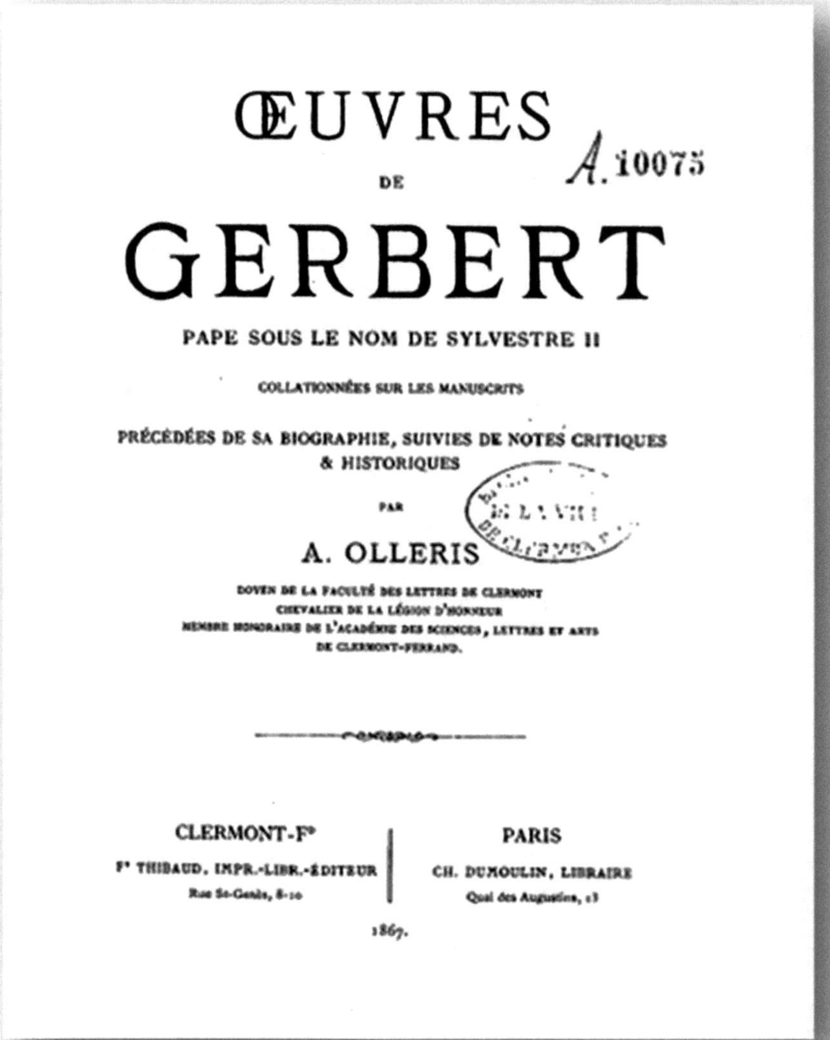
ŒUVRES

DE

GERBERT

PAPE SOUS LE NOM DE SYLVESTRE II

COLLATIONNÉES SUR LES MANUSCRITS

PRÉCÉDÉES DE SA BIOGRAPHIE, SUIVIES DE NOTES CRITIQUES & HISTORIQUES

PAR

A. OLLERIS

DOYEN DE LA FACULTÉ DES LETTRES DE CLERMONT
CHEVALIER DE LA LÉGION D'HONNEUR
MEMBRE HONORAIRE DE L'ACADÉMIE DES SCIENCES, LETTRES ET ARTS DE CLERMONT-FERRAND.

CLERMONT-Fd	PARIS
Fd THIBAUD, IMPR.-LIBR.-ÉDITEUR	CH. DUMOULIN, LIBRAIRE
Rue St-Genès, 8-10	Quai des Augustins, 13

1867.

A. 10075

Fig. 2 A. Olleris, Oevres de Gerbert, 1867

representing for all apprentices the reference for any note, the teaching based on the monochord would instead provide an independent and reliable means for tuning.

Unfortunately, only through indirect evidence can we affirm the usage of this object by Gerbert, but this hypothesis is in our opinion at least a quite reasonable one. But the future Pope's reputation in the field of music is mostly related to the construction in Reims of an outstanding pipe organ: an instrument with incomparable steady and harmonious sound thanks to the constant pressure provided by

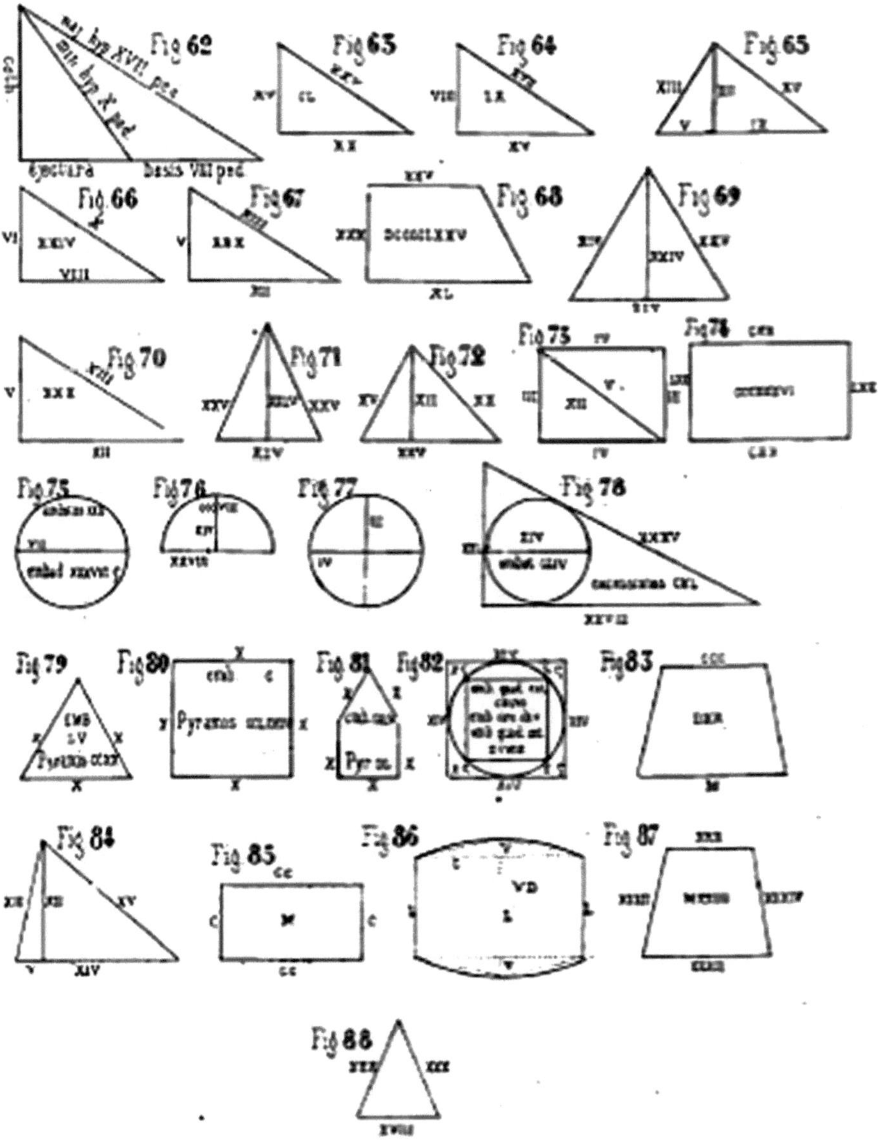

Fig. 3 A. Olleris, Oevres de Gerbert, 1867

water power and the perfect geometry of the pipes. Again it is not a technical problem for Gerbert: better, the technical problem is the final part of a complex process in which solutions are progressively tuned according to theoretical investigations and practical calculations. In the *mensura fistularum*, the work devoted to pipes dimensioning, the Scholastic of Reims discusses the applicability to the organ

Fig. 4 A. Olleris, Oevres de Gerbert, 1867

cylinders of the numerical ratios used for the monochord. There is evidently no automatic correspondence between the two: for a string, its length is the only variable to be considered in order to produce a certain sound, while in a cylinder its

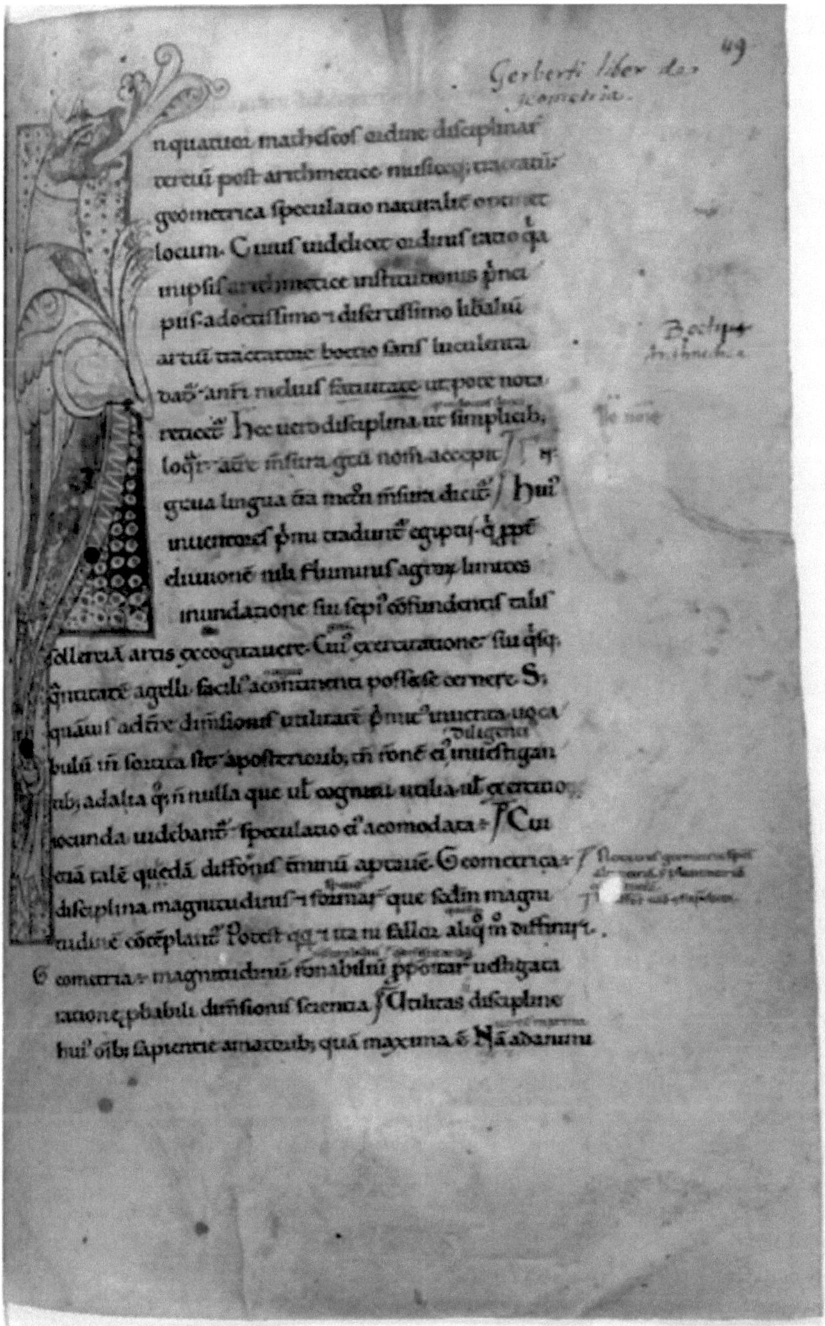

Fig. 5 Gerbert d'Aurillac, Liber des geometria

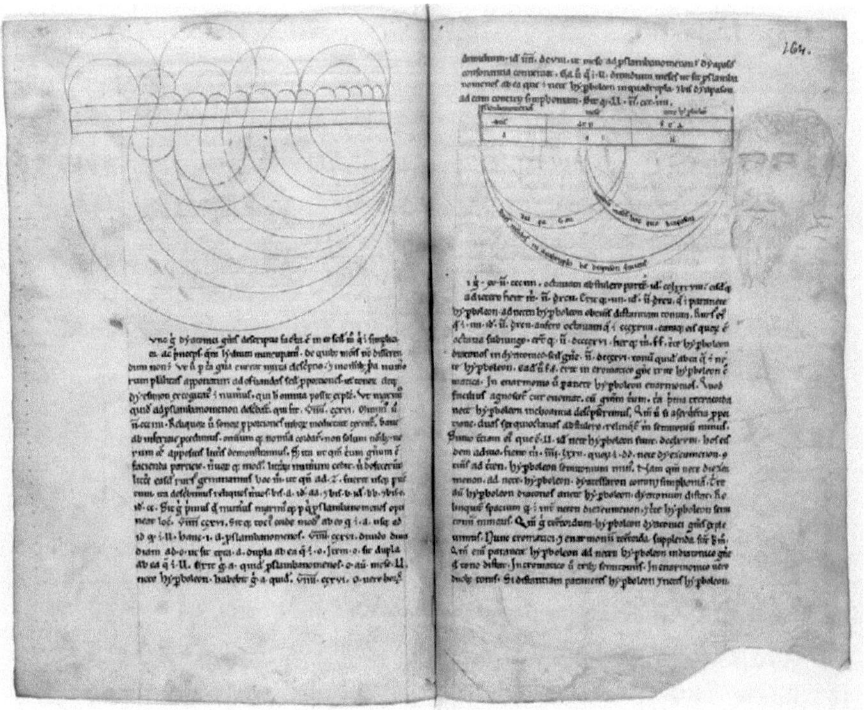

Fig. 6 Gerbert d'Aurillac, De Geometria

volume too must be taken into account. The "practical" rules of medieval *organa* constructors are thus completely bypassed and revised according to this rigorous vision. But Gerbert pushes his research far beyond, demonstrating with complicated calculations that the proportions suitable for cylinders receive the same ratios of the monochord and, finally, of the Universe itself.

Nam tempore nocturno ardentibus stellis operam dabat; agebatque ut eas in mundi regionibus diversis obliquatas, tam in ortu quam in occasu notarent… [In fact, he dedicated himself to the stars twinkling in the night and labored because (students) would notice, both in their rise that in their sunset, those declined in different regions of the sky]. Gerbert's astronomy is then essentially an observation activity. But in a period in which astronomy was often tangled with astrology, the future Pope addresses the discipline coherently with its "scientific" program in which theory and practice are faces of the same medal and instruments are the means to pass from one to the other. In this framework many objects have been traditionally reported as used in Gerbert's teaching. First of all the *spherae*: a *sphere of the stars* (made in solid wood and covered with leather, showing on its surface all constellations); the *armillary sphere*, one of the most famous astronomical objects since Greek antiquity (a spherical framework of rings, centred on Earth, that represent lines of celestial longitude and latitude and other astronomically relevant

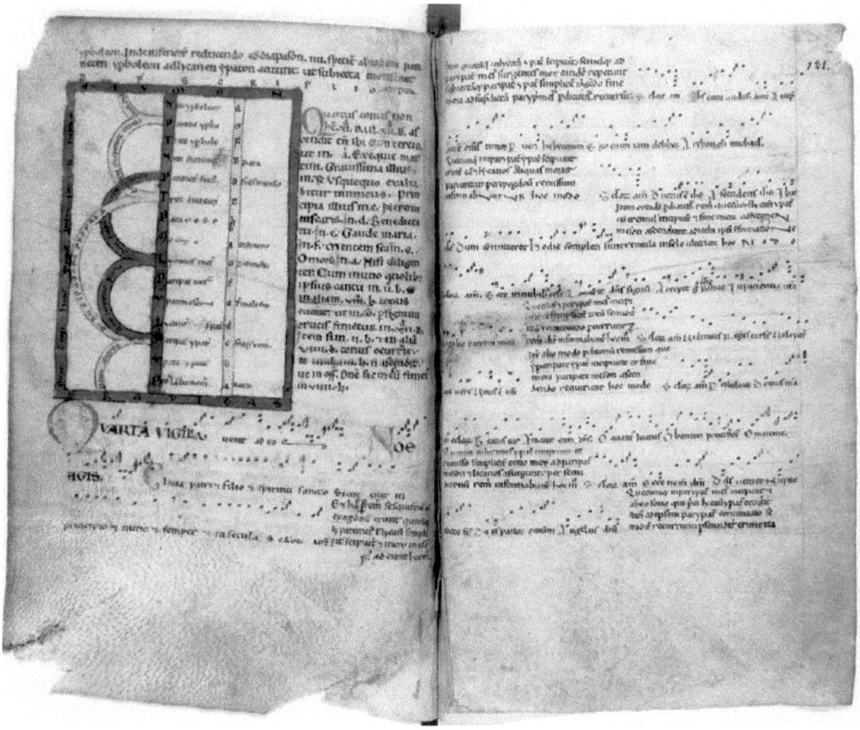

Fig. 7 Gerbert d'Aurillac, De Geometria

elements such as the ecliptic plane); the so-called *half-sphere* (made in solid wood showing on its surface poles, equator, tropics and one meridian) used for observing from its center the celestial bodies through holes (*fistulae*) positioned at the intersection between the parallels and the meridian; and finally a fourth sphere being a sort of fusion between an armillary sphere and the sphere of stars that, thanks again to a pointing system (from the poles to the Ursa Major) could lead to an alignment of the object coherent with the Earth inclination.

Much more debated is what role Gerbert could have played in the adoption and diffusion of the astrolabe in the western European culture. This fundamental astronomical instrument had again a Hellenistic origin but only with the Arabs did it reach its full maturity. At the end of X century it was quite common across the Arab world and it is thus probable that Gerbert too could have seen or even owned one during his stay in the *marca hispanica* or later. Nevertheless the simple possession of such an instrument does not mean the possibility of its comprehensive use: in fact without knowing at least the fundamentals illustrated in the Ptolemy's Planisphere an astrolabe would result in a mysterious object practically useless for most astronomical purposes. As all this information (both on the usage of the astrolabe itself and the Planisphere) would reach Europe only in the XII century

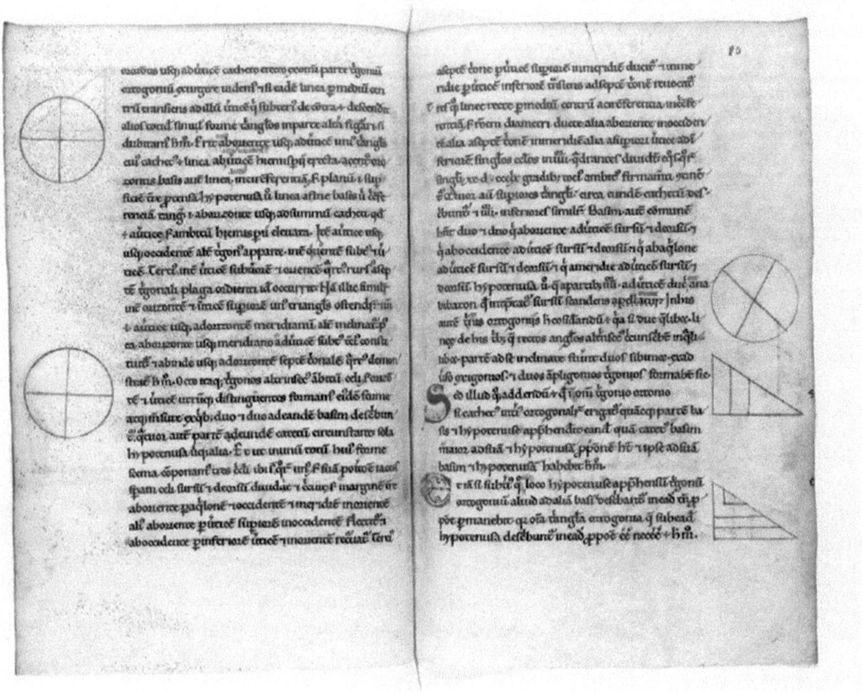

Fig. 8 Anicii Manlii Boëtii, De Arithmetica libri duo

with some systematic translation from the arabic, it is reasonable to imagine that the instrument was known by Gerbert but not completely understood unless for some geometrical application less relevant though for astronomy (i.e. indirect measuring). This hypothesis is also reinforced by the wide use of the *half-sphere* by Gerbert in his teaching activity. This object will in fact rapidly become obsolete with the introduction of the astrolabe (more precise and handy) and it would have been illogic for Gerbert to keep on using it as an alternative to the astrolabe.

Geometry completes the picture of the *quadrivium* disciplines. The content of Gerbert's teaching program actually coincides with the so-called *geometria practica*, an ensemble of geometric principles and propositions addressing the solution of some practical problems. Even if the term appears only in the XII century in the *Didascalicon* written by Hugh de Saint Victor aiming at distinguishing within the tree of medieval geometry two different branches, *geometria theorica* and *geometria practica,* (For more details refer to Bianchini 1994, 1995a, 1995b), nevertheless Gerbert's work represents an important step towards this crucial result. On one side, in fact, he contributes to the conservation and transmission of the roman geometric knowledge in the form established by Boethius to the Christian western culture; on the other, he sows the seeds of a new geometric vision that, trying again

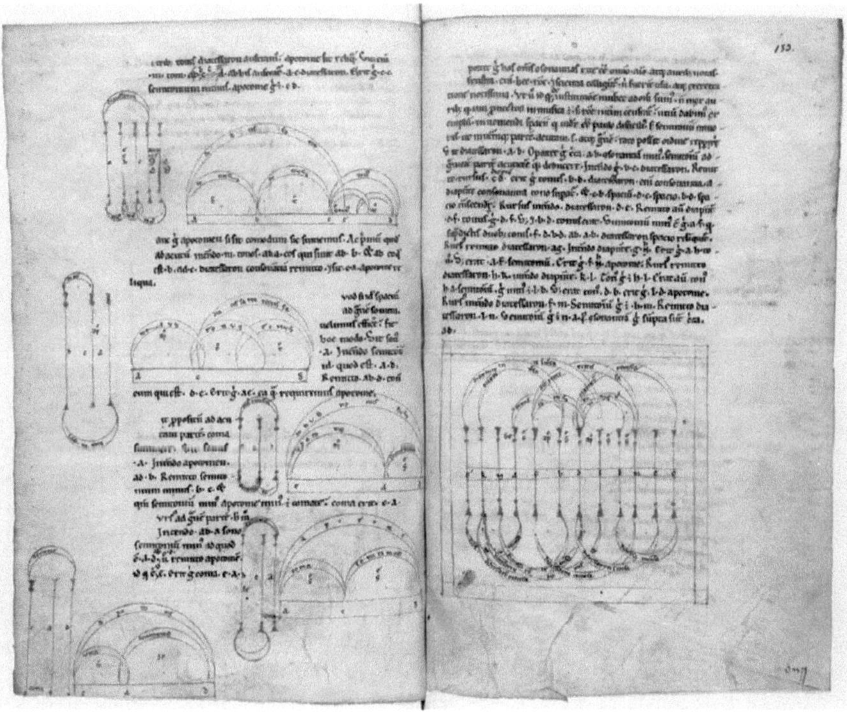

Fig. 9 Gerbert d'Aurillac, De Geometria

to merge together theory and practice, will finally disclose the possibility of repossessing all ancient greek knowledge and especially Euclid's Elements.

Quite apart from the still debated attribution of Gerbert's geometric writings (the so-called *Geometria Gerberti* (Materni 2008) and its probable "collective" character, still this work represents a significant example of that *sub-euclidean geometry* "in use among scholars who have not studied Euclid himself as their immediate source" (Evans 1976, p. 114). The structure of the treatise, quite original for the time, included: an introductive section listing some postulates, definitions and theorems mainly coming from the first four books of the Elements; a metrology section defining a collection of units of measure; methods for calculating the areas of polygons and circles; finally a part—probably apocryphal—concerning solutions to some practical geometric problems (i.e. the direct and indirect measuring of lengths and areas). Many different sources seem thus to have merged together in the *Geometria Gerberti*: certainly the Boethian tradition (cited by Gerbert itself); likely also many of the notions and problems coming from the *corpus agrimensorum romanorum* included in the *codex arcerianus* that the future pope could have come in touch with in the Bobbio Abbey; but certainly the new information he had learned during his spanish stay and, later on, from arabic sources. Actually the treatise reveals the very advanced geometric knowledge of its author: for instance,

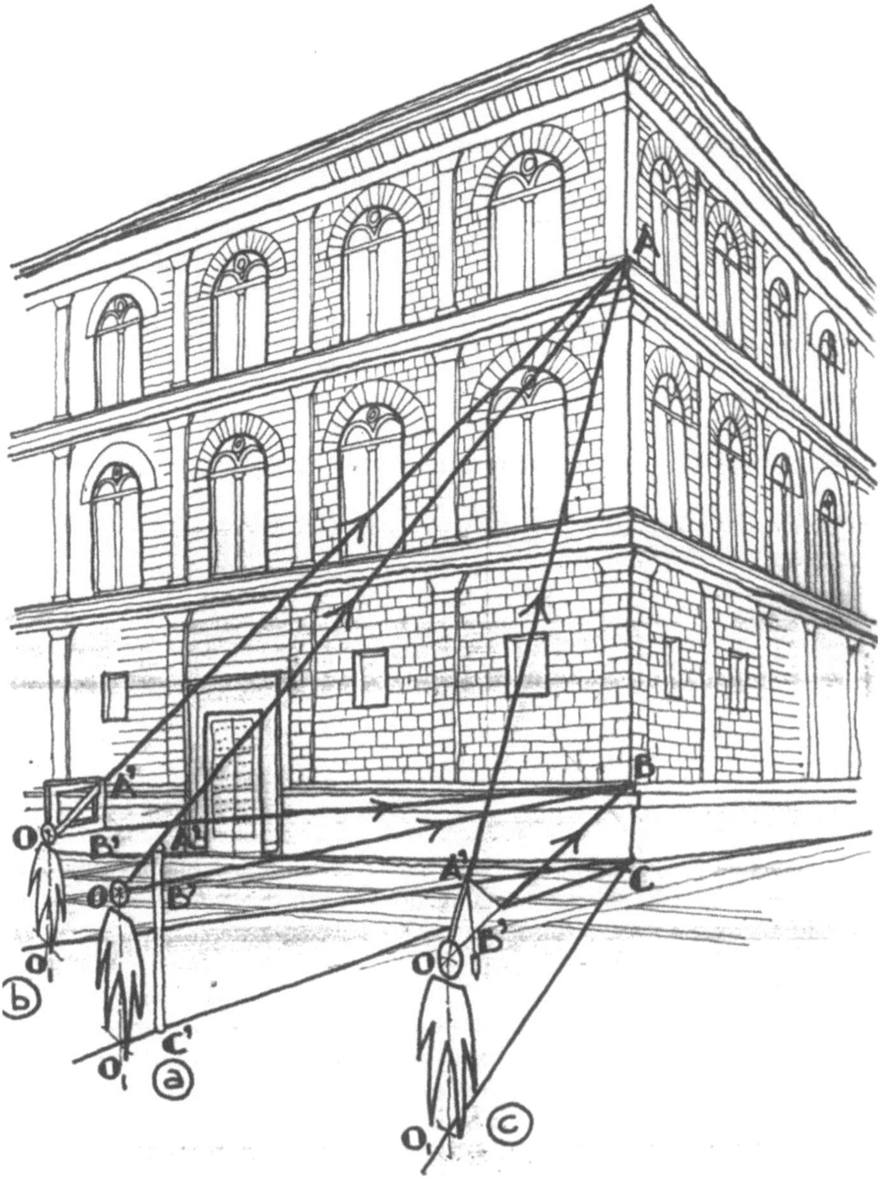

Fig. 10 The instrumental measure of a height according to Gerbert

Gerbert can easily deal with triangles using the Pythagorean theorem (not receding even when fractions are involved in calculations) when his contemporaries had difficulties even in understanding simple geometric definitions like "internal" and "external" angle of a triangle.

But also the probably apocryphal part of the *Geometria*, that recent investigations anyway ascribe to the closer circle of Gerbert's disciples (Jaquemard 2000), shows an original character in comparison with other medieval works. The practical problems illustrated in the different propositions are in fact solved, bending the abstraction of geometric principles to the needs of real life. And this transposition is put in practice thanks to the use of specific measurement instruments and procedures that, as expression of a rigorous geometric background, are capable of leading to more reliable and general results.

From this standpoint, the *Geometria Gerberti* is not only a catalogue of problems, but also an inventory of measurement instruments and techniques. Based on the properties of proportions between similar triangles, some typical survey problems are in fact solved in different ways. More than 20 propositions in *caput IV*, in fact, refer to the "instrumental" measurement of lengths, heights or depths. Instruments, though, can be very simple objects (a stick, arrow and string, a mirror) but also more refined ones, like isosceles, rectangles and magic (3, 4, 5) triangles, squares; finally we find the *horoscopus* (Probably an astrolabe or a quadrant) and the astrolabe.

Quite apart from the clear differences pertaining to instruments, the theory that ensures the correctness of results is more or less the same: the possibility of constructing a proportional relationship between two similar triangles, one on the instrument and the other "on site". In Fig. 10 we can see in fact in O an observer aiming at point A situated, for example, at the top end of the AC height to be measured. The visual ray OA identifies on the instrument the segment A'B' that, together with OB (the horizontal distance between O and the height AC conventionally measured) defines the proportion OB: O′B′ = AB:A′B′ and finally allows the calculation of AB. The final length of AC is then the result of AB + BC, height of point O from the reference plane (the ground). It is worth mentioning the procedure in which a reflecting surface is used (a mirror, a bowl with water). In this case, by choosing a specific point of the mirror to serve as a marker, the observer O reaches a position from which the upper A point is centered on the repair. Two similar triangles are thus constructed also in this situation: the former is made by the unknown height, its distance from the repair on the mirror and by the ray BA; the latter by the height of the observer, its distance from segment BA and the reflected ray. Quite apart from the construction (indeed a little naïve) the procedure implies some knowledge concerning optics (i.e. the linearity of light rays and the law of reflection) not quite usual during the Middle Ages.

Conclusions

Gerbert has clearly devoted his life to the praise of God: in this the enlightening of the perfect mathematical structure of the Creation represents an outstanding component. But his activity was not self-referencing but on the contrary constantly oriented to the spreading of knowledge actuated through the education especially of young generations.

In continuity with the ancient Latin tradition and the Boethius lesson, he devised several means (also practical) for teaching the fundamental *quadrivium* disciplines: the abacus; celestial globes; a hemisphere for observing stars and visualizing imaginary celestial circles; armillary spheres; the astrolabe, even if probably used only for measurements.

With his writings on geometry he tended to merge together the available Euclid's fragments, the knowledge of roman *gromatici* and some new notions acquired from the Arabs.

He showed a very profound acquaintance with music enlightening the close connection between numbers and notes and devising new ways to conceive and build instruments to demonstrate it.

Finally, he coherently went through all disciplines convinced of the substantial unity of knowledge based on mathematics.

For these reasons (probably more than for his ecclesiastic and political career), Gerbert remains a key figure of late 10th century and one of the most relevant scholars in all medieval culture.

Bibliography

Beaujouan G (1971) L'enseignement du quadrivium. In: La scuola nell'Occidente latino dell'Alto Medioevo, Spoleto, CISAM 1971, pp 639–667

Bianchini C (1994) Conservazione e sviluppo delle conoscenze geometriche durante il medioevo: il ruolo della geometria pratica. In: XY dimensioni del disegno, 21–22/8, Officina Edizioni, Roma, pp 55–59

Bianchini C (1995a) Conservazione e sviluppo delle conoscenze geometriche durante il medioevo: il ruolo della geometria pratica. Ph.D. thesis

Bianchini C (1995b) Teoria e tecnica del rilevamento medievale. In: Disegnare idee immagini, nn° 9–10, Gangemi editore, Roma

Bobnov N (1898) Gerberti Opera Mathematica, Berlin (ried. Hildesheim 1963), pp 48–97

Charbonnel N, Iung JE (ed) (1997) Gerbert L'européen, Actes du colloque d'Aurillac (Aurillac, 4–7 juin 1996) (Société des lettres, sciences et arts "La Haute Auvergne", Mémoires 3), Aurillac

Cigola M, Ceccarelli M (1995) On the evolution of Mechanisms drawing. In: Proceedings of IXth IFToMM world congress, vol. 4, pp 3191–3195, Politecnico di Milano

Cigola M, Ceccarelli M (2001) Trends in the drawing of mechanisms since the early middle ages. J Mech Eng Sci 215:269–289. Professional Engineering Publishing Limited, Suffolk

Cigola M (2012) In praise of parallel theories: Descriptive geometry and applied mechanics. In Carlevaris L, Filippa M (eds) In praise of theory. The fundamentals of the disciplines of representation and survey pp 39–46. Roma Gangemi editore

Evans G (1976) The 'Sub-euclidean' Geometry of the earlier middle ages up to the mid-twelfth century. Arch Hist Exact Sci 16(2):105–118

Flusche AM (2005) The life and legend of Gerbert of Aurillac: the Organbuilder who became Pope Sylvester II, New York

Frova C (1974b) Trivio e Quadrivio a Reims: l'insegnamento di Gerberto d'Aurillac, Bullettino dell'Istituto storico italiano per il Medio Evo n. 85, 1974–1975, pp 53–87

Hock KF (1846) Silvestro II Papa ed il suo secolo, Milano
Levet J-P(1997a) Gerbert. Liber Abaci I (Cahiers d'histoire des mathématiques et d'épistémologie), Poitiers
Materni M (2008) Attività scientifiche di Gerberto d'Aurillac. In: Archivum, I
Migne JP (ed) (1853a) Gerbertus, Geometria Gerberti. In: Patrologia Latina CCCXXXIX, Paris
Migne JP (ed) (1853b) Gerbertus, De rationale et ratione. In: Patrologia Latina CCCXXXIX, Paris
Nuvolone FG (2001) Gerberto d'Aurillac da abate di Bobbio a papa dell'anno 1000, Atti del Congresso Internazionale (Bobbio, Auditorium di S. Chiara, 28–30 settembre 2000) (Archivum Bobiense - Studia 4), Bobbio
Nuvolone FG (2008) Nuvolone, Zh/sej, he tu viva!. Dall'eredità scientifica pluriculturale della Catalogna, ai risvolti contemporanei (Archivum Bobiense 29), Bobbio
Olleris A (1867) Olleris, Oeuvres de Gerbert, pape sous le nom de Sylvestre II… / précédées de sa biographie, suivies de notes critiques, historiques par A. Olleris, Paris
Pez (1721) Gerbertus, Geometria Gerberti. In: Thesaurus, III/2
Riché P, Callu JP (ed) (1993) Gerbert, Correspondance (Les Classiques de l'Histoire de France au Moyen Age 35–36), Paris
Riché P (1984) Riché, Le scuole e l'insegnamento nell'Occidente cristiano dalla fine del V secolo alla metà dell'XI secolo, Roma
Riché P (1985) L'enseignement de Gerbert à Reims dans le contexte européen. In: Tosi 1985a, pp 51–69
Riché P (1987) Gerbert d'Aurillac. Le pape de l'an Mil, Paris 1987 (ultima ristampa parzialmente riveduta, Paris 2006) (trad. italiana P. Riché, Gerberto d'Aurillac. Il papa dell'anno Mille, Cinisello Balsamo 1988
Riché P (2000) Le *Quadrivium* dans le haut moyen âge. In: Freguglia, pp 14–33
Sachs KlJ (1970–1980) Mensura fistularum. Die Mensuriering der Orgelpfeifen in Mittelalter, tomo I, Stuggart-Murrhardt 1970–1980, pp 59–72
Segonds APh (ed) (2008) Gerbert, Lettres scientifiques. In: Gerbert, Correspondance, II, pp 662–708
Tosi M (1985) Gerberto. Scienza, Storia e Mito. Atti del Gerberti Symposium. Bobbio 25–27 Luglio 1983, Archivum Bobiense Studia II, Bobbio
Zimmermann M (1997) La Catalogne de Gerbert. In: Charbonnel 1997, pp 79–101

Francesco Feliciano De Scolari (1470–1542)

Arturo Gallozzi

Abstract Francesco Feliciano De Scolari, also known as "Francesco Lazesio" or simply Lazisio (or Lasezio) in his native Lazise, a master of mathematics and expert surveyor who worked in Verona and other parts of the Italian peninsula in the late fifteenth and the first half of the sixteenth century, owes his popularity mainly to the famous treatise known by its original title, "Scala Grimaldelli". Possessing a wide range of technical skills, he covered many aspects of the engineering disciplines in his work. In addition to some brief biographical notes, this study will explain a few salient aspects of his published works, with particular attention to the use of the "surveyor's cross", which is described for the first time in print.

Biographical Notes

One of the major proponents of the principles of arithmetic, algebra and geometry put forward by Leonardo Fibonacci (1170–1240) and Luca Pacioli (1445–1517) at the turn of the fifteenth and sixteenth centuries, Feliciano De Scolari is remembered for his work as an arithmetic master and land surveyor and for the extraordinary success of his published treatise known as the "Scala Grimaldelli".

De Scolari was born around 1470, at Lazise on Lake Garda, in the province of Verona. There is little information about his family, of which only the name of his father, Domenico, is known, and the biographical details supplied by the author himself in his first work (Fig. 1), the "Libro de Abbacho nuovamente composto per magistro Francescho da Lazesio veronese" [Book of Abacus newly compiled by master Francescho da Lazesio Veronese], edited in 1517 and published in Venice on behalf of Nicolò Aristotele de' Rossi (1478/1480-active until 1544), known as Zoppino and "mister Vincentio his partner" in which the final pagereads:

A. Gallozzi (✉)
DICeM—Department of Civil Engineering and Mechanics, University of Cassino and Southern Lazio, via G. Di Biasio, 43, 03043 Cassino, Italy
e-mail: gallozzi@unicas.it

© Springer International Publishing Switzerland 2016
M. Cigola (ed.), *Distinguished Figures in Descriptive Geometry and Its Applications for Mechanism Science*, History of Mechanism and Machine Science 30, DOI 10.1007/978-3-319-20197-9_3

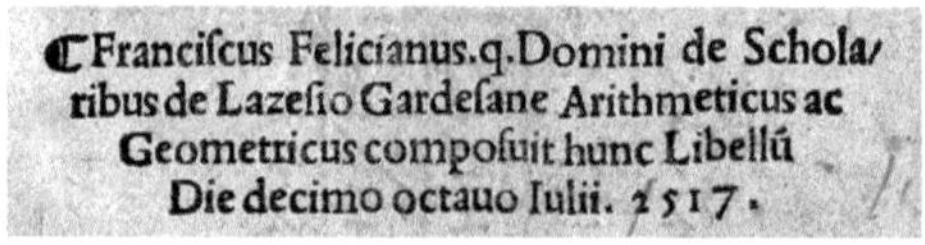
Francifcus Felicianus.q.Domini de Schola/
ribus de Lazefio Gardefane Arithmeticus ac
Geometricus compofuit hunc Libellũ
Die decimo octauo Iulii. 1517.

Fig. 1 Information in the colophon of the "Libro de Abbacho" 1524 edition, where the full name of Francesco Feliciano De Scolari da Lazesio appears, and the original date that the book was written, 1517

"Franciscus Felicianus, q. Dominici de Scholaribus de Lazesio Gardesane Arithmeticus ac Geometricus composuit hunc Libellum Die decimo octavo Iulii 1517". This is the only explicit reference that Master Francesco Feliciano makes to his lineage.

His childhood was spent in Lazise, after which he moved to Verona, where he remained for the rest of his life. He first took lodgings in the San Marco district, later moving to Santa Cecilia.

In 1533 he was appointed master of arithmetic in Verona, where he taught at the Palazzo della Ragione, in Piazza dei Signori. Although not documented at the time, it is likely that he occupied this position from the early sixteenth century. It was during this time, in fact, that De Scolari presumably succeeded to the teaching role vacated by Master Battista. The latter had succeeded his father Baldassarre, who founded the tradition of Veronese teachers, replacing the former Tuscan teachers who had long been in tenure (Belloni Speciale 1991, p. 347).

Following the established tradition, the role was passed on a few years later to De Scolari's nephew, Giacomo or Giacobbe (Garibotto 1923, p. 324). It was the nephew who cited the work of his uncle in an application submitted to the City Council of Verona, in which he requested that his teaching mandate be renewed following the incident in 1541 that destroyed Piazza delle Erbe and much of nearby Palazzo della Ragione, reminding them that it was his uncle who had been the previous incumbent of the post: "…voles levar scola nel palazzo … si fece un seraglio di asse e travi a tutte sue proprie spese, et convene di pagarle fino da sei lire l'anno & ivi insegnava leggere, scrivere, sumare, sottrarre e far conti d'ogni sorte, et altro che apertiene alle mathematice, cosa certo utilissima et necessaria in una città …" [he had a school in the palace…he built a space out of boards and beams at his own expense…and there taught reading, writing, addition and subtraction, all kinds of accounting and everything related to mathematics; a useful and essential resource for a city…] (Cavazzocca Mazzanti 1909, p. 16).

In addition to his principle activity of teaching, Feliciano, like many of his "master of arithmetic" colleagues, performed other professional duties to supplement his teacher's salary, making full use of his technical skills. This work mainly took the form of valuations, estimates and accounting checks for private and commercial companies, as well as frequent work in surveying and measurement. In relation to the latter, as De Scolari noted in the preface to his "Libro di aritmetica", he had the opportunity to travel: "… quasi per tuttal'Italia a misurare in compagnia, e senza compagnia de misuratori, terre, feni, biave, vini, muri, boschi, paludi,

&livelar acque, e simil cose…" […almost throughout the whole of Italy to measure, either alone or in the company of other technical surveyors, land, hay, corn, wines, walls, woods, marshes, drainage plateaux and such like…] (Feliciano 1536, c 1 verso).

A leading member of the cultural milieu of the time, Francesco Feliciano did not shrink from routinely pitting himself against his colleagues on scientific issues, not only to test their skills and potentially enhance his own, but also to gain publicity for his activities, which resulted in cash prizes and the acquisition of new students or new professional assignments. In any case it was expected, since provision was made in a section of the ancient statutes of the Veronese city: "Every professor of any Liberal Arts salaried by the City of Verona, is constrained to and must enter a challenge every month of the winter to determine it" (Zagata 1745, p. 216). The mathematical and geometric problems set were often actual scientific challenges that reflected the rules of chivalry. None of the contenders, in fact, could set his opponent any question that he himself was not able to answer. Testimony of this activity is found in the confrontations he had with Niccolò Tartaglia (1499–1557), who was also active in the first half of the sixteenth century in Verona, where he taught in the other public school of arithmetic in the city, located in Piazza delle Erbe.

Nicolò Fontana, who was called Tartaglia because of his stammer, which was caused by a throat injury that he sustained during the sacking of Brescia in 1512, is among the leading scholars of cubic equations. The contributions he made to the application of ballistic science are also significant. In his work "Quesiti, et inventioni diverse" [Queries and various ideas], the first edition of which was published in Venice in 1546, and in which he collected all the scientific issues related to more complex equations, among the many that were proposed, Tartaglia includes—in the ninth edition—three questions submitted by De Scolari (Fig. 2), the first of which dates back to 1521; the other two to 1526 (cc 98–100). He also cites a question by a certain "Master Antonio Veronese, son-in-law of master Francesco Feliciano", proposed in 1527. This fact is interesting because it adds another element to Feliciano's family history, since he must have fathered a daughter who married a man with whom, presumably, the scholar shared an interest in mathematics.

The problems put forward by De Scolari in Tartaglia's Book IX form part of a line of research that was quite popular at the time, which falls into the controversy about the priority of the resolution of the cubic equation, the start of which can be traced back to the disputes that occurred in the early sixteenth century among Tartaglia, Giovanni de Tonini da Collio (16 C.), Antonio Maria del Fiore (15–16 C.) and Girolamo Cardano (1501–1576). In particular, the questions posed by Feliciano concern three problems related to quadratic equations while his son-in-law proposes a geometrical problem that leads to a quartic equation reducible to a second-degree equation.

Tartaglia also takes up the theme in his "General trattato di numeri et misure" [General treatise on numbers and measures] (Venice 1556–1560), probably in the wake of Feliciano's treatise on the surveyor's cross—the first printed essay on the

LIBRO NONO DELLI QVESITI, ET INVENTIONI DIVERSE DE NICOLO TARTALEA BRISCIANO SOPRA LA

QVESITO PRIMO FATTO DA MAESTRO Francesco Feliciano l'anno. 1521. in Verona.

QVESITO SECONDO FATTO DA VN fra Raphaelle de. S. Zorzi de Verona l'anno. 1524.

NONO 100

QVESITO OTTAVO FATTO DA MAESTRO Francesco Feliciano l'anno. 1526. in Verona.

QVESITO NONO FATTO DAL MEDESIMO maestro Francesco Feliciano l'anno. 1526.

Fig. 2 Tartaglia, Quesiti, et invenzioni diverse, Venice 1546, carte 98 and 100 of the ninth book. Questions one, eight and nine proposed by Francesco Feliciano, one in 1521, eight and nine in 1526

use of this instrument. He covers it in his third book, "Dell'istromento materiale, necessario a' misuratori di terreni, chiamato squadro, come se fabbrichi e si conosce se sia giusto" [On the physical instrument necessary for the measurement of land, called the suryeyor's cross, how to use it and know whether it is right]. He reaffirms the importance of angles about 30 years after the publication of "Scala Grimaldelli", which have not been satisfactorily referenced by scholars of applied mathematics, and offers a vast case history of practical applications, providing directions on how to construct them.

A register of living and dead persons in St. Cecilia, which is preserved in the St. Anastasia archives in Verona, is usually referred to for the date of Feliciano's death, recorded as October 10, 1542 and referred to as the death of "Maestro Francesco de abacho" (Belloni Speciale 1991, p. 349).

Major Published Works

The need to dissect the complexity of the mathematical and algebraic procedures that underpin trade, and to codify the principles and promote the instruments governing land surveying, form the basis of the two publications for which Feliciano achieved fame in the field of applied mathematics: his "Libro de abbacho", the first edition of which dates back to 1517, and in particular his "Libro di arithmetica & geometria speculative & praticale … intitulato Scala Grimaldelli" [Book of arithmetic and practical and experimental geometry entitled Scala Grimaldelli] published from 1526/27.

The two works were highly successful, as evidenced by the numerous reprints and editions that were released, especially for the "Libro di arithmetica" which was also published posthumously a number of times with contributions from other authors and circulated over a period of about 200 years.

The first edition of "Libro de Abbacho" was published in Venice and is traceable back to 1517, although the date printed on the endpage is 1518; later editions were reprinted in 1519 and 1524, and a final edition, revised and expanded by the author in 1532, was published posthumously in 1550. All editions were published in octavo and printed in Venice, a major publishing centre at that time with about two hundred active printers.

Feliciano's second work was essentially a review and significant expansion of the first. The editions of the "Libro di arithmetica & geometria…" published in the sixteenth century were also printed in Venice, in 1526, 1527, 1536, 1545, 1550, 1560, 1561, 1563 and 1570. All volumes were published in "quarto" with the exception of the 1563 edition which was published in "ottavo". Other editions were also published in the seventeenth century, notably: 1602, printed in Verona, 1629 in Padua and finally 1669 and 1692 both in Venice (Smith 1908, pp. 146–148). The seventeenth century editions are marked by a change in the position of the book's title. The term "Scala Grimaldelli", by which the treatise was now commonly known, appears at the top of the page, and not at the bottom. Additionally, all of the later editions (Fig. 3) were supplemented and expanded by Master Filippo Macario Veronese (16–17 C.), "Rasonato pubblico" [well-known accountant]. Macario, as stated on the title page, belonged to the "Collegio dei Rasonati" (the accountants of the day) founded in Venice in 1581. The Rasonati were the principle users and popularisers of the double-entry accounting method invented by Luca Pacioli in 1494, which was used to record business operations in two sets of accounts simultaneously, becoming the most reliable and sought-after professionals in the management of public affairs. In "Scala Grimaldelli", in addition to reviewing and correcting some errors in the text, Macario enhances "Libro primo" with the addition of eight problems, among which is the so-called "Catain" rule—as in the Arabic usage—or rule of the false, which he developed, as in Fig. 5.

Finally, starting with the 1669 edition, they also contain a further addition of twenty-three "dottissimi problemi" [most learned problems] of algebra, placed at the end of the second book, and signed with the initials BB. Riccardi speculates that this may refer to the famous mathematician, engineer and poet Bernardino Baldi (1553–1617), considering that an ascription such as this would constitute recognition of an unprecedented contribution by the illustrious native of Urbino (Riccardi 1873, p. 23). No acknowledgment of this hypothesis is made in any literature, however, and it seems even more unlikely in view of the chronological disparity between the date of publication and the period of Baldi's life.

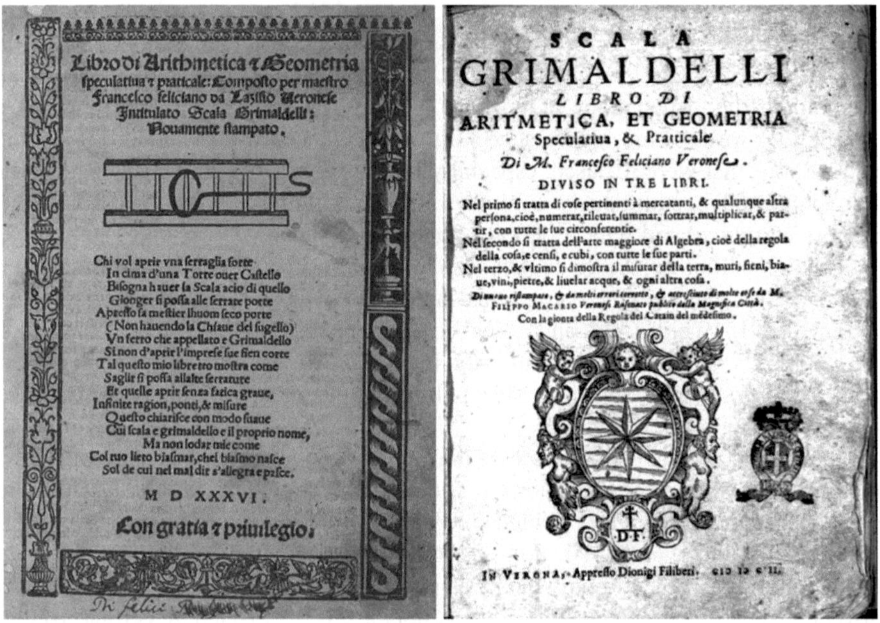

Libro di Arithmetica & Geometria speculatiua & praticale: Composto per maestro Francesco feliciano da Lazisio Veronese Intitulato Scala Grimaldelli: Nouamente stampato.

Chi vol aprir vna serraglia forte
In cima d'una Torre ouer Castello
Bisogna hauer la Scala acio di quello
Gionger si possa alle serrate porte
Apresso fa mestier lhuom seco porte
(Non hauendo la Chiaue del sugello)
Vn ferro che appellato e Grimaldello
Si non d'aprir l'imprese sue sien corte
Tal questo mio libretto mostra come
Saglir si possa allalte serrature
Et quelle aprir senza fatica graue,
Infinite ragion, ponti, & misure
Questo chiarisce con modo suaue
Cui scala e grimaldello e il proprio nome,
Ma non lodar mie come
Col tuo lieto biasmar, chel biasmo nasce
Sol de cui nel mal dir s'allegra e pasce.

M D XXXVI.

Con gratia & priuilegio.

SCALA GRIMALDELLI

LIBRO DI ARITMETICA, ET GEOMETRIA

Speculatiua, & Pratticale

Di M. Francesco Feliciano Veronese.

DIVISO IN TRE LIBRI.

Nel primo si tratta di cose pertinenti à mercatanti, & qualunque altra persona, cioè, numerar, tileuar, summar, sottrar, multiplicar, & partir, con tutte le sue circonferentie.

Nel secondo si tratta dell'arte maggiore di Algebra, cioè della regola della cosa, e censi, e cubi, con tutte le sue parti.

Nel terzo, & vltimo si dimostra il misurar della terra, muri, fieni, biaue, vini, pietre, & liuelar acque, & ogni altra cosa.

Filippo Macario Veronese

Con la gionta della Regola del Catain del medesimo.

D.F.

In Verona, Appresso Dionigi Filiberti.

Fig. 3 Frontispiece from the "Libro di Arithmetica et Geometria speculativa et praticale" published in 1536 in Venice by the Bindoni and Pasini presses (*left side*) and the expanded edition by Filippo Macario, printed in Verona in 1602 by the Angelo Tamo press (*right side*)

The "Scala Grimaldelli" was very popular as a textbook for arithmetic schools; produced in a classic type-set, it was well-established both in its construction and its content. Feliciano himself makes several quotes from "Libro de abacho" by Pietro Borghi (15 C.-1494) as well as explicit reference to "Frate Luga de Borgo" (Luca Piacioli di Borgo San Sepolcro) mostly in corroboration of these theories, but sometimes to refute them.

In the third volume of his "Bibliotheca mathematica" (Leipzig 1803), Murhard attributes a work entitled "Nuovo Libro di Statica e Meccanica" [New Book of Statics and Mechanics] to "M. Francesco Feliciano" published in octavo in Venice in 1667 (Murhard 1803, p. 10). There is no other reference to this treatise in any other record, and only Ricciardi quotes this reference (Riccardi 1873, p. 23).

The "Libro de Abbacho" and the "Libro Di Arithmetica E Geometria Speculativa et Praticale"

The books of arithmetic and the subsequent works that were inspired by them make up a significant section of the Italian mathematics output from the thirteenth century until well into the seventeenth, edited largely in imitation of the "Liber Abaci"

(1202) written in Latin by Leonardo Pisano—the aforementioned Fibonacci—who propounded the algebraic thinking that forms the basis of modern science and engineering. These introduced the positional system into the West, repeating many of the teachings of al-Khwarizmi (780–850) and other Arab mathematicians, wherein the value of a number is based on the position occupied by each digit and its Indo-Arabic numeral that finally supplanted the cumbersome Roman numeral system. The works that ensued, mainly written in the vernacular, formed general background reading for the abacus [arithmetic] schools, where techniques for performing arithmetic operations, rules for calculating area and volume, the basics of practical geometry and the resolution of business and finance issues were studied, and to which a large number of future "technicians" were drawn (Devlin 2013, pp. 21–39).

On the title page of "Nova Scientia" by Niccolò Tartaglia, published in 1537 in Venice, there is a symbolic representation of the concept of knowledge as perceived in the first half of the sixteenth century, in which mathematics has an important role. The throne of the kingdom of knowledge, occupied by philosophy, is guarded by Aristotle and Plato, waving the motto of his Academy ("Nemo huc geometrie expers ingrediat"—Do not enter, those who are not mathematicians), while at the door of the underlying broad outer limits of knowledge is Euclid, the author of "Elements". Below this, in the enclosure of knowledge, in addition to firearms symbolising the new science of the motion of projectiles, there are the four arts of the quadrivium, with arithmetic and geometry at the forefront of the other disciplines, including architecture, mechanics, geography, etc. (Fig. 4). Similarly, a previous symbolic representation contained in the work "Aepitoma omnis phylosophiae alias margarita phylosophica tractans de omni genere scibili" (Strasbourg 1504) by Gregor Reisch (1470 c.-1525), and significant to the understanding of mathematics, illustrates the argument between abachists, represented by Pythagoras, and algorithmists represented by Boethius (often cited by Feliciano—Scala Grimaldelli 1563—c 2 recto) in allegorical form. Arithmetic is personified as a female figure, and by the look she is casting at Boethius (475–525) on her right, one can argue the preference for calculation using the new Indo-Arabic numerals, which is also confirmed by the sad expression of Pythagoras (570 BC–495 BC), compared to the composure of Boethius.

In this context, the abacus schools are situated in an environment combining two distinct cultural traditions of different degree and order. The first is the knowledge of "scholars": this uses the Latin language and is found mainly in universities and in the Renaissance courts; the second is the "intermediate cultural stratum" to which the "technicians" belong (masters of arithmetic, surveyors, artisans, merchants, architects, engineers, plumbers, mechanics, etc.); consumers and often authors, in the vernacular, of many treatises on mathematics and practical geometry (Maccagni 1996, pp. 279–292), primarily related to teaching in abacus schools. The latter, which were by now well-established, were the major producers of mathematical works in the fifteenth and much of the sixteenth century and disseminated the basics of understanding in the field of arithmetic geometry and algebra at the "technician" level, allowing tradesmen, surveyors, engineers and the military to quickly solve practical problems relating to their professions without the need for a profound

Fig. 4 On the *left*, 1537 frontispiece from the "Nova Scientia" by Niccolò Tartaglia, a Renaissance portrayal of the concept of knowledge relating to mathematics—on the *right*, the 1504 "Typus Arithmeticae" engraving in "Margarita philosophica" by Gregor Reisch (comparing abachists and algorithmists)

understanding of mathematics. But it was precisely this fundamental characteristic of the schools—their strong orientation towards solving practical problems—that was one of their greatest limitations, encouraging rote learning of certain operations and procedures at the expense of a proper theoretical approach to the problem itself. The awareness of this fact defined the way in which mathematical disciplines were characterised as an area of independent knowledge from the start of the sixteenth century.

Feliciano's place, then, falls within a context of scholastic transformation which began two centuries earlier with Leonardo Pisano, and we can perceive in his works, therefore, his origins as a master of an abacus school and expert land surveyor.

The "Libro de Abbacho"

Analysis of the editions of "Libro de Abbacho" reveals that this was drawn up not only for use in schools, but was aimed primarily at the mercantile milieu, with its many different types of trade. The book, the first edition of which was printed by

Giacomo Penzio (operating between 1495 and 1527), Niccolò Zoppino and Vincenzo di Paolo (XV–XVI C.) in Venice in 1517, was reprinted on many occasions, as mentioned before, in 1518, 1519 and 1524; the first two editions published in Venice, the city of Zoppino and Vincenzo di Paolo, and the third by Francesco Bindoni and Maffeo Pasini, who also edited the numerous editions of "Scala Grimaldelli". Mazzanti (1909, p. 8) also refers to the 1550 edition, probably the final one, which was published under the title "Libro de Abaco Nuovamente Composto per maestro Francesco da Lazesio Veronese, intitolato Spasso de mercadanti, il quale insegna a fare molte ragioni, & come respondeno le monete, con alcuni giochi piacevoli per via de numeri, & il modo di mesurar le terre, & il modo de laureo numero e la pata, & la inditione delli nodari in perpetuo"; the volume, as reflected in the Boncompagni (1866, p. 275), is extended by four carte as opposed to the original twenty of the previous editions and was revised and expanded by the author in 1532.

The first edition of the "Libro d'abaco", as mentioned, as well as providing information on the author's family background, states the main purpose of the work, which was aimed chiefly at merchants, in a brief prologue, "Queste sono alcune regule breve facitile da imparare & utilissime & laudabile ad ogni persona: le quale acadeno a tuto homo de giorno I giorno come se pole in esse considerare & vedere &c..." [These are some easy rules to learn, useful and advisable to every person, for events that happen to all men every day, and how they can be viewed and considered etc.] (Feliciano 1524, c 6 verso).

On the title page of the work the same engraving is reproduced that was used for another famous treatise of that era: "Itinerario" (first edition 1510) by Ludovico de Varthema (1465/70–1517), in which the author describes his marvelous trips to Egypt, Syria, Arabia, Persia, India and Ethiopia (Fig. 5). The engraving shows the traveller marking the places on the globe that he has visited. The use of the same engraving on Feliciano's cover is confirmed by the fact that the engraver, identified by his initials (ZA) in the bottom left corner, was probably Giovanni (Zoan) Andrea Valvassore, called Guadagnino (active 1510–1573), at the Zoppino printshop in Venice, where the first edition of "Libro de Abbacho" was printed. In his professional life, Guadagnino took care of the illustration requirements for several Venetian printers, including the printing house of Giorgio Rusconi who was the editor of "Itinerario", and who had a close working relationship with Zoppino (Atzeni 2013, pp. 299–310). Additionally, the image borrowed by Feliciano was well-suited to his libro d'abaco, both for the main purpose of the manual, which was aimed mainly at merchants with transnational trade, and as an implicit reference to the dissertation on positional notation which had its origins in Indo-Arabic countries.

The typesetting for the treatise, which originally comprised twenty pages, as mentioned, follows classic manual styles for use in schools and for handbooks of a mercantile nature. Feliciano begins the book with a description of the positional Indo-Arabic numbers; four carte of multiplication tables follow, and a reminder of the casting out sevens and nines test. Then after this brief prologue he starts setting out a number of examples, mainly of a commercial nature, appropriately arranged

Fig. 5 The same engraving used in the "Libro de abacus" by Feliciano published in 1519 (*left*) and in the 1517 edition of "Itinerario de Ludovico de Varthema…" (*right*)

in order of increasing difficulty. He also describes a unique "Rota dominicale" [Sundays rota] for determining Easter Sundays, illustrated by an engraving that shows an alphanumeric circle superimposed on a numbered hand. Finally there are two rules for guessing numbers known only by the interlocutor, using an arithmetical sequence, as used by accountants to include recreational mathematics problems in their works. These, in general, were problems that could be solved with simple mathematical reasoning on various subjects and formulations "…se puol sapere quanti denari ha uno in borsa…" [if you want to know how much money there is an a bag], inserted as an amusement in order to intrigue and rouse the attention of students.

"Scala Grimaldelli"

De Scolari's reputation is, however, primarily linked to the treatise known as the "Scala Grimaldelli". The author explains the original title with a drawing on the front depicting an actual ladder and a key, accompanied by a sonnet. Like all Renaissance scholars, De Scolari uses poetry to depict the content of his work, in tribute to the widespread custom of the time, in this case suggesting that you need a ladder [scala] to attack a castle and a key [grimaldello] to open the lock, just as you need a book to gain access to knowledge. (Fig. 6).

The book of arithmetic and geometry was developed as an extension to the "Libro d'abaco", following the themes and lessons covered in school.

The work is divided into three "Libri"; the first is of a commercial nature, in the second the author covers roots and algebra, also known as the "Regola della cosa" [rule of the thing], whilst the third book is dedicated to practical geometry and solving the key problems of measurement.

There are many themes dealt with in "Scala Grimaldelli", of varying interest. Since it is not possible -in this context- to analyse them systematically as a whole, and omitting any examination of content related to the first two books, on which much has already been written, we will only focus on some of the most significant aspects of "Libro terzo" namely, the essay on the use of the surveyor's cross.

The 1526 edition of Feliciano's "Libro di arithmetica & geometria speculativa e praticale" is in fact the first printed work indicating the usage and applications of the surveyor's cross. In the first half of the sixteenth century the "arte di misurare con la vista" [art of measuring by sight] that had always had a large following in the field of surveying and was still popular at the time of De Scolari, was becoming increasingly outmoded, with the publication of a great number of essays on the

"Chi vuol aprir una serraglia forte
In cima d'una Torre over Castello,
Bisogna haver la Scala, acciò di quello
Gionger si possa alle serrate porte.
Apresso fa mestier l'huom seco porte
(Non havendo la Chiave del sugello)
Un ferro, che apellato è Grimaldello.
Se non d'aprir l'imprese sue, sien corte.
Tal questo mio libretto mostra come
Salir si possa a l'alte serrature,
E quelle aprir senza fatica grave.
Infinite ragion, ponti, e misure
Questo chiarisce, con modo suave,
Cui Scala e Grimaldello è il proprio nome.
Ma non lodar mie come,
Co'l tuo lieto biasmar, che 'l biasmo nasce
Sol di cui nel mal dir s'allegra, e pasce. "

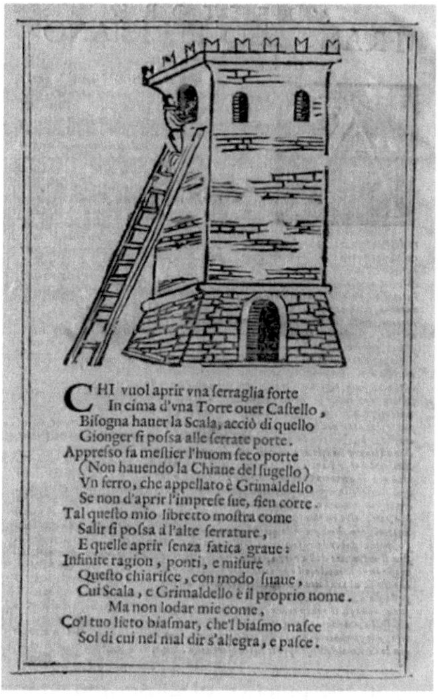

Fig. 6 Illustration accompanying the sonnet of "Scala Grimaldelli" in the Venice edition of 1669, expanded by Macario

subject. Feliciano highlights the importance of the correct use of angles in the numerous editions of his "Libro di Arithmetica", starting from 1526. The surveyor's cross had been in use for some time, often erroneously, for all types of measurement. In fact, in the prologue to his third book De Scolari writesabout the "...difficulta de larte mesuratoria ... molti sono che tale arte fanno, e pochi che siano boni ... ogni ignorante a tal impresa si metteno e dimostrano el negro per il bianco quando sono inter cecos ... tali alcuni ce ne pur che collo squadro o altro istrumento lavorano assai bene ... alcuni sono ancora che portano lo squadro .. e poco o niente lo sanno adoperare, hor che diremo de quelli che misurano senza alcuno istrumento, ma solu cosi coll'occhio dicendo, va de qua e va de la, e questi sono quasi un numero infinito de homini da villa & scriveno in su li legni ... e questi sono totalmente quelli che hanno guastato larte che più non se studia..." [... the difficulty of the art of measuring... many follow this art, but few are capable... those who undertake it in ignorance only distinguish black from white when operating amongst blind people... there are some who work well with the surveyor's cross and with other tools... others still carry the cross knowing little or nothing about its usage, except to say to those who measure without any tool and only with the eye, "go here and go there", and these are an almost infinite number of field technicians who write books... and these are the ones who have all ruined the art by not studying any more...] (Scala Grimaldelli 1536, c 65 recto).

The main concepts of surveying procedures were in fact relics of ancient times, and were generally limited to the simple operation of measuring surfaces, defining and adjusting boundaries and divisions, etc. In surveying as in other fields, these methods had no specific theoretical basis, but instead were carried out -in many cases- by the measurement of distances and empirical estimations to obtain an approximate quantification of the area, often following creative procedures for expediency put forward by self-proclaimed experts. Moreover, the agrarian institutions based on Roman geometric rules were already suffering from a gradual decline in the Middle Ages. Consequently, problems related to the form and dimensions of the earth seem to have disappeared completely, to surface again with the rediscovery of the concepts of geometry, which were frequently being applied in surveying. An important finding, subsequent to the essays of Leonardo Fibonacci ("Liber abaci" 1202 and "Practica Geometriae" 1220), is found in "Summa de arithmetica, geometria, proportioni e proportionalità" (Venice 1494) by Luca Pacioli, in which he draws extensively on the texts of Pisano with many notions on practical geometry applied particularly to the measurement of land. Thus we arrive in the fifteenth century, when knowledge of mathematics was fairly widespread and of a good standard, both on an experimental level and even more so in practice, as a result of instruction in applied arithmetic and geometry in the abacus schools. These disciplines, designed for the training of merchants, engineers, technicians, mechanics and artisans by teaching them the principal procedures for resolving issues, devised essentially without recourse to theory or demonstration and useful only in the practice of these trades, had made great progress.

In this context Feliciano addresses his treatise on measurement to all other areas of practice in daily life beyond that of measuring land, including the issues of

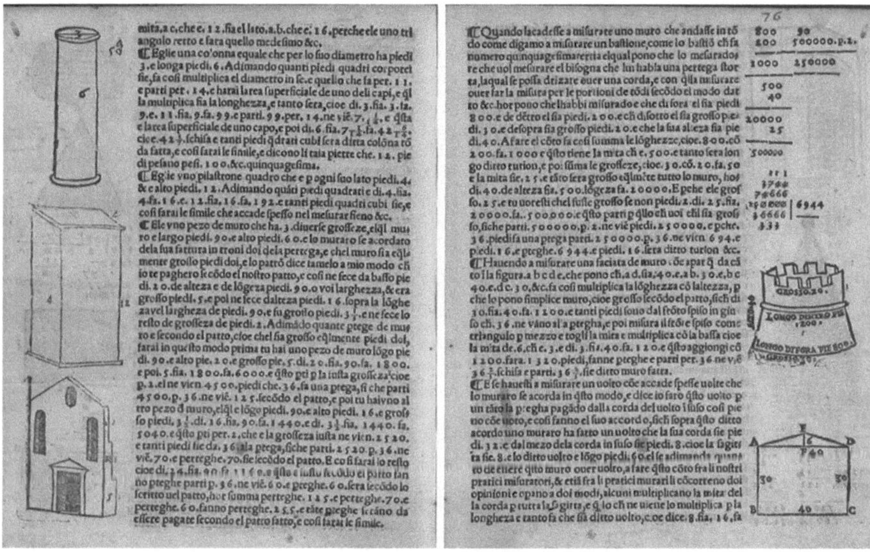

Fig. 7 Libro di Arithmetica & Geometria speculative & praticale…, Venice 1536 carte: 75 verso 76 recto, Libro terzo Problems for the measurement of columns, walls, towers, etc

measuring of chests, vats and barrels, columns, the stone pillars of buildings and ramparts, haystacks, etc. (Fig. 7). However, in the “Terzo libro” of his Scala Grimaldelli he concerns himself mainly with demonstrating the ease of use of the surveyor’s cross and the accuracy of the result compared to other methods in use at the time. The work is enhanced with an interesting overview of some case studies relating to the measurement of land which are considered difficult to resolve, providing the solutions to them. To do this he sets out a description of three different ways of measuring for each of the problems posed, supported by drawings, in order to highlight the procedures that lead to correct results compared to those that lead to errors or inaccurate results.

The author hypothesises that the measurement is carried out by three different technicians: one operates without any instruments, measuring only “by sight”; another carries out the measurement with the aid of the surveyor’s cross; the third uses numerical calculation, measuring triangles (Fig. 8). He demonstrates that two of the measurements are always consistent and in accordance with each other, whilst the other is obviously wrong (De Scolari 1536, c 65 recto-verso).

Feliciano was the first to provide a description of how the surveyor’s cross is used in a printed text; the structure of it is not explained, but it is referred to as an instrument that has already been well-known for some time by more accomplished surveyors. The instrument, which is essentially based on the groma used by the Romans, most likely consisted initially of a horizontal disc with two pairs of cross pieces mounted at right-angles (Cigola 2011, pp. 23–29), or of a circular wooden

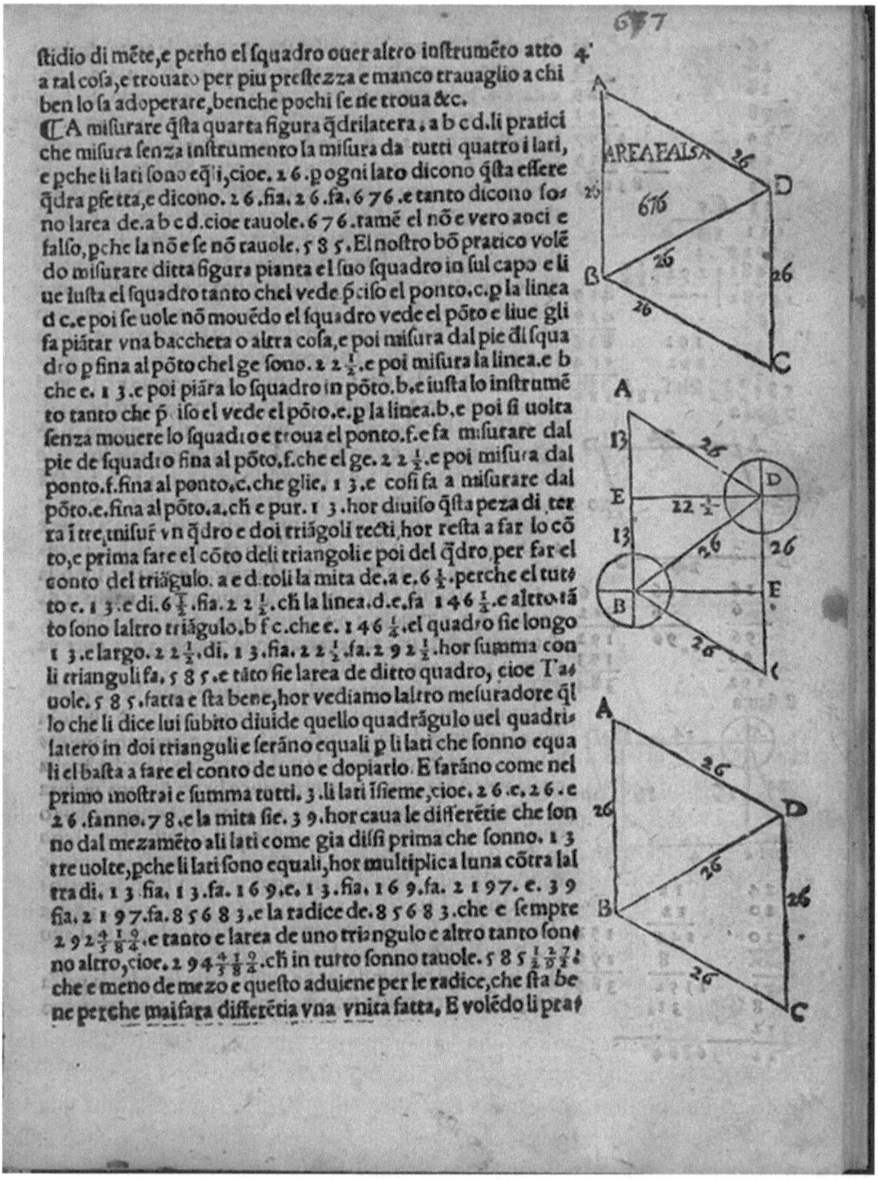
ſtidio di mēte,e perho el ſquadro ouer altro inſtrumēto atto
a tal coſa,e trouato per piu preſtezza e manco trauaglio a chi
ben lo ſa adoperare,benche pochi ſe ne troua &c.
¶A miſurare q̄ſta quarta figura q̄drilatera.a b c d.li pratici
che miſura ſenza inſtrumento la miſura da tutti quattro i lati,
e pche li lati ſono eq̄li,cioe.26.p ogni lato dicono q̄ſta eſſere
q̄dra pfetta,e dicono.26.fia.26.fa.676.e tanto dicono ſo-
no larea de.a b c d.cioe tauole.676.tamē el nō e vero anci e
falſo,pche la nō e ſe nō tauole.585.El noſtro bō pratico volē
do miſurare ditta figura pianta el ſuo ſquadro in ſul capo e li
ue iuſta el ſquadro tanto chel vede p̄ciſo el ponto.c.p la linea
d c.e poi ſe uole nō mouēdo el ſquadro vede el pōto e liue gli
fa piātar vna baccheta o altra coſa,e poi miſura dal pie di ſqua
dro p fina al pōto chel ge ſono.22½.e poi miſura la linea.e b
che e.13.e poi piāra lo ſquadro in pōto.b.e iuſta lo inſtrumē
to tanto che p̄ciſo el vede el pōto.e.p la linea.b.e poi ſi uolta
ſenza mouere lo ſquadro e troua el ponto.f.e fa miſurare dal
pie de ſquadro fina al pōto.f.che el ge.22½.e poi miſura dal
ponto.f.fina al ponto.c.che glie.13.e coſi fa a miſurare dal
pōto.e.fina al pōto.a.ch e pur.13.hor diuiſo q̄ſta peza di ter
ra i tre,miſur̄ vn q̄dro e doi triāgoli recti,hor reſta a far lo cō
to,e prima fare el cōto deli triangoli e poi del q̄dro,per far el
conto del triāgulo.a e d.toli la mita de.a e.6½.perche el tut-
to e.13.e di.6½.fia.22½.ch la linea.d.e.fa 146¼.e altro tā
to ſono laltro triāgulo.b f c.che e.146¼.el quadro ſie longo
13.e largo.22½.di.13.fia.22½.fa.292½.hor ſumma con
li triangoli fa.585.e tāto ſie larea de ditto quadro, cioe Ta-
uole.585.fatta e ſta bene,hor vediamo laltro meſuradore q̄l
lo che li dice lui ſubito diuide quello quadrāgulo uel quadri-
latero in doi triangoli e ſerāno equali p li lati che ſonno equa
li el baſta a fare el conto de uno e dopiarlo. E farāno come nel
primo moſtrai e ſumma tutti.3.li lati iſieme,cioe.26.e.26.e
26.fanno.78.e la mita ſie.39.hor caua le differētie che ſon
no dal mezamēto ali lati come gia diſſi prima che ſonno.13
tre uolte,pche li lati ſono equali,hor multiplica luna cōtra lal
tra di.13.fia.13.fa.169.e.13.fia.169.fa.2197.e.39
fia.2197.fa.85683.e la radice de.85683.che e ſempre
292 4/5 1/8 1/4.e tanto e larea de uno triangulo e altro tanto ſon-
no altro,cioe.294 4/5 1/8 1/4.ch in tutto ſonno tauole.585 ...
che e meno de mezo e queſto aduiene per le radice,che ſta be
ne perche maiſara differētia vna vnita fatta, E volēdo li pra-

Fig. 8 Libro di Arithmetica & Geometry speculative & praticale…, Venice 1536—carta: 67 recto, Libro Terzo. Application of the three measurement procedures for a rhomboid shaped area; in the second figure a surveyor's cross is used, placed first of all at point D and then at point B

plate with two incisions at 90°. This is how it appears, for example, in Part III of the third book of the "General trattato de numeri et misure" (Venice 1560) by Nicolò Tartaglia (Fig. 9).

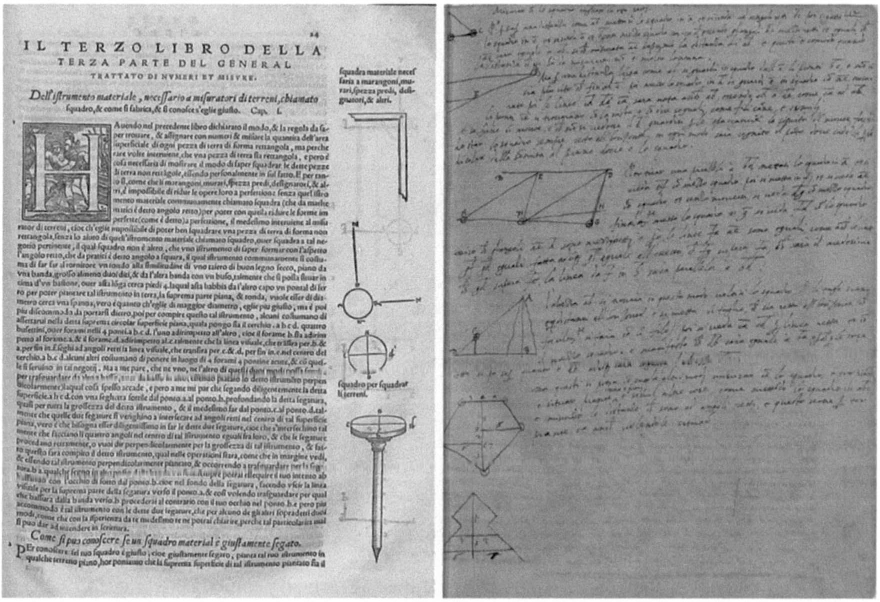

Fig. 9 *Left* NicolòTartaglia, “La terza parte del General trattato de numeri et misure” [Part three of the general treatise on numbers and measurements] (Venice 1560) carta 24 recto: “Dell’istrumento materiale, necessario a misuratori di terreni, chiamato squadro, & come si fabrica…” [On the material instrument required for measuring land, called surveyor’s cross and hoe it is made]. *Right* Guidobaldo dal Monte, “Misurar con lo squadron tagliato in otto parti” [Measuring with the surveyor’s cross cut into eight parts], p. 12 (carta 8 recto), in: Meditatiunculae de rebus mathematicis, original paper manuscript (1587–1592), Fonds Latin 10246, Bibliothèque nationale de France. This paragraph returns in the “Del Misurare” treatise by Guidobaldo in the vernacular on pages 6–12, and addresses some measurement problems solved by the use ofthe surveyor’s cross

The technical instrumentation of the time reflects the limited technology of the early sixteenth century, particularly as regards precision working. The rarity of sheet metal in brass and copper, the absence of cylindrical screws and the complexity of using a lathe all contributed to the inherent difficulty of devising and constructing instruments.

Subsequently, Guidobaldo dal Monte (1545–1607), in his signature manuscript of about 1589, “Meditatiunculae de rebus mathematicis” describes various problems of measurement by using a surveyor’s cross “tagliato in otto parti” [cut into eight pieces]. It is not specified whether the shape is cylindrical or octagonal, but it is clear that the “squadron” [cross], which has lines at 90° and 45°, goes beyond the simple orthogonality of previous tools enabling it to achieve fixed angles of 45, 90, 135, 180, 225 and 270 degrees.

Giovanni Pomodoro (1550–1599) also presents a “Squadro” in his “Geometria prattica” (Rome 1599), for the measurement of land (Fig. 10) and illustrates its usage from Fig. 10 left. The instrument shown is still a simple cross, cylindrical

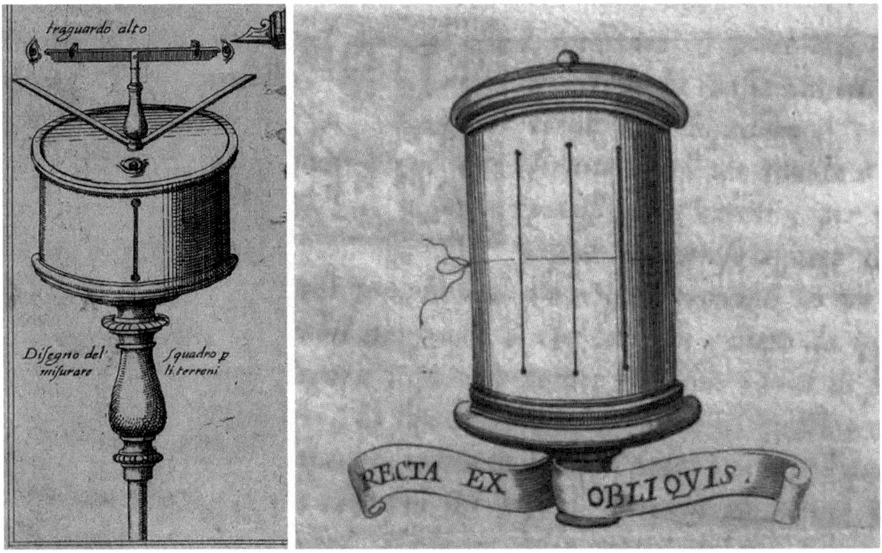

Fig. 10 *Left* Giovanni Pomodoro, "Geometria prattica" [Practical geometry] (Rome 1599), Table I—"Squadro" Surveyor's cross for measuring land. *Right* Muzio Oddi, "Dello Squadro" [On the surveyor's cross] (Milan 1625) from the frontispiece

in form with four slots set at ninety degrees, and still appears in the application tables as a circular plate with four incisions. We only arrive at a complete explanation of the instrument with Muzio Oddi (1569–1639) in his treatise "Dello Squadro" (Milan 1625), in which he describes how it is constructed and the usage of it. Oddi outlines the shape of the cross, saying: "…alcuni sono quadri, altri a otto & più faccie; ma per la maggior parte sogliono farli a foggia di Cilindro…" [some are square, others have eight or more sides, but most are made in the shape of a cylinder] (Fig. 10). In fact, the cylindrical or octagonal shape of the cross with vertical incisions, unlike the flat disc type, provided further advantages for taking measurements, giving line plans rather than visual lines, and even allowing the marking out of corners on uneven ground, as already stated by Guidobaldo in his "Meditatiunculae" (Gamba, Mantovani 2013, p. 219).

In fact, most traditional surveying practices were based mainly on the ability to solve only right-angled triangles, applying Pythagoras' theorem (570 BC–495 BC). First with the groma and later with the cross (technical gauges), surfaces were divided into right-angled triangles of which the sides were measured. It is only in the late sixteenth century and the beginning of the seventeenth, in the wake of the "De triangulis omnimodis" (Norimberga 1533) by Regimontano (1436–1476) who first enunciated the law of sines, that formulas for resolving any triangle begin to be followed for land measurement (Maccagni 1987, pp. 47–52). However, the work of Feliciano, who was the first to systematise the use of the instrument, together with the work of Oddi, dedicated exclusively to the cross and instrumental in its perfection, can be considered as the beginning of modern surveying.

After explaining the various problems of surveying, De Scolari goes on to review a number of other cases, referring to "…misurar biave, vini, muri, e tutte quelle cose che sono corporee…" [the measurement of fodder, liquids, masonry, and everything else] (Fig. 11), according to Veronese custom, referring to the units of measurement described at the beginning of his third book, where he had explained the system used in his treatise.

In particular, both for land and for smaller things, Feliciano refers to the Veronese measurement known as the "Pertica", which is made up of six feet, each of which comprises twelve ounces, scoring the figure in the margin as the exact measurement of half a foot and an ounce so as to allow the reader to reconstruct a perch for measurements in the field (Libro di Arithmetica… 1536, carta 65 verso and recto). In seventeenth century editions of Scala Grimaldelli, the description of the system is enhanced by the additions of Filippo Macario, who specifies, "per maggiore intelligenza" [for greater intelligence], submultiples of ounces, referring to "rotti" of an ounce into twelve points, a point into twelve atoms and an atom into twelve minutes.

In describing the problem of measuring a hay wagon, De Scolari sets out the qualities that a good consultant must possess: "…bisogna che'l misuratore habbia buona prattica, e sopra ogni cosa buona coscienza, e non voler servir a uno per parole, e manco per fatti, perche no puol servire a uno che'l non disserva all'altro, e perche bisogna che'l misuratore sia homo da bene, e che l'habbia buona prattica

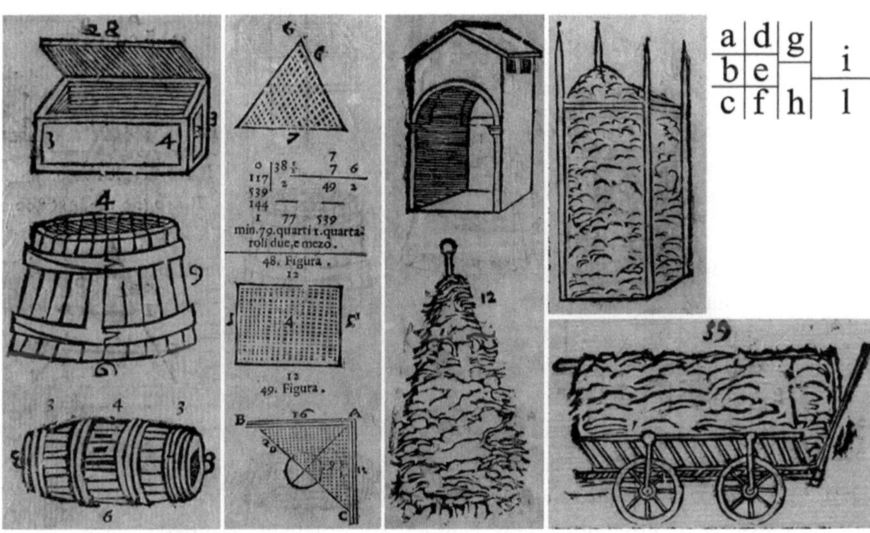

Fig. 11 Scala Grimaldelli, Venice 1669—Book Three, pp. 228–234. Illustrations of the various measurement problems proposed. **a** Trunk. **b** Vat. **c** Barrel. **d–f** Piles of grain. **g** Vault. **h** Haystack. **i** "Barcottoquadro". **l** Wagon loaded with hay

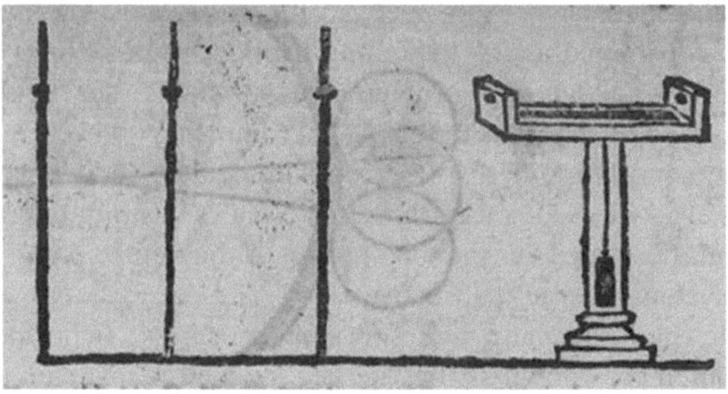

Fig. 12 Libro di Arithmetica & Geometria speculativa & praticale ..., Venezia 1536—carta 79 verso, Libro Terzo. Sketch of the levelling instrument, with bacchette [batons] for sights

..." [...the expert measurer should have a good practice and above all honesty, not making words and lacking in facts, not serving one and displeasing another...] (Scala Grimaldelli 1669, p. 234).

Finally, at the end of the third book, Feliciano describes, with examples of applications, the use of another tool for professional surveying: "...molte volte accade al vero misuratore livellare, overo condur acque da un luogo all'altro, per far fontane, over per adaquare prati, o campi..." [...it often happens to the true measurer that he must level off, or carry water from one place to another, to create fountains or to level meadows or fields...] (De Scolari 1536, carta 79 recto-verso, carta 80 recto). This is the "livello" [level] which, in the text, is given a schematic portrayal with levelling poles. The instrument, from how it looks in the picture, is equipped with a central plumb line to ensure it is perfectly horizontal, with two holes on the ends of the level to detect a target altitude that appears in the line of sight (Fig. 12).

On the Circulation of Work

As we have seen, the reputation of Francesco Feliciano is essentially linked to the particular scholastic and publishing success of his treatise, which became known as the "Scala Grimaldelli" over the span of about 200 years bridging the sixteenth and seventeenth centuries. This, according to its primary application as a manual intended for the teaching of the main processes of arithmetic, algebra and measurement of the earth, is repeatedly mentioned -and not always accredited to the author- in the many and various accounts of the history of mathematics.

Among these is the aforementioned Pietro Riccardi (1828–1898), engineer, bibliographer of mathematics and scholar of practical geometry, who in his "Biblioteca Matematica Italiana" (Riccardi 1873) lists the numerous editions and reprints of the two major works of the Veronese author.

In addition, a detailed description, with some specific peculiarities relating to the early editions of Feliciano's "Libro d'abbaco", can be found in the work "Intorno ad un trattato d'aritmetica stampato nel 1478" (Boncompagni 1863) by Prince Baldassarre Boncompagni (1821–1894), a high-profile scholar of mathematics, founder and editor of the Bulletino di Bibliografia e Storia delle Scienze Matematiche e Fisiche [Bulletin of the Bibliography and Historyof Mathematical and Physical Sciences], one of the first reviews devoted entirely to the history of the exact sciences. In his well-structured dissertation, an analysis of the anonymous treatise on abacus in Treviso of 1478, published by the Accademia de' Nuovi Lincei, the author recalls, examines and compares various editorial features and numerous calculation methods present in the volumes he examined.

The published works of De Scolari are thus acknowledged in the growing and widespread demand of his time, characterised by an increasing need for an understanding of mathematics and practical geometry. So it appears that the publication of his "Libro de Abbacho" follows on chronologically from four earlier, similar publications: the first one, which was anonymous, printed in Treviso in 1478 (Boncompagni 1866), followed by Pietro Borghi's publication, printed in Venice in 1484, then the one by Pellizzati Francis (1450–1500) printed in Turin in 1492 and finally the work of Girolamo and Giannantonio Tagliente published in Venice in 1515 (Cavazzocca Mazzanti 1909, p. 20—Smith 1908, p. 50, 114). However, Feliciano's treatise is more complete than the "Libro di Treviso", more modern than the "Libro de abacho" by Pietro Borghi and more concise and practical than Pacioli's treatise (Devlin 2013, pp. 14–17). This certainly justifies the success of the text that has typified, over the many years that have passed since it was first published, the teaching of elementary mathematics and the applications of practical geometry.

Below are the title pages of several major reprints and editions of the "Libro d'abbaco" and the "Libro di aritmetica" e "Geometria pratica" that have been released over the years.

1519

1524

Libro de Abbacho novamente composto per magistro Francesco da Lazesio veronese, il quale insegna a fare molte ragione mercantile, & come rispondeno li precii, & monete novamente stampato
Venetia, per Nicolo Zopino e Vincentio suo compagno, 1519;
Venezia, per Francesco I Bindoni & Maffeo Pasini, 1524

1526

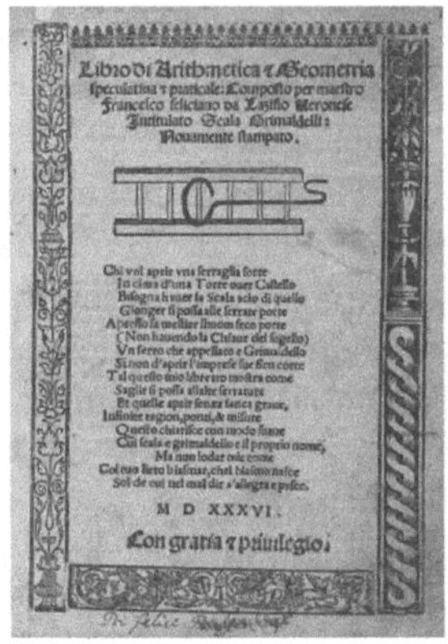

1536

1545

1550

1560

1563

Libro di Arithmetica & Geometria speculativa & praticale, Composto per maestro Francesco Feliciano da Lazisio Veronese Intitulato Scala Grimaldelli, Novamente stampato
Venezia, per Francesco I Bindoni & Maffeo Pasini, 1527(1526), 1536, 1545, 1550;
Venezia, per Francesco Di Leno, 1560, 1563

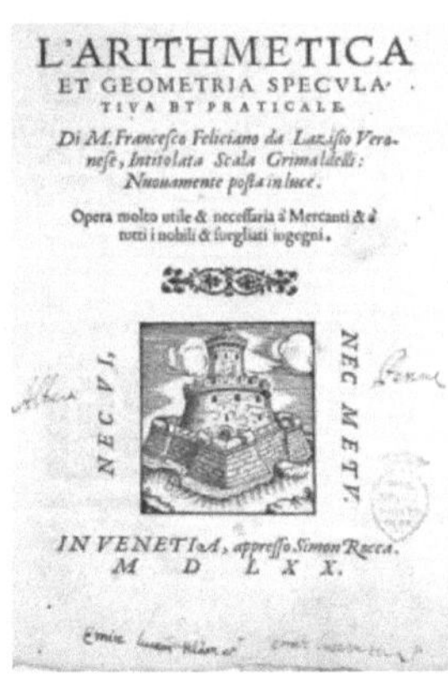

L'ARITHMETICA
ET GEOMETRIA SPECVLA-
TIVA ET PRATICALE.
Di M. Francesco Feliciano da Lazisio Veronese, Intitolata Scala Grimaldelli:
Nuouamente posta in luce.
Opera molto utile & necessaria à Mercanti & à tutti i nobili & suegliati ingegni.
NEC VI, NEC METV.
IN VENETIA, appresso Simon Rocca.
M D L X X.

L'arithmetica et geometria speculativa et praticale. Di m. Francesco Feliciano da Lazisio Veronese, Intitolata Scala Grimaldelli, Nuovamente posta in luce. Opera molto utile & necessaria a Mercanti & a tutti i nobili & svegliati ingegni
Venezia, per Francesco Di Leno ad instantia de Simone Rocca, 1570

SCALA
GRIMALDELLI
LIBRO DI
ARITMETICA, ET GEOMETRIA
Speculatiua, & Pratticale
DIVISO IN TRE LIBRI.

1602

1602

Scala Grimaldelli
Libro di aritmetica, et geometria speculatiua, & pratticale di M. Francesco Feliciano Veronese. Diviso in tre libri.
Di nuovo ristampato, & da molti errori corretto, & accresciuto di molte cose da M. Filippo Macario ...
In Verona : appresso Dionigi Filiberi, 1602
(Nella stamparia di Angelo Tamo, 1602)

SCALA
GRIMALDELLI
LIBRO DI
ARITMETICA, ET GEOMETRIA
Speculatiua, & Pratticale
DIVISO IN TRE LIBRI.
In Padoua, Per Donato Pasquardi, & compagni. 1629.

1629

Scala Grimaldelli
Libro di aritmetica, et geometria Speculativa, & Pratticale
Di M. Francesco Feliciano Veronese.
Diviso in tre libri.
Di nuovo ristampato, & da molti errori corretto, & accresciuto di molte cose da M. Filippo Macario Veronese Rasonato publico della Mag. Città.
In Padova : per Donato Pasquardi, & compagni, 1629

SCALA
GRIMALDELLI
LIBRO DI
ARITMETICA, E GEOMETRIA
Speculatiua, e Pratticale
DI M. FRANCESCO FELICIANO
VERONESE.
GIO. BATTISTA SORER.
VENETIA. MDCLXIX.
Presso Gio. Giacomo Hertz.

1669

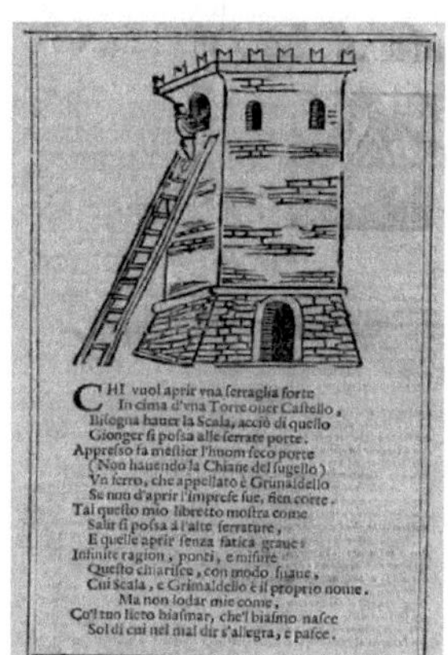

1669 (c 4 verso)

Scala Grimaldelli
Libro di aritmetica, e geometria speculativa, e pratticale di M. Francesco Feliciano ...
... Accresciuto in quest'ultima impressione di dottissimi problemi da B.B.
Di nuovo ristampato, e da molti errori corretto, & accresciuto di molte cose da M. Filippo Macario...
Venezia : presso Gio. Giacomo Hertz, 1669

SCALA
GRIMALDELLI
LIBRO DI
ARITMETICA, E GEOMETRIA
Speculatiua, e Pratticale
DI M. FRANCESCO FELICIANO
VERONESE.
DIVISO IN TRE LIBRI.
AL MOLT' ILLVSTRE SIGNOR
GIO. BATTISTA SORER.
VENETIA MDCXCII.
Presso Gio. Giacomo Hertz.
CON LICENZA DE' SVPERIORI.

1692

CHI vuol aprir vna ſeraglia forte
In cima d'vna Torre, ouer Caſtello,
Biſogna hauer la Scala, accio di quello
Giongere ſi poſſa alle ferrate porte.
Appreſſo fa meſtier l'huom ſeco porte
(Non hauendo la Chiaue del ſugello)
Vn ferro, che appellato è Grimaldello
Se non d'aprir l'impreſe ſue, ſien corte.
Tal queſto mio libretto moſtra come
Salir ſi poſſa à l'alte ferrature,
E quelle aprir ſenza fatica graue:
Infinite ragion, ponti, e miſure
Queſto chiariſce, con modo ſuaue,
Cui Scala, e Grimaldello è il proprio nome.
Ma non lodar mie come,
Co'l tuo lieto biaſmar, che'l biaſmo naſce
Sol di cui nel mal dir s'allegra, e paſce.

1692 (c 4 verso)

Scala Grimaldelli
Libro di aritmetica, e geometria speculativa, e pratticale di M. Francesco Feliciano …
… Accresciuto in quest'ultima impressione di dottissimi problemi da B.B. Di nuovo ristampato, e da molti errori corretto, & accresciuto di molte cose da M. Filippo Macario …
Venezia : presso Gio. Giacomo Hertz, 1692

Conclusions

Teacher of arithmetic, respected expert of surveying and a man of his time, De Scolari, Francesco Feliciano da Lazesio, contributed to the training of generations of merchants, technicians and experts, especially through his principal published work, at the turn of the sixteenth and seventeenth centuries, and combined technical information with the ethical qualities that a practitioner should apply to his professional practice.

His high profile as a technician is also due to his dissemination of the correct methodology for measuring land by the accurate use of the "Squadro agrimensorio" [surveyor's cross]. Contributing through his work to the transition from the practice of "by sight" measurement to procedures based on scientific rigour, he was the founder of new surveying methods which became fully developed in succeeding years.

Finally, it is as yet uncertain whether Feliciano was the author of the treatise entitled "Nuovo libro di Statica e Meccanica" published in Venice in 1667, as examined by Murhard (Murhard 1803, p. 10). The treatise, which would in any case have been published posthumously, may be consistent with the additions to the 1669 edition of "Scala Grimaldelli", initialled BB by the alleged Bernardino Baldi and demonstrating an increased interest in the work of Lazesio by scholars of the new mechanical disciplines that were becoming more widespread in the early seventeenth century.

Bibliography

Atzeni G (2013) Gli incisori alla corte di Zoppino. In ArcheoArte, n. 2, UNICA, Cagliari, pp 299–328

Belloni Speciale G (1991) De Scolari, Francesco Feliciano. Dizionario biografico degli italiani, Istituto della Enciclopedia Italiana fondata da Giovanni Treccani, Roma 39:347–349

Boncompagni B (1863) Intorno ad un trattato d'aritmetica stampato nel 1478, Estratto dagli Atti dell'Accademia Pontificia de' Nuovi Lincei, Anno XVI, Tomo XVI, Sessioni 1^, 2^, 3^, 4^, 5^, 6^ e 7^, 1862–63, Tipografia delle belle arti, Roma

Cavazzocca Mazzanti V (1909) Un matematico di Lazise (Francesco Feliciano De Scolari). Stab. Tip. M. Bettinalli e C, Verona

Cigola M, Ceccarelli M, Rossi C (2011) La Groma, lo Squadro agrimensorio e il corobate. Note di approfondimento su progettazione e funzionalità di antiche strumentazioni. In Disegnare Idee Immagini, anno XI n. 42/2011, Cangemi, Roma, pp 22–33

Docci M, Maestri D (2009) Manuale di rilevamento architettonico e urbano. Laterza, Bari, pp 334–343

Devlin K (2013) I numeri magici di Fibonacci. RCS Libri, Milano

Gamba E, Mantovani R (2013) Gli strumenti scientifici di Guidobaldo del Monte. In: Guidobaldo del Monte (1545–1607), Theory and Practice of the Mathematical Disciplines from Urbino to Europe. Proceedings 4, Edition Open Access, Max Planck, Berlino, pp 209–239

Garibotto E (1923) Le scuole d'abbaco a Verona. In: Atti e memorie dell'Accademia di agricoltura, scienze e lettere di Verona, s. 4, XXIV, pp 315–328

Maccagni C (1987) Evoluzione delle procedure di rilevamento: fondamenti matematici e strumentazione. In: Cartografia e istituzione in età moderna, Atti della Società Ligure di Storia Patria, Vol XXVII (CI), Fasc I, Istituto Poligrafico e Zecca dello Stato, Roma, pp 43–57

Maccagni C (1996) Cultura e sapere dei tecnici nel Rinascimento. Piero della Francesca: tra arte e scienza, a cura di Dalai Emiliani M e Curzi V. Marsilio, Venezia, pp 279–292

Murhard F W A (1803) Bibliotheca mathematica, Breitkopf und Härtel, Lipsia, Tomo III, parte Prima

Riccardi P (1873) Biblioteca matematica italiana *dall'origine della stampa ai primi anni del sec. XIX*, Società Tipografica, Modena, Vol. II, Fasc. I, pp 19–23

Smith DE (1908) Rara arithmetica; a catalogve of the arithmetics written before the year MDCI with a description of those in the library of George Arthvr Plimpton of New York. Ginn and Company Publishers, Boston and London

Niccolò Tartaglia (1500c.–1557)

Alfonso Ippolito and Cristiana Bartolomei

Abstract The article presents Niccolò Tartaglia as a mathematician active in various fields of science such as mathematics, arithmetic, mechanics, geometry as well as ballistics and military architecture. Although he won general recognition for the Tartaglia's Triangle and his solution to cubic equations, he made important discoveries in ballistics, geometry and military architecture. Among them were calculations of the trajectory of cannon balls, the volume of complex figures and requirements for constructing fortifications able to resist enemy attacks. But his activity remains of interest today mainly because he knew how to fuse theoretical knowledge with practical experience—the fundamental principle of modern science.

Introduction

In the late Middle Ages Italy underwent a commercial revolution which made Italian merchants the most important intermediaries between Europe and the Middle East in the trade of textiles and spices.

The phenomenon reached such a scale that Italian merchants had to get organized into societies and became involved in developing instruments and methods of dealing with goods and the proceeds they generated. Efficient methods of counting, of calculating rates of exchange, loans and interests had to be devised.

A. Ippolito (✉)
Department of History Representation and Restoration of Architecture, Sapienza University of Roma, Piazza Borghese 9, 00186 Rome, Italy
e-mail: alfonso.ippolito@uniroma1.it

C. Bartolomei
Department of Architecture, University of Bologna, Viale Risorgimento 2, 40136 Bologna, Italy
e-mail: cristiana.bartolomei@unibo.it

© Springer International Publishing Switzerland 2016

M. Cigola (ed.), *Distinguished Figures in Descriptive Geometry and Its Applications for Mechanism Science*, History of Mechanism and Machine Science 30, DOI 10.1007/978-3-319-20197-9_4

All these methods assumed some knowledge—at least elementary—of mathematical concepts and notions. Since most Italian commercial societies were family businesses where the trade passed from father to son, it became necessary to provide the successors with education in this field. To satisfy this demand there appeared the so-called 'abacus' schools or studios, where 'masters of the abacus' taught mathematics for a fixed salary. From the beginning of the XVI century these masters spread their knowledge in two ways: they either published the solutions to mathematical problems so that they could be used for self-teaching or they collected a series of mathematical problems into a kind of catalogue in order to teach their solutions orally upon payment: understandably, as long as the methods and solutions had not yet become public knowledge, since the latter would have diminished the likelihood of winning mathematical challenges which defined the market value or 'masters of the abacus'. Naturally, this value depended also on the knowledge of as many methods and solutions as possible in various domains of mathematics.

Niccolò Tartaglia and his activity can be considered representative of this new social setting.

Biographical Notes

Niccolò Tartaglia (Fig. 1) was born in Brescia around the year 1500 (the precise year of his birth remains unknown) in a poor family. In 1506 he was orphaned by his father and in 1512—during the 'sack of Brescia' he was gravely wounded in the face—so much that he almost lost the ability to speak, hence his nickname Tartaglia (meaning "stammerer", his real name seems to have been Fontana) that he himself adopted to demonstrate the power of spirit in adversity.

In 1518 he moved to Verona where he got married probably around the year 1530. He lived in modest conditions and supported himself and his family teaching commercial mathematics publicly and privately. He was consulted on the problems of mensuration, calculations, currencies and rates of exchange. We know for certain that he taught himself Latin to be able to study great mathematicians like Euclid and Archimedes. In 1534 he moved to Venice, where he worked as a mathematics teacher and published his works. He returned to his home town where he spent the years 1548 and 49. There he continued to teach before moving back to Venice for good to complete his studies in mechanics and mathematics. He died on December the 13th in Venice.

List of (Main) Works

The list contains the fundamental publications of Niccolò Tartaglia

Fig. 1 A portrait of Niccolò Tartaglia (1500c.–1557)

1. Nova Scientia, 1537. (the second version including book III was published in 1550, and then posthumously in 1558 and 1583. It was translated into several languages).
2. Euclide Megarense, 1543. (the first italian translation of Euclid's Elements, the first printed publication in a modern European language)
3. Opera Archimedis, 1543 (a collection—in the Latin translation of William of Moerbecke—of three works of the Siracusan mathematician Archimedes: De centersi gravium libri duo, Tetragonismus or De quadratura circuli and De insedentibus aquae).
4. Quesiti et inventioni diverse, 1546 (a variety of problems presented in dialogue form concerning problems of physico-mathematic sciences, re-edited in 1554, posthumously in 1562 and translated—though partially—into English, French and German).
5. Cartelli di sfida matematica, 1547–1548 (concerns numerous mathematical problems regarding the mathematical disputes he had with Ludovico Ferrari in defence of Gerolamo Cardano).
6. Travagliata inventione (con la regola generale, i ragionamenti e il supplemento), 1551. (sets forth procedures for raising sunken ships. Included in it is also a partial English version with Tartaglia's comments on his own work, the first book of Archimedes' De insedentibus acque.)
7. General trattato di numeri et misure, 1556–1560 (an important treatise published in part posthumously. It concerns arithmetic, geometry and algebra. The part on

arithmetics was re-edited posthumously in 1592–93 and translated—partially—into other languages).

He was profoundly interested in subjects which embrace such fields of study as mathematics, arithmetics, mechanics, geometry as well as ballistics and military architecture.

Review of Main Works

'Nova scientia' (New Science) (Fig. 2), the first printed work of Niccolò Tartaglia —his debut in fact—appeared in Venice in 1537 and was devoted to ballistics, and in particular to the trajectory of cannon balls.

The book heralds a new approach to mechanics, and particularly to kinematics. Even though Tartaglia had no practical experience in this field, he applies and develops scientific knowledge and proposes constructive indications, making use of sketches. He inaugurated new ballistics unequivocally rejecting Aristotelian principles. In the letter dedicating the work to the Duke of Urbino, which opens the volume, Tartaglia includes details of his enquiry into mathematics applied to the problems of ballistics (Fig. 3).

Through this enquiry he discovered that the howitzer has the longest range when elevated to 45°, i.e. the projectile launched with a given force covers the longest distance when the angle of the initial launch is 45°. As he writes in his letter: it was an 'expert bombardier' who introduced him to this field of endeavour while the empirical discovery that the optimum angle for the longest distance of a cannonball is 45° convinced him to continue his research in ballistics (Fig. 4).

Taking Aristotle's distinction between natural and violent motion as a point of departure, Tartaglia envisaged a curved trajectory assuming that the two motions are super imposable, a notion rejected by Aristotle (Nova scientia, Libro II). Tartaglia claims natural motion is the movement of a body generated solely by gravity. He considers all other kinds of motion to be unnatural/violent/(Nova scientia, Libro I). In a horizontal and oblique shot the superimposition of both movements create a trajectory which cannot be a perfectly straight line at any part.

The trajectory can approximately be said to be composed of the initial, linear phase caused by the propulsion force, followed by an arch of a circle when the violent motion gives way to its natural counterpart; the latter is finally substituted by a straight vertical part corresponding to natural motion. Tartaglia calculated that within his ballistic theory the angle of maximum range is that of 45° which he demonstrated by quoting shots launched at the angle of 30° and of 45° (Fig. 5).

He also established that every given range can be achieved with two angles of elevation. He claimed, moreover, that with a single measurement of the elevation angle and its consequent range, it is possible to calculate the range for each and everyone angle of elevation. However, he does not describe the method of doing it —either in 'Nova scientia' or in any successive writings. When this element is

Fig. 2 Nova Scientia, 1537. (its second version was published in 1550 with libro III added, and then—in 1558 and 1583—the posthumous version translated into various languages)

found, the next step—as Tartaglia himself put forward—is to establish the distance of the aim to be struck. In Libro III of 'Nova scientia' he describes the method to evaluate 'a vista' the distance between the observer and the inaccessible object to

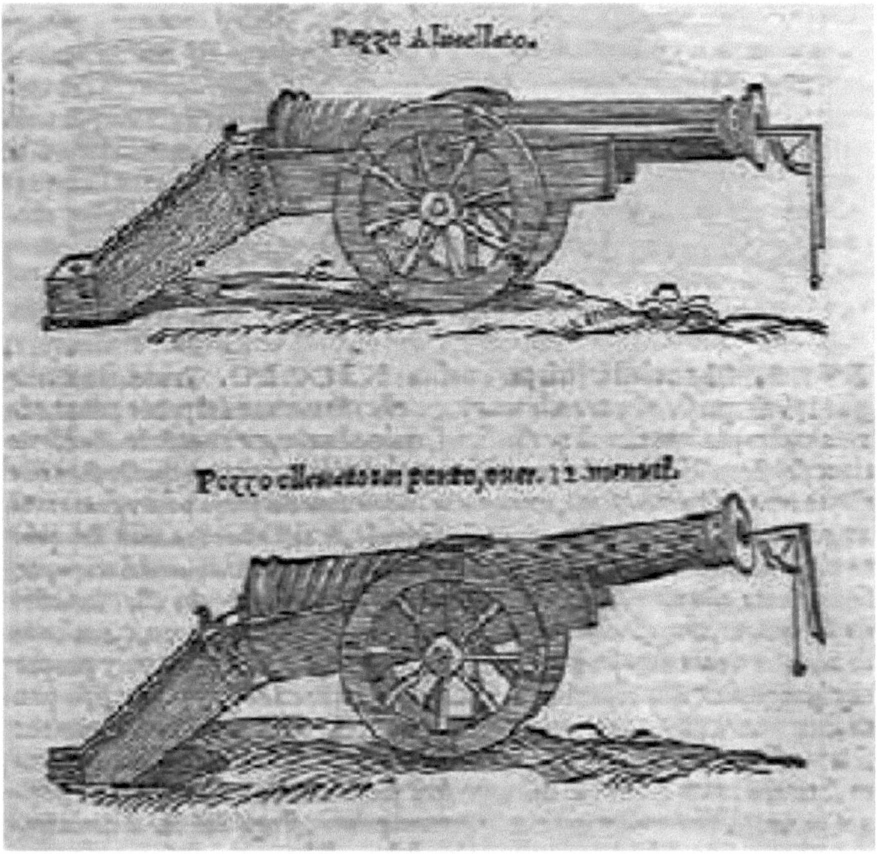

Fig. 3 Inclination of pieces of artillery verified with proper set square

hit. This is done by using two different kinds of cross-sights equipped with quadrants. The first kind is a cross-sight with a gnomon in whose internal right angle is placed a square with the side shorter than that of the cross-sight. The external vertical side of the cross-sight is furnished with two viewfinders (Fig. 6).

From the vertex of the right angle is suspended a plumb line which ensures that the instrument is in vertical position or serves to measure the inclination. The square is used to measure altitudes. The other instrument for measuring horizontal distances is the cross-sight with mobile viewfinders formed by a square placed horizontally on an axis and with sides divided into 12 equal parts. Two alidades move in their centre. They are placed at the right angle and have notches and fissures. Understandably these instruments were rather heavy for artillerymen who needed to establish quickly and only approximately the distance between them and the aim to be hit.

Fig. 4 Survey of distance performed with proper instrument

Land surveyors need more precise measurements, so Tartaglia invented two other instruments which he mentions in his 'Quesiti et inventioni diverse' and 'General trattato di numeri et misure'.

Ballistics as conceived by Tartaglia strikes a compromise between empirical experience and traditional explanations of motion in terms of physics and introduces new elements into the techniques of using firearms, namely mathematics and geodesy. During his stay in Verona, between 1521 and 1533, Tartaglia was approached with about 15 'quesiti' (queries) of the 42 that compose the volume 'Quesiti et inventioni diverse' published in Venice in 1546 (Fig. 7).

One of the first cases he was supposed to deal with concerned the game of dice. A ludic matter, no doubt, but they inaugurated the very first research in combinatory. Tartaglia was especially well known for the combinations with repeating six elements from 1 to 1 and from 8 to 8. As he himself writes in his 'General trattato di

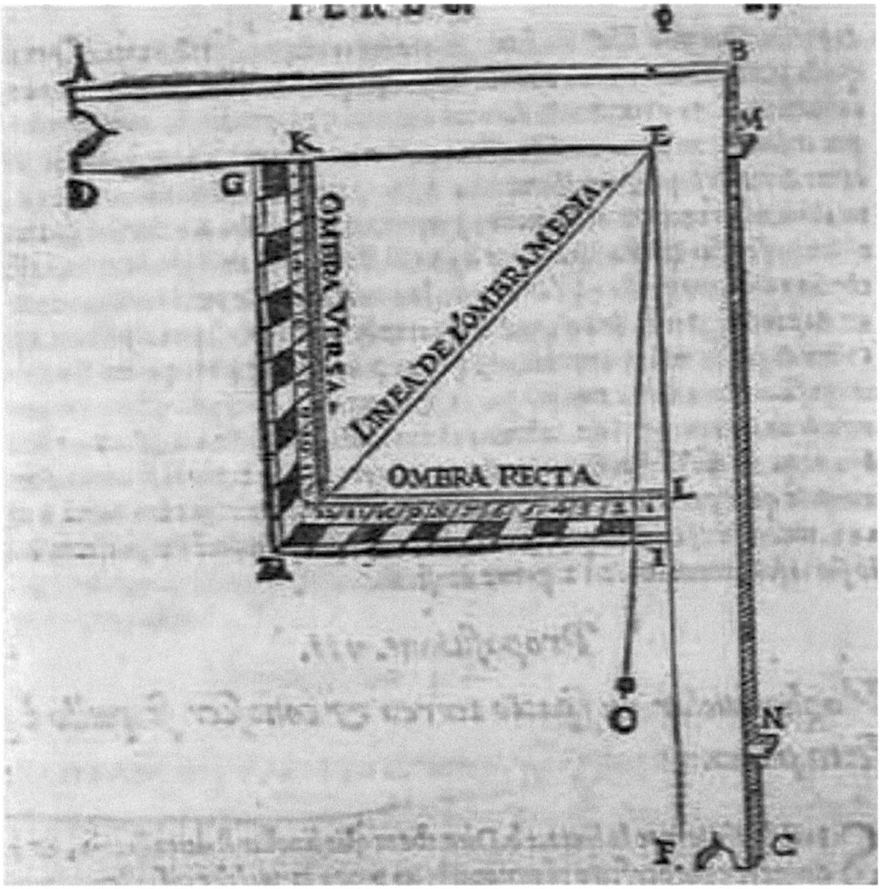

Fig. 5 Detail of instrument used for surveying distance

numeri et misure' (libro I, Cap XVI, c.17r: "the general rule discovered by the author on the first day of lent of 1523 in Verona to be able to find in how many ways it is possible to vary the throw of the dice and how many dice are necessary to do it".) Even though he does not say it explicitly, the structure of the combinatorial calculus is exactly the same in the so-called 'Tartaglia's Triangle'.

Thanks to the queries (Quesiti et inventioni diverse, libro IX, quesito 14, cc. 101rv) posed in Verona in 1530, he started to deal with cubic equations. Their solutions—considered impossible so far—belong with his major mathematical achievements. They were so important that in the libro IX, quesito 34, he expresses the formulae in a poetic form. When in 1534 he moved to Venezia, the very first quesito (Quesiti et inventioni diverse, libro IX, quesito 19, cc. 103rv) posed to him by a Venetian mathematician Giovanni Battista Memmo (1503–1579) regarded the squaring of the circle. In the church of San Zanipolo in Venezia he started to teach

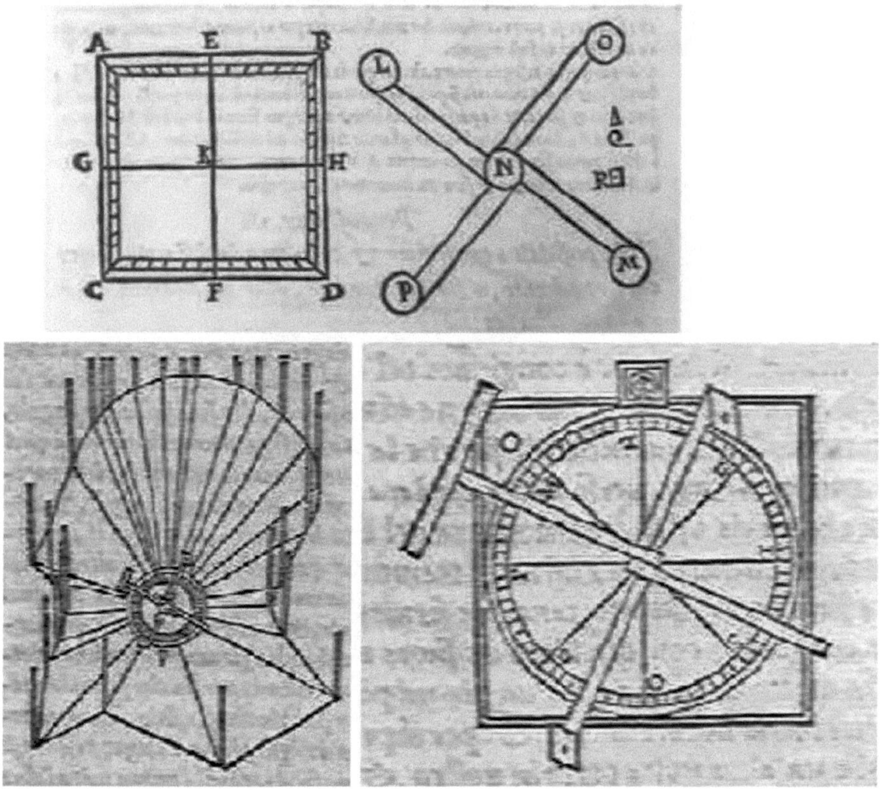

Fig. 6 Drawings from Tartaglia's treatises illustrating topographic surveying

Euclid, explaining various books and chapters of Euclid's manual. He will resume this activity in Brescia between 1549 and 1549. Tartaglia was the first to publish in print and in a modern language—i.e. the vernacular Italian—Euclid's manual 'Elements' published in 4 different editions (1543–65/1566–1569–1585/1586) (Fig. 8). Euclid's handbook seems to be so important to Tartaglia that he makes reference to it in all his works, particularly in the last one, 'General trattato' (Fig. 9). There he claims that Euclidean problems can be solved with the help of a ruler and a fixed compass. Euclid used the compass with a variable aperture. The problem was of enough significance to make him enter into a mathematical dispute with his rival L. Ferrari (1522–1565).

Even G. Cardano (1501–1576) was interested in the matter. However, their methods of solving these problems seemed too complicated to Tartaglia who provides his convincing solution in 'General trattato' (part V—Libro III, c. 64r). Another important discovery he made in relation to Euclid was to add two figures to the 12 discovered by the Greek mathematician, that can be inscribed into regular solids, namely the icosahedron into the cube and the octahedron into the

QVESITI, ET INVENTIONI DI-
VERSE DE NICOLO TARTALEA
BRISCIANO.

Con gratia, & priuilegio dal Illustrissimo Senato Veneto, che niuno ardisca ne presuma, di stampare la presente opera, ne stampate altroue uendere ne far uendere in Venetia, ne in alcuno altro luoco, o terra del Dominio Veneto, per anni diece sotto pena de ducati trecento, & perdere le opere, el terzo della qual pena immediate che sia denontiata, si applica al Arsenale, & un terzo sia del magistrato, ouer rettore del luoco doue se fara la essecutione, & laltro terzo sara del denuntiante, ouer accusatore, & sara tenuto secreto, come nel priuilegio appare.

Fig. 7 Quesiti et inventioni diverse, 1546 (queries of various kinds presented in dialogue form on matters concerning diverse physical-mathematical problems, re-edited in 1554, then posthumously in 1562, translated in part into English, French and German)

icosahedron. Apart from republishing Euclid in 1543, he also published the works of Archimedes based on the version of the Flemmish dominican William of Moerbeke (1215–1286) (Fig. 13).

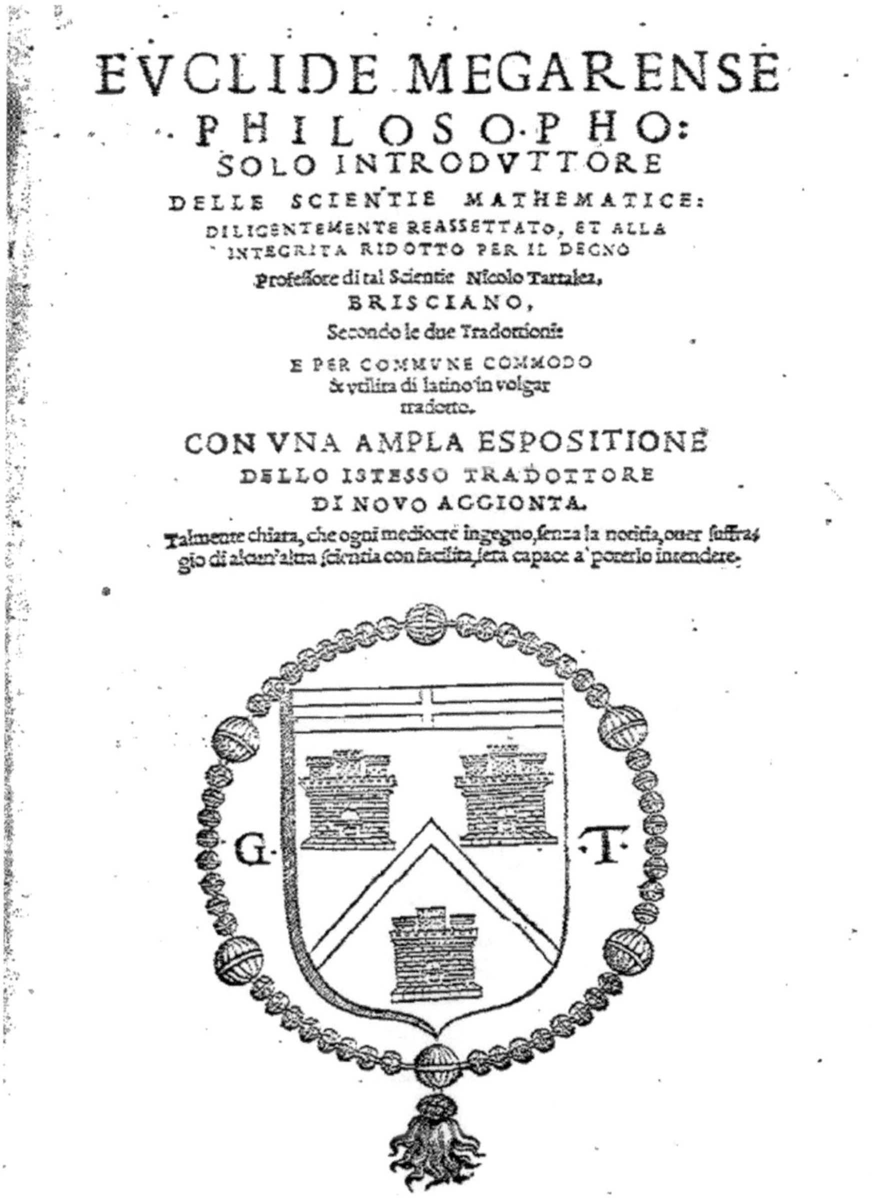
EVCLIDE MEGARENSE
PHILOSOPHO:
SOLO INTRODVTTORE
DELLE SCIENTIE MATHEMATICE:
DILIGENTEMENTE REASSETTATO, ET ALLA
INTEGRITA RIDOTTO PER IL DEGNO
Professore di tal Scientie Nicolo Tartalea,
BRISCIANO,
Secondo le due Tradottioni:
E PER COMMVNE COMMODO
& vtilita di latino in volgar
tradotto.
CON VNA AMPLA ESPOSITIONE
DELLO ISTESSO TRADOTTORE
DI NOVO AGGIONTA.
Talmente chiara, che ogni mediocre ingegno, senza la notitia, ouer suffragio di alcun'altra scientia con facilita, sera capace a' poterlo intendere.

Fig. 8 Euclide Megarense, 1543. (the first Italian translation of Euclid's Elements and the first printed version in a modern European language)

Archimedes' treatise was referred to in many of his works: in 'Travagliata Inventione' (Fig. 12) where he presents a complicated method of recovering a sunken ship (Fig. 10), based on Aristotle's principles, consisting in a gradual

LA PRIMA PARTE DEL GENERAL TRATTATO DI NVMERI, ET MISVRE DI NICOLO TARTAGLIA,
NELLAQVALE IN DIECISETTE LIBRI SI DICHIARA TVTTI GLI ATTI OPERATIVI, PRATICHE, ET REGOLE NECESSARIE NON SOLAmente in tutta l'arte negotiaria, & mercantile, ma anchor in ogni altra arte, scientia, ouer disciplina, doue interuenghi il calculo.

MALIGNITA'
NOIAR NON PVO
A' FORTEZZA

CON LI SVOI PRIVILEGII.
In Vinegia per Curtio Troiano dei Nauò.
M D LVI.

Fig. 9 General trattato di numeri et misure, 1556–1560 (a powerful treatise on arithmetic, geometry and algebra, published posthumously. The chapters on arithmetic were reedited posthumously in 1592–93 and translated in parts into other languages)

emptying out of two supporting ships (in the very same work he presents his version of a bell which allows man to stay underwater—it was never realized but stimulated future development) and in 'General trattato' (libro III, Chapters I and II,

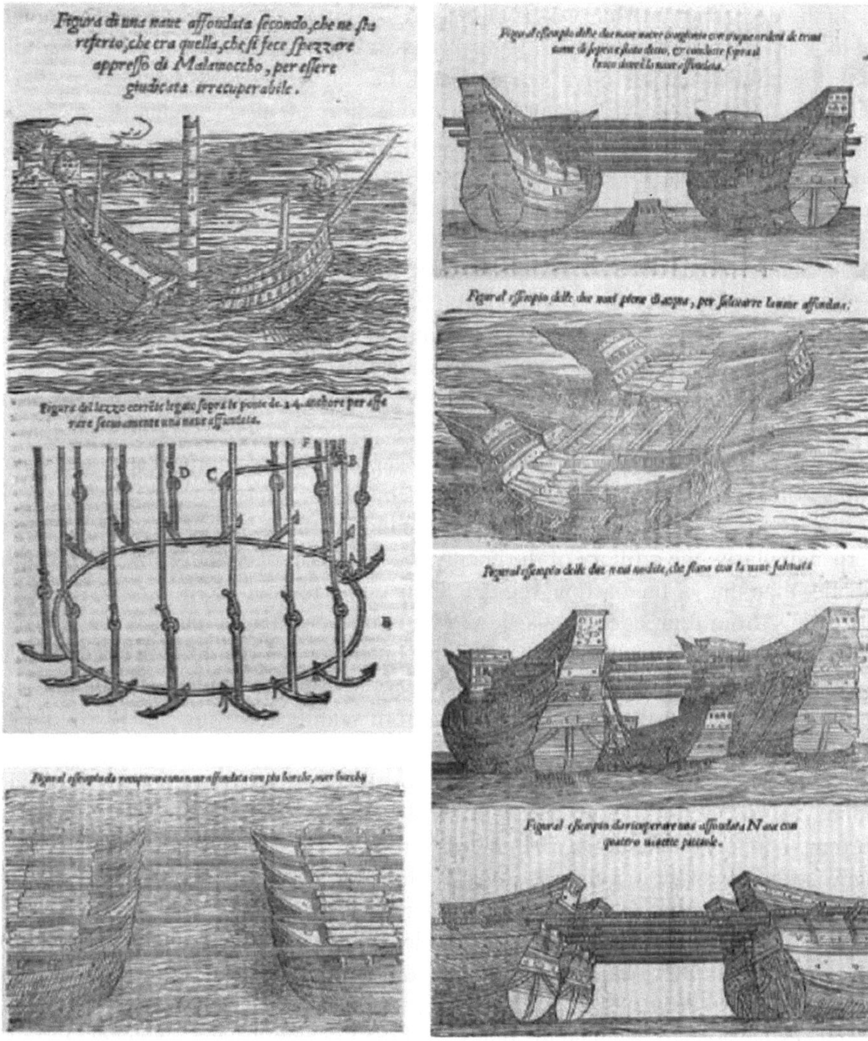

Fig. 10 Method of recovering sunken ships, from "Travagliata Inventione" (1551)

cc. 43v–60v and chapter III cc. 60v–63v) in which he describes 11 examples concerning the area and the volume of the sphere as well as 7 examples that pertain to parts of the area and of the volume, making reference to the famous Archimedes' work on measurement or squaring the circle.

Yet his most significant work which presents his discoveries with scientific rigor is undoubtedly the 'Quesiti et inventioni diverse', first published in Venice in 1554 and then, posthumously in 1562. For the first time the frontispiece contains a lithographic portrait of the author.

In the dedication to Henry VIII (1491–1547), king of England, he explains why he took up his first research into ballistics as well as why he continued his enquiry with the publication of 'nova scientia' and the 'quesiti et inventioni diverse'. He states that the latter work constitutes a collection of all the scientific-technical questions posed to him by various interlocutors while also being a compendium of the results of his discoveries. The first was the invention of an instrument for land surveyors (libro V, quesito primo, cc. 54v and foll.). He called it a "bossolo" as it is partly a compass. It is built of a large metal circle with a scale and a small compass in its centre which is also equipped with two alidades with "pinnule" moving around the centre and placed at the right angle to each other. Most likely, they are meant not to provide exact measurements but to enable orientation. Libro VI (book vi) is devoted to the ways of building fortifications of cities. All the 8 queries were posed to Tartaglia by Gabriele Tadino (1476–1543), Knight of Rhodes, prior of Barletta, who asked him if it was possible to fortify the city of Turin better than it was fortified at that time. In response Tartaglia—having pointed out the weak points of the Torino fortifications—enumerates six innovative requisites: it is necessary to avoid the situation where the cannonballs of the enemy hit the walls perpendicularly; the artillery of the enemy must not approach the walls at a distance smaller than that of the bastion; the defenders ought to have the possibility to fire at the enemy from four sides; walls should be built in such a way that when they fall into ruins under artillery attack, they will become even a more difficult obstacle than they were when they stood erect; it is necessary to construct additional (supporting) works, like moats, to prevent the enemy from scaling the walls; there should be some cultivable land within city walls.

These observations provided the basis for the construction of star forts or bastioned fortifications built in the form of a polygon with bastions at each apex forming support for cannons. Their sides were composed of tiers where cannonballs were stored. Even though the bastioned fortification was weaker at the front sides, it enabled the defenders to fight in all directions while reducing the possibility of enemy attack thanks to the wide moat filled with water. Its embankment (called glacis) protected the structure from direct cannon fire and at the same time met the requirement of symmetry. It is interesting to observe that while enumerating the requisites he stops to evaluate the efficacy of their design through sketches. In point of fact, sketches are for Tartaglia an important tool for evaluating the designs or for realizing physical models for studying some phenomena. In its initial form this methodology provides the basis for an important development.

Also in the 'Quesiti et inventioni diverse'—in libro IX, he explains how to inscribe a square into an equilateral triangle and provides a simple and original demonstration.

The successive work of 1547–1548, 'Cartelli di sfida matematica' (Fig. 11) is a response to the attack of Ludovico Ferrari who took upon himself to defend his master Gerolamo Cardano, at whom Tartaglia had launched several unfavourable epithets in libro IX of the 'Quesiti et inventioni diverse', connected with the solution of cubic equations.

I SEI CARTELLI

DI MATEMATICA DISFIDA

PRIMAMENTE INTORNO ALLA GENERALE RISOLUZIONE DELLE EQUAZIONI CUBICHE

DI

LODOVICO FERRARI

COI SEI CONTRO-CARTELLI IN RISPOSTA

DI

NICOLÒ TARTAGLIA,

COMPRENDENTI

LE SOLUZIONI DE' QUESITI DALL'UNA E DALL'ALTRA PARTE PROPOSTI

RACCOLTI, AUTOGRAFATI E PUBBLICATI

DA ENRICO GIORDANI,

Bolognese.

Premessa notizie bibliografiche ed illustrazioni sui Cartelli medesimi,
estratte da documenti già a stampa ed altri
manoscritti favoriti dal Comm. Prof. SILVESTRO GHERARDI, Preside dell'Istit. Tecn. Prov. di Firenze.

MILANO, 1876.

R. STABILIMENTO LITOGRAFICO DI LUIGI RONCHI
E TIPOGRAFIA DEGL' INGEGNERI.

Fig. 11 Cartelli di sfida matematica, 1547–1548 (analysis of various mathematical problems—testimony to the mathematical dispute he got involved in with Ludovico Ferrari in defense of Gerolamo Cardano)

This work is based on a number of written blows and repartees and is articulated into six files and six counter files which pose and partly solve 62 mathematical problems.

Particularly interesting is the one which demands a calculation of the volume of a semi-regular polyhedron with all its sides in the form of regular, though different polygons whose number is limited to 62 sides inscribable into a sphere.

To this number Ferrari limited the calculation of the altitudes drawn from the centre of the circular sphere to the various sides, suggesting that the solution can be provided by constructing a physical model. Tartaglia dealt with many problems from the files in his last work, the 'General trattato' of 1556–1560, which however he never completed or published in an integral form. Its very title bears significant meanings. On the one hand the term 'trattato' (treatise) refers to a research and a didactic tool in itinere, hence open to later development.

On the other hand, the term 'general' seems to express a wish to consider both theoretical aspects and those applicable in geometry, algebra and arithmetic's. The treatise is devoted mainly to the "indigitatio", i.e. the possibility to represent with the hands the rule of three (also called the golden rule) that allows us to solve problems of simple and complex interest and the discount.

A part of the work deals with mathematical riddles, i.e. mathematical brain-teasers and paradoxes. The best known among them is that of a wolf, a goat a cabbage and crossing the river (General trattato, parte I, libro XVI, n.141, c.257r) which is the source of the Italian saying "to save the goat and the cabbage".

He also writes about arithmetic problems connected with divination, that is, how to guess a number that the interlocutor has in his mind by making him communicate it underhand during the game. Presented in the General trattato is also the "triangolo di Tartaglia", an array of binomial coefficients, i.e. of coefficients serving to develop a binomial raised to any power. As a matter of fact, the triangle was not his invention. It had been known for a long time. The merit of Tartaglia was to popularize it. As has been said above, Tartaglia also invented various measurement instruments, the last of them is described in the General trattato and can be said to have been the precursor of the land-surveyor's cross-sight.

In geometry he was mainly interested in calculating the volume of the tetrahedron when the sides of the triangle at its base are unequal. He found the solution through the length of the six apices. He also provided the solution to the problem later termed as "the problem of Malfatti", i.e. that of how to inscribe two or three equal circles into an equilateral or isosceles triangle (parte V, Libro I, cap. X cc. 18v–19r). The final, sixth, part of the General trattato presents problems of algebra. Here Tartaglia does not explain anything new that has not been dealt with in the writings of 'masters of the abacus'. His argumentation, and especially the exercises, are so complex that one must have a profound knowledge of mathematics to grapple with them.

It is necessary to bring out yet another important aspect of Tartaglia's work, namely the language. He makes use of its two different forms: the Italian vernacular, a modern European language which ensures that knowledge be accessible to all, and the language of images. Some of the frontispieces preceding the text of his works provide a metaphor of his conception of various branches of mathematics. The sketches on the cover of Nova scientia convey different meaning depending on the level of interpretation. On the fundamental level they show that the knowledge of mathematics is the key to mastering other disciplines like geometry, arithmetics, astronomy, music, etc. On the higher level the picture of Plato holding a written motto: "Nemo huc geometriae expers igrediatur" (Let no one ignorant of geometry enter) seems to express that the Platonic science comes closer to philosophy than the Aristotelian approach which is closer to physical-mathematical sciences.

On the Circulation of Works

Tartaglia's views on the importance of knowing mathematics, on the relation between the theoretical and operative functions of knowledge based on mathematics, inspired an epistemological debate during the XVI century.

The mathematical rivalry between Tartaglia and Ferrari to demonstrate certain Euclidean theorems with the help of a ruler and a fixed compass became the issue for quite a number of mathematicians resulting in multiple publications on the subject. The most eminent mathematician who analyzed the problem was G.B. Benedetti (1530–1590). The solution of the equation of the third degree (x3 + q = px) stimulated the study of complex numbers guaranteeing that Tartaglia's formula could be used even when the solution was a negative number, an idea never even considered before Tartaglia's times. His explanations concerning the economy of loans as well as his calculus of simple and complex interest are common practice to this very day.

The volume of the tetrahedron was the problem taken up also by scholars of later date, like Euler (1707–1783), Lagrange (1736–1813), Monge (1746–1818), who found in it a number of exceptional properties. The same can be said about the problem of inscribing three circles into a triangle, which became an object of remarkable studies in the XIX century and gave rise to the science of stereotomy. Only in 1803 was the problem solved analytically by Gian Francesco Malfatti (1731–1807), for which it started to be called "the problem of Malfatti".

As far as fortifications are concerned, Tartaglia demonstrated that their weakest points were the surfaces of the bastion's front sides because they are covered only by the side of the nearest bastion and thus constitute the favourite aim of enemy attack. These weaknesses can be overcome in many ways—the basis of a number of fortification systems which evolved between the end of the XVI century and the XVIII and can approximately be grouped into three main systems: Italian, Dutch and French.

The most significant and fruitful, however, were his studies in ballistics. Until his times ballistics was not even considered a branch of science but an aggregate of non-formalized ways of acting made applicable through the personal experience of people. Tartaglia's studies provided an impulse for further studies which started in the XVI century (Figs. 12 and 13).

RAGIONAMENTI DE NICOLO
TARTAGLIA SOPRA LA SVA TRAVAGLIATA
INVENTIONE.
Nelli quali se dechiara uolgarmente quel libro di Archimede Siracusano Intitolato.
De insidentibus aquae, Con altre speculatiue pratiche da lui ritrouate sopra le
materie, che stano, et chi non stano sopra lacqua, Vltimamente se asse-
gna la ragione, et causa naturale di tutte le sottile, et oscure
particularità dette, et dechiarate nella detta sua
trauagliata inuentione cō molte altre
da quelle dependenti.

Apresso di Lautore.
Cō gratia, et priuilegio del Illustriss. Senato Veneto che niun possa stāpare ne far stāpa
re la presente operina ne parte di quella, ueder ne far uedere in Venetia, ne in alcun al
tro loco, o terra del dominio Veneto per anni dieci senza consentimento del Autore
[illegible] perder le opere. Come che nel preuilegio appare.

Fig. 12 Travagliata inventione (con la regola generale, i ragionamenti e il supplemento), 1551. (presents procedures to recover sunken ships and contains also a partial Italian version with Tartaglia's commentary on the first book De insidentibus acque di Archimede)

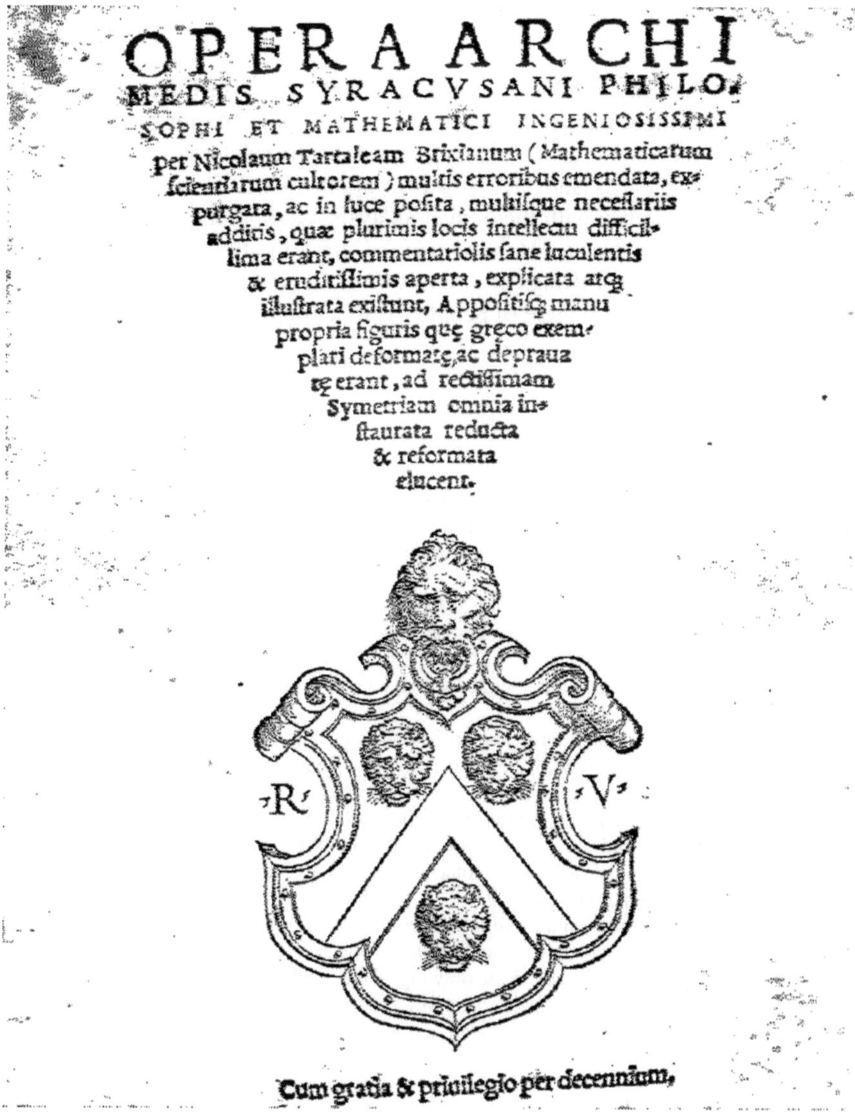
OPERA ARCHIMEDIS SYRACVSANI PHILOSOPHI ET MATHEMATICI INGENIOSISSIMI
per Nicolaum Tartaleam Brixianum (Mathematicarum scientiarum cultorem) multis erroribus emendata, expurgata, ac in luce posita, multisque necessariis additis, quae plurimis locis intellectu difficillima erant, commentariolis sane luculentis & eruditissimis aperta, explicata atq; illustrata existunt, Appositisq; manu propria figuris quę greco exemplari deformatę, ac deprauatę erant, ad rectissimam Symetriam omnia instaurata reducta & reformata elucent.

Cum gratia & priuilegio per decennium.

Fig. 13 Opera Archimedis, 1543 (the collection of three works of Archimedes, the mathematician of Siracusa: De centri gravium libri duo, Tetragonismus ovvero De quadratura circuli e De insedentibus aquae, translated into Latin by William of Moerbecke)

Conclusions

To Nicolò Tartaglia goes the historical merit to have preserved and spread much fundamental knowledge, mainly in mathematics, indispensable for practical applications. On the other hand, however, he invented instruments and apparatus necessary for various sciences and their practical applications. There he manages to unite experimental enquiry with theoretical analysis—the procedure that lies at the foundations of modern science. He also was the first to publish and translate with his commentary scientific writings of antiquity and made the results of his research accessible to a much greater circle of people by disseminating them in printed form.

Bibliography

Amodeo F (1908) Sul corso di storia delle matematiche fatto nella Università di Napoli nel biennio 1905/1906-1906/1907, B. G. Teubner Berlino

Baldi B (1707) Cronica de matematici: overo Epitome dell' istoria delle vite loro, Angelo Antonio Monticelli Urbino

Bassi G (1666) Aritmetica e geometria pratica, Piacenza

Benedetti GB (1585) Diversarum speculationum mathematicarum & physicarum liber, Beuilaquae Torino

Biggiogero G (1936) La geometrica del tetraedro, Enciclopedia delle matematiche elementari, vol II, pp 220–245

Bittanti L (1871) Di Nicolò Tartaglia matematico bresciano, Tip. Apollonio Brescia

Boffito G (1929) Gli strumenti della scienza e la scienza degli strumenti. Multigrafica Editrice Roma

Boncompagni B (1881) Testamento inedito di Nicolò Tartaglia, Hoepli Milano

Bortolotti E (1926) I contributi del Tartaglia, del Cardano, del Ferrari, e della Scuola matematica bolognese alla teoria algebrica delle equazioni cubiche, Studi e memorie per la storia dell'Università di Bologna, vol IX, p 89

Capobianco A (1618) Corona e palma militare di artiglieria, Bariletti Venezia

Capra A (1717) La nuova architettura civile e militare, Ricchini Cremona

Ceccarelli M, Cigola M (2001) Trends in the drawing of mechanisms since the early Middle Ages, Journal of Mechanical Engineering Science, vol 215, pp 269–289

Cerasoli M (1981) Two matrices constructed in the manner of Tartaglia's triangle (Italian), Archimede,vol 33, pp 172–177

Chirone E, Cambiaghi D (2007) Meccanica e Macchine nella rappresentazione grafica fra Medioevo e Rivoluzione Industriale, Atti del Convegno Internazionale XVI ADM/XIX Ingegraph, Perugia

Chirone E, Pizzamiglio P (2008) Niccolò Tartaglia matematico e ingegnere, Atti del 2°Convegno di Storia dell'Ingegneria, Napoli, pp 1051–1060

Cigola M (2012) In praise of parallel theories: Descriptive Geometry and Applied Mechanics. In: Carlevaris L, Filippa M (eds) In praise of theory. The fundamentals of the disciplines of representation and survey, Gangemi editore Roma, pp 39–46

Cigola M, Ceccarelli M (1995) On the evolution of Mechanisms drawing. In: Proceedings of IXth IFToMM World Congress Politecnico di Milano, vol 4, pp 3191–3195

Costabel P (1973) Vers une mécanique nouvelle. In: Roger J (eds) Sciences de la renaissance, Paris, pp 127–142

Demidov SS (1970) Gerolamo Cardano and Niccolo Tartaglia, Fiz.-Mat. Spis. Bulgar. Akad Nauk, vol 13, pp 34–47
De Pace A (1993) Le matematiche e il mondo. Ricerche su un dibattito in Italia nella seconda metà del Cinquecento, Franco Angeli Milano
Di Pasquale L (1957) Le equazioni di terzo grado nei "Quesiti et inventioni diverse" di Nicolò Tartaglia, Periodico di matematiche Ser. IV, vol 35 n°2, p 79
Drake S, Drabkin IE (1969) Mechanics in Sixteenth-Century Italy: Selections from Tartaglia, Benedetti. The University of Wisconsin Press Madison, Guido Ubaldo & Galileo, pp 61–143
Dugas R (1955) Histoire de la Mecanique, Griffon Neuchatel
Favaro A (1883) Notizie storico-critiche sulla divisione delle aree, Memorie del R. Istituto veneto di scienze, lettere ed arti 22:151–152
Gabrieli GB (1986) Nicolo Tartaglia: invenzioni, disfide e sfortune. Università degli studi di Siena
Gabrieli GB (1997) Nicolò Tartaglia. Una vita travagliata al servizio della matematica, Biblioteca Queriniana Brescia
Geppert H (1929) Sulle costruzioni geometriche che si eseguiscono colla riga ed un compasso di apertura fissa, ibid., vol 9, pp 303–309, pp 313–317
Giordani E (1876) I sei cartelli di matematica disfida primamente intorno alla generale risoluzione delle equazioni cubiche di Lodovico Ferrari coi sei contro-cartelli in risposta di Nicolò Tartaglia comprendenti le soluzioni de' questi dall' una e dall' atra parte proposti, Ronchi Milano
Keller A (1971) Archimedean Hydrostatic Theorems and Salvage Operations in 16th-Century Venice. Technol Cult 12:602–617
Kiely ER (1947) Surveying Instruments, New York, pp 210–11
Klemm F (1966) Storia della tecnica, Feltrinelli Milano
Koyré A (1960) La dynamique de N. Tartaglia, La science au seizième siècle - Colloque international de Royaumont 1957, pp 91–116
Libri G (1840) Histoire des sciences mathématiques en Italie: depuis la renaissance des lettres jusqu'à la fin du 17e siècle, J. Renouard Paris
Mackay JS (1887) Solutions of Euclid's Problems. With a Ruler and One Fixed Aperture of the Compasses, by the Italian Geometers of the Sixteenth Century, Proceedings of the Edinburgh Mathematical Society 5:2–22
Manzi P (1976) Architetti e ingegneri militari italiani: dal secolo XVI al secolo XVIII, Istituto Storico e di Cultura dell'Arma del Genio Roma
Maracchia S (1979) Da Cardano a Galois. Momenti di storia dell'algebra. Feltrinelli Milano
Marcolongo R (1919) Lo sviluppo della meccanica sino ai discepoli di Galileo, Tip. Accademia dei Lincei Roma
Masotti A (1962) Quarto centenario della morte di Niccolo' Tartaglia, Atti del convegno di storia delle matematiche 30–31 maggio 1959, La Nuova Cartografica Brescia
Masotti A (1962) Atti del convegno in onore del Tartaglia, Ateneo di Brescia Brescia
Masotti A (1963) N. Tartaglia, Storia di Brescia vol II, Giovanni Treccani degli Alfieri Brescia, pp 597–617
Masotti A (1963) Gli atti del Convegno di Storia delle Matematiche tenuto in commemorazione di Nicolò Tartaglia: parole di presentazione nell'adunanza del 16 giugno 1962, F.lli Geroldi Brescia
Masotti A (1974) Gabriele Tadino e Niccolo' Tartaglia, Atti dell'Ateneo di scienze, lettere ed arti, vol 38, pp 363–374
Montagnana M (1958) Nicolò Tartaglia quattro secoli dopo la sua morte, Archimede. Rivista per gli insegnanti e i cultori di matematiche pure e applicate, vol X, nn. 2–3, pp 135–139
Natucci A (1956) Che cos'è la "Travagliata inventione" di Nicolò Tartaglia, in Period. Mat. vol 34, pp 294–297
Olivo A (1909) Sulla soluzione dell'equazione cubica di Nicolò Tartaglia: Studio storico-critico, Tip. A. Frigerio Milano
Oddi M (1625) Dello squadro, Milano

Piotti M (1998) Un puoco grossetto di loquella. La lingua di Niccolò Tartaglia (la "Nova scientia" e i Quesiti et inventioni diverse). LED Milano

Pizzamiglio P (2010) Atti della Giornata di Studio in memoria di Niccolò Tartaglia nel 450° anniversario della sua morte, Supplemento ai Commentari dell'Ateneo di Brescia, Ateneo di Brescia Brescia

Pizzamiglio P (2012) Nicolò Tartaglia nella storia con antologia degli scritti, EDUCatt Milano

Pizzamiglio P (2005) Niccolò Tartaglia (1500ca.–1557) nella storiografia, Atti e Memorie—Memorie scientifiche, giuridiche, letterarie, ser. VIII, vol VIII fasc. II, pp 443–453

Pizzamiglio P (2000) Niccolò Tartaglia, Tutte le opere originali riprodotti su CD, Biblioteca di Storia delle Scienze Carlo Viganò Brescia

Pollack MD (1991) Turin 1564–1680, University of Chicago Press

Rossi G (1877) Groma e squadro, ovvero storia dell' agrimensura italiana dai tempi antichi al secolo XVII°, Loescher Torino

Sansone G (1923) Sulle espressioni del volume del tetraedro, Periodico di matematiche, vol 3, pp 26–27

Toscano F (2009) La formula segreta. Tartaglia, Cardano e il duello matematico che infiammò l'Italia del Rinascimento, SIRONI Milano

Wauvermans H (1876) La fortification de N. Tartaglia, Revue belge d'art, de sciences et de technologie militaires, vol 1, pp 1–42

Zilsel E (1945) The Genesis of the Concept of Scientific Progress. J Hist Ideas 6(3):325–349

Federico Commandino (1509–1575)

Ornella Zerlenga

Abstract During the sixteenth century, Federico Commandino was drawn to the attention of the scientific and cultural community for his role as an erudite scholar, as well as his contributions to the disciplines of Mechanics and Descriptive Geometry. To Commandino can be attributed important Latin translations of Greek texts as well as the furthering of scientific knowledge on determination of the centre of gravity and the concept of perspective.

Introduction

Fourteenth century Humanism led to the search for study and circulation of the works of classical poets, philosophers and historians, which over time also reached the field of exact sciences. However, only with the invention of movable type and printing did the rediscovery of classical texts in mathematics and geometry exert all its influence to the benefit of a wider scientific community.

In fact, during the sixteenth century, the works of the great Greek mathematicians, along with several minor ones, were published. In 1505, the Venetian mathematician and humanist Bartolomeo Zamberti (XV–XVI century) edited the first Latin translation of the Greek work "Elementi" by Euclide (III–II century B.C.) with the title "Euclidis Megarensis philosophi platonici mathematicarunt disciplinarum janitoris". It is worth highlighting the erroneous identification by Zamberti of the Greek mathematician Euclide (III–II century B.C.) with the Socratic philosopher Euclide of Megara (V–IV century B.C.). In 1533, "Elementi" was published in Greek in Basel and in 1544 in Arabic in Rome, while in 1570, the first English translation was edited by Sir Henry Billingsley (XVI century–1606), who made the

O. Zerlenga (✉)
Department of Architecture and Industrial Design "Luigi Vanvitelli",
Second University of Naples, Via San Lorenzo ad Septimum,
81031 Aversa, Caserta, Italy
e-mail: ornella.zerlenga@unina2.it

© Springer International Publishing Switzerland 2016

M. Cigola (ed.), *Distinguished Figures in Descriptive Geometry and Its Applications for Mechanism Science*, History of Mechanism and Machine Science 30, DOI 10.1007/978-3-319-20197-9_5

same mistake of confusing the mathematician Euclide of Alexandria with the philosopher Euclide of Megara. Similarly, in 1560, the first Latin version of the "Commento al I Libro degli Elementi di Euclide di Proclo Licio Diadoco" (412–485) by Francesco Barozzi (1537–1604) was published in Venice. In 1518 and 1557, the first versions of the Latin "Sferica" by Teodosio Tripolita (I century B.C.) were produced. In 1575, the first version of "Arithmetica" by Diofanto of Alexandria (II–III century) was published. In 1537, the first Latin edition of the first four books of the "Coniche" by the Greek mathematician Apollonio of Perga (approx. 262–180 B.C.) was translated by Giovan Maria Memmo (1503 or 1504–1579).

However, the publication of some of these early versions (often taken from original manuscripts that were incorrectly copied by scribes) diffused imperfections and errors and led to misunderstandings. This subsequently led to the mathematicians of the time not only fully reviewing the published texts, correcting and commenting on them, but also undertaking original research on unfinished mathematical determinations. Thus, this also constituted the primary reason for the revival of mathematics and geometry research throughout the sixteenth century. In this scientific and cultural context of the dissemination and reinterpretation of classic mathematical thinking, Italy played a particularly prominent role through the work of two eminent figures, Francesco Maurolico (1494–1575) and Federico Commandino (1509–1575). To these two mathematicians, who studied classical texts while maintaining a fruitful correspondence, can be attributed the correct translation into Latin of the most important works of Greek mathematics and geometry as well as numerous original studies for the advancement of scientific investigations in these fields.

Specifically, in this scientific-mathematical retelling of classic (and late-medieval) works, Federico Commandino translated into Latin and/or Italian some of the writings of Archimede (287–212 B.C.), Euclide (III–II century B.C.), Pappo (320–? B.C.), Aristarco Samio (310–230 B.C.), Tolemeo (100–175), Giordano Nemorario (XIII century), Erone (I century?). Among his original scientific studies, there is the determination of the centre of gravity of a pyramid, the building in the perspective of a point and the circles of a sphere. In particular, the mathematical writings of Commandino on linear perspective laid the foundation for studies in an area that would be purely pictorial to a geometrician-mathematician.

Biographical Notes

Federico Commandino (1509–September 3, 1575) was born and died in Urbino. His family, of noble origins, oriented the young man towards the study of humanities and sciences, continuing a family tradition strongly marked by a scholarly and masterful education (Fig. 1).

His humanistic education was therefore detailed and consolidated, with not only the study of Greek and Latin under the guidance of Giacomo Torelli da Fano (XVI

Fig. 1 Federico Commandino in a sixteenth century Baroque school portrait

century), but also the learning of scientific subjects, such as mathematics, with Commandino being introduced to determinations by Gian Pietro de'Grassi (XVI century). He lived in Rome for a few years, starting from 1530 when Gian Pietro de'Grassi entered the service of Cardinal Niccolò Ridolfi (1501–1550) and took the twenty year old Commandino with him. In Rome, Commandino furthered his mathematical studies and had the opportunity—through Cardinal Ridolfi's library—to consult some Greek manuscripts, and perhaps became personal secretary to Pope Clement VII (1478–1534) participating in scientific debates on the Copernican system then in place.

After 1534, the year of the death of Clement VII, Commandino moved to Padua to study philosophy, medicine and mathematics, the latter under the guidance of Federico Delfino (1477–1547). In 1544, Commandino moved to Ferrara where, according to his disciple and biographer Bernardino Baldi (1553–1617), in 1546 he was awarded a doctorate in medicine even if there seems to be no trace in the archival documents of the University of Ferrara. In the same year, Commandino returned to Urbino, where he married and had children, but the untimely death of his son and wife was a source of distrust in medical studies to the point of him abandoning medicine forever. A meeting with Cardinal Ranuccio (1530–1565) who, like Commandino, cultivated a passion for the humanities and mathematics led him to being employed by the family until the death of the cardinal in Rome, where he now had an intense interest in the culture of mathematics and the rediscovery of Greek manuscripts. In addition, frequent communications with Cardinal Farnese and a subsequent meeting with Cardinal Marcello Cervini (1501–1555), allowed Commandino to access the consultation versions of mathematical texts in Greek, which would have been otherwise impossible.

Due to a good knowledge and understanding of Greek and Latin, Commandino had an acute critical and decisive mathematical sense, thus enabling him to devote his entire life to the publication of classical geometry texts, correctly translated and commented with shrewd and advanced scientific considerations. Due to his excellent translations skills, it is possible to recognise Federico Commandino as one of the greatest translators of mathematical works of the Greek-Hellenistic period. At the same time, his knowledge of Greek, Latin and mathematics allowed him to cover an important scientific role in the revival of mathematical speculation in Europe during the sixteenth century, offering the scholars of the time the opportunity to access the most important works on mathematics and geometry that had long remained unexplored. In this sense, his careful translations laid the groundwork for the more mature reflection of Galilean thought, even if, apart from the translations of Euclid and Apollonius, his works were never widely disseminated. Commandino was therefore a precursor of the rediscovery of the great classics of ancient mathematics, promoting them and paving the way to the masters of modern mathematics, who through his translations took inspiration from ancient mathematical thought. His profound knowledge of the humanities and mathematics allowed him to be the focus of an intense network of scientific and cultural speculation. After Cardinal Farnese, Commandino had the opportunity to meet the mathematicians Niccolò Tartaglia (1499–1557) and Giambattista Benedetti (1530–1590) in Venice and during his second stay in Rome, he came into contact with the writer and mathematician Annibale Caro (1507–1566). Commandino's active dedication to mathematical studies is also demonstrated by the "Dedications" with which he began his works. In them, he always proclaimed the superiority of the "mathematicae disciplines" over all the other sciences, including philosophy, while praising the great mathematicians of antiquity including Archimedes of Syracuse (287–212 B.C.). In fact, when considering Archimedes, he claimed that only those who engaged in the study can be considered true mathematicians.

Federico Commandino is also attributed with the invention of the polymetric compass in 1568, the forerunner of the geometric and military compass upon which the controversy between Galileo Galilei (1564–1642) and Baldassare Capra (1580–1626) is based. In the last years of his life, Commandino also held mathematics classes in his native town, Urbino, and in addition to the son of the Duke of Urbino, his pupils also included Guidobaldo del Monte (1545–1607), Bernardino Baldi (1553–1617) and Torquato Tasso (1544–1595).

List of Works

"Ptolemaei Planisphaerium. Iordani Planisphaerium. Federici Commandini Vrbinatis in Ptolemaei Planisphaerium commentarius. In quo uniuersa scenographices ratio quam breuissime traditur, ac demonstrationibus confirmatur". Venetiis: Paolo Manuzio, 1558.

"Archimedis Opera non nulla à Federico Commandino Vrbinate nuper in Latinum conuersa, et commentariis illustrata". Venetiis: Paulum Manutium, Aldi f., 1558.

"Clavdii Ptolemaei Liber de analemmate: a Federico Commandino Urbinate instauratus, & commentariis illustratus, qui nunc primum eius opera e tenebris in lucem prodit. Eiusdem Federici Commandini liber de Horologiorum descriptione". Romae: Apud Paulum Manutius Aldi F., 1562.

"Archimedis De iis quae vehuntur in aqua libri duo. A Federico Commandino Urbinate in pristinum nitorem restituti, et commentariis illustrati". Bononiae: ex officina Alexandri Benacii, 1565.

"Federici Commandini Vrbinatis Liber de centro gravitatis solidorum". Bononiae: ex officina Alexandri Benacii, 1565.

"Apollonii Pergaei Conicorum libri quattuor. Una cum Pappi Alexandrini Lemmatibus, et commentariis Eutocii Ascalonitae. Sereni Antinsensis philosophi Libri duo nunc primum in lucem editi. Quae omnia nuper Federicus Commandinus Urbinas mendis quamplurimis expurgata è Graeco convertit, & commentariis illustravit. Sereni Antinsensis libri duo". Bononiae: ex officina Alexandri Benatii, 1566.

"De superficierum diuisionibus liber Machometo Bagdedino ascriptus… Nunc primum Ioannis Dee Londinensis, & Federici Commandini Vrbinatis opera in lucem editus. Federici Commandini De eadem re libellus". Pisauri: apud Hieronymum Concordiam, 1570.

"Euclidis Elementorum libri XV. Unà cum Scholijs antiquis. A' Federico Commandino Vrbinate nuper in latinum conversi, commentarijsque quibusdam illustrati". Pisauri: Iacobus Chriegher German, 1572 (Pisauri: apud Camillum Franceschinum, 1572).

"Aristarchi De Magnitudinibus, et Distantiis Solis, et Lunae, Liber, cum Pappi Alexandrini explicationibus quibusdam. A' Federico Commandino Urbinate in latinum conversus, ac commentarijs illustratus". Pisauri: apud Camillum Francischinum, 1572 (Pisauri: apud Camillum Francischinum, 1572).

"De gli elementi d'Euclide libri quindici. Con gli scholii antichi. Tradotti prima in lingua latina da M. Federico Commandino da Vrbino, & con commentarij illustrati, et hora d'ordine dell'istesso trasportati nella nostra vulgare, & da lui riueduti". Urbino: appresso Domenico Frisolino, 1575 (In Vrbino: In casa di Federico Commandino, 1575).

"Heronis Alexandrini. Spiritalium Liber, a Federico Commandino urbinate, ex graeco, nuper in latinum conversus", Urbini 1575 (postumo).

"Pappi Alexandrini Mathematicae collectiones à Federico Commandino Vrbinate in latinum conversae, et commentariis illustratae" (Pisauri, apud Hieronymum Concordiam, 1588). Bononiae: ex typographia HH. de Duccijs, 1660 (Bononiae: ex typographia HH. de Ducijs, 1658).

Review of Main Works

In 1558, Commandino published in the printing works of Paolo Manuzio in Venice his Latin translation of the work by Claudius Ptolemy (100–175), whose full title reads: "Ptolemaei Planisphaerium. Iordani Planisphaerium. Federici Commandini Vrbinatis in Ptolemaei Planisphaerium commentarius. In quo uniuersa scenographices ratio quam breuissime traditur, ac demonstrationibus confirmatur".

Ptolemy, an astronomer and geographer, was a leading figure in the world of ancient sciences. His cosmological system (based on a mathematical model which placed the Earth at the centre of the universe around which the Sun, the Moon and the other planets moved) was described by Ptolemy in the book "Almagest" and remained unsurpassed for centuries. Ptolemy not only wrote a text entitled "Geografia" (an equally important work that contained maps with the coordinates of many places of antiquity, that had never been discovered) but he also devoted himself to the discussion of the projections of points on the celestial sphere in the book "Analemma" as well as solid projections in "Planisphaerium". In the latter, Ptolemy described stereographic projection in which the points of the celestial sphere are projected along lines that go from one pole to a plane (from the South Pole to the plane of the equator). Ptolemy sensed that in this transformation, a circle not passing through the centre of projection became a circle in the plane and a circle passing through the pole was projected in a straight line. Furthermore, the transformation was consistent and the angles were maintained. This projection was recognised as the mathematical basis for the construction of the astrolabe, one of the most popular medieval astronomical instruments, of which Ptolemy described the operations for surveying surfaces as well as how to calculate the times of sunrise.

Federico Commandino dedicated his book to Cardinal Ranuccio Farnese. In the dedication, Commandino introduces the usefulness of the translation, but, above all, expresses his personal opinion about a possible use in architecture of some of the determinations made by Ptolemy and, more specifically, as the theoretical-operative basis to develop linear perspective. After the dedication, Commandino places in order "Claudii Ptolomei. Sphaerae a planetis proietio in planum" (page 26) and Iordanus de Planisphaerii figurazione" (page 23), contribution, the latter, of Giordano Nemorario (XIII century), mathematician and physicist, founder of the medieval school of mechanics that was interested in issues related to the displacement of masses, along with the motion of weights on inclined planes as well as the study of astronomy, including the work of Ptolemy. Commandino finally concluded the work with his own contribution entitled "Federici Commandini urbinatis in Planisphaerium Ptolomei Commentarius" (page 64). The entire volume is amply illustrated with geometric designs (Fig. 2).

It is worth noting "Commentarius" where, as it can be deduced from the title "In quo uniuersa scenographices ratio quam breuissime traditur, ac demonstrationibus confirmatur", Commandino felt that the work of Ptolemy could also be helpful for the architects with the use of "Scenographia". In fact, the work by Ptolemy contained the theoretical foundations of stereographic projection, i.e. a method of

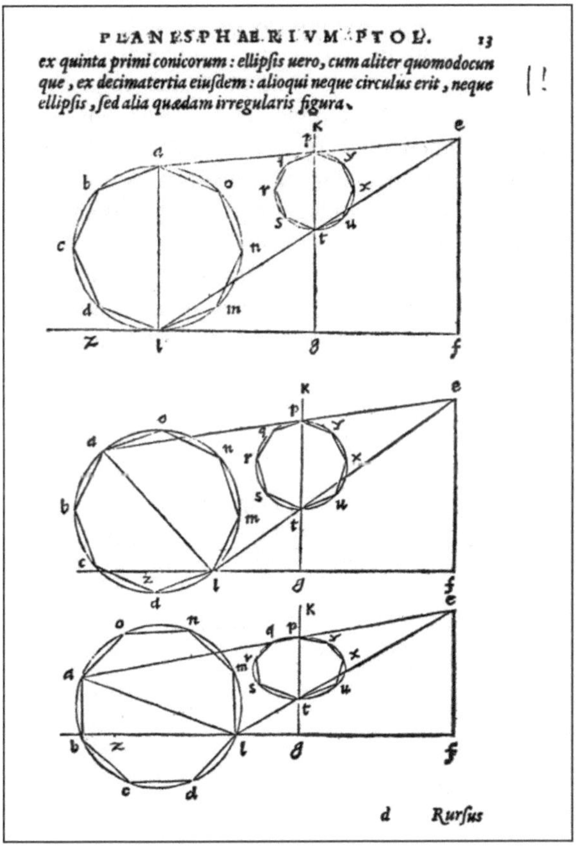

Fig. 2 Federico Commandino, "Ptolomaei Planishaerium" (Venetiis, 1558)

geometric representation of the figures in the plane. Therefore, Commandino commented the work developing considerations that led him to take some interesting positions on linear perspective. Commandino referred to figures considered to be on two mutually perpendicular planes (in vertical and horizontal positions and to which to attribute the plane and elevation of a body) and assumed for purposes of the representation, a third plane (perpendicular to the previous) and for perspective, a point on the vertical plane. Overturning the framework on the vertical plane and thus arranging the fixed elements, Commandino described the case in which the figures to be represented in perspective belonged to a horizontal plane and illustrated two simple geometric constructions for any point of the plane. From this assumption, he continued with the application of the perspective of the circles of a sphere. However, the construction was so theoretical that the painters of the time did not apply it.

In 1558 and dedicated to Cardinal Ranuccio Farnese, among the writings reserved to Archimedes of Syracuse (287–212 B.C.), Commandino published in Venice "apud Paulum Manutium, Aldi F.", the Latin version (translated from the

Greek) entitled “Archimedis Opera non nulla à Federico Commandino Vrbinate nuper in Latinum conversa, et commentariis illustrata”. The work included five treaties (in the order: “Circuli dimensio”; “De lineis spiralibus”; “Quadratura paraboles”; “De conoidibus, et sphaeroidibus”; “De arenae numero”) followed by the “Commentarii” by Commandino (Fig. 3). Archimedes, the greatest physicist-mathematician of Hellenism, was a scholar of geometry and mechanics and his scientific production showed brilliant and original research skills to the point that his works could only be understood by a public that was admittedly well-versed on the subject. However, the legacy of Archimedes cannot be read chronologically, due to the lack of a reliable dating but rather thematically, as collected by Commandino in his translation of the five books.

The book “Circuli dimensio” (a booklet to be used in study of a larger work, the original of which has yet to be discovered) constituted an important contribution by Archimedes to the problems of squaring the circle and the rectification of the

ARCHIMEDIS
OPERA NON NVLLA
A FEDERICO COMMANDINO
VRBINATE
NVPER IN LATINVM CONVERSA,
ET COMMENTARIIS
ILLVSTRATA.
Quorum nomina in sequenti pagina leguntur.

CVM PRIVILEGIO IN ANNOS X.
VENETIIS,
apud Paulum Manutium, Aldi F.
M D LVIII.

Fig. 3 Federico Commandino, “Archimedis Opera non nulla” (Venetiis, 1558)

circumference. Unlike the mathematicians who had preceded him, Archimedes did not show the area of the circle with the geometric constructions that they had used (as usual) ruler and compass but resorted to the determination of an issue of "measure", arriving with satisfactory approximation to the value of the relationship π of the circumference. Similarly, the contribution "De lineis spiralibus" constituted a step forward in the definition of the fundamental properties of plane curves commonly called "Archimedean spiral". It seems that these curves were introduced for the first time by Conon of Samos (280–220 B.C.), but it is without doubt Archimedes to whom can be attributed their analytical study (including, rectification, tangent, quadrature) and especially the hypothesis that these curves could be represented by equations in polar coordinates.

In the book "Quadratura paraboles", Archimedes obtains for the first time the original determinations for the squaring of a flat area not bounded by straight or circular lines. These come from his studies on the equilibrium of planes, which are the first scientific treatment of static and mechanical behaviour with a method consistent with that adopted by Euclid (III–II century B.C.) and where every point was characterized by the three spatial coordinates plus its weight. The general results of this study were then applied for the first time in the field of geometry to the parabola. The contribution of "De conoidibus, et sphaeroidibus" was equally innovative, where Archimedes excellently dealt with second-order rotation surfaces, i.e. those surfaces that are generated by the rotation of an ellipse around one of its axes, of a parabola around its axis, of a branch of hyperbola around its transverse axis (with the exception of an hyperboloid to a flap rotation). In relation to these remarkable surfaces, Archimedes studied the expression of the volumes of those figures that are generated by intersecting them with planes perpendicular to the axis, anticipating the integral calculus and arriving at results that can be extended to any quadric surface.

Commandino completed the Latin translation of Archimedes works with a final contribution entitled "De arenae numero", a real gem of scientific literature. Archimedes surpassed the limits of the mathematical thinking of his time and tried to give a name to numbers that cannot only count the grains of sand that surrounded Syracuse and the then known universe, but also to an even bigger multiplicity. Archimedes' aim was to be able to represent the biggest number imaginable. He showed that the sequence of numbers could be extended indefinitely, thus anticipating a numbering system equivalent to the positional system.

Already in that time, the translation into Latin of the remarkable works of Archimedes by Commandino constituted a contribution of great excellence for the advancement of scientific thought. In his work as a translator, Commandino integrated the gaps and commented on the Archimedean demonstrations, defining Archimedes as the greatest mathematician of all time. Commandino added to his translations and corresponding notes, the translated version by Byzantine mathematician Eutocius of Ascalon of Ascalon (480–540?) of "De circuli dimensione", which constituted an important source of the unknown aspects of the history of Greek mathematics.

In 1562, in the absence of the original Greek version and thanks to the code given to him by Cardinal Marcello Cervini (1501–1555), Commandino published in Rome with Paolo Manuzio "Claudii Ptolemaei liber de analemmate, a Federico Commandino Urbinate instauratus, et commentariis illustratus, qui nunc primum eius opera e tenebris in lucem prodit", followed by "Liber de Horologiorum descriptione" (Fig. 4).

In Greek literature, and by Ptolemy, three different methods were conceived to represent the earth's surface on a plane, the model of which was assimilated to a sphere. Ptolemy developed these methods in three of his various works, "Geografia", "De Plaenisphaerium" and "De analemmate" where, in order, he anticipated the criteria formulated by Gerardus Mercator (1512–1594), discussing stereographic projection, introducing the orthogonal projection to a sphere on three mutually orthogonal planes to support the construction of sundials. In the third work, "De analemmate", Ptolemy formulated a graphical method to calculate the angles of sundials as well as to obtain the projection on a plane of sundials. This graphical method (which poses in relation the possibility of determining the angles

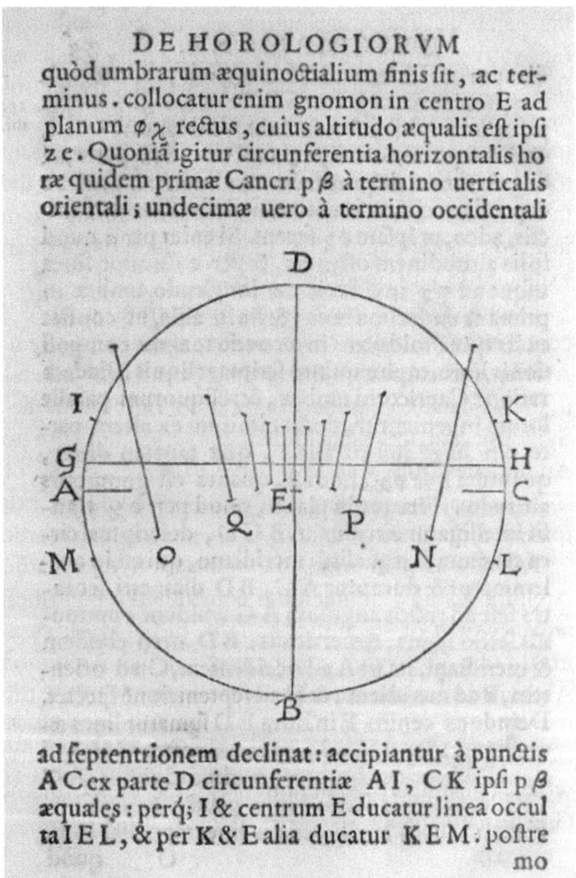
DE HOROLOGIORVM
quòd umbrarum æquinoctialium finis sit, ac ter-
minus. collocatur enim gnomon in centro E ad
planum φχ rectus, cuius altitudo æqualis est ipsi
z e. Quoniã igitur circunferentia horizontalis ho
ræ quidem primæ Cancri p β à termino uerticalis
orientali; undecimæ uero à termino occidentali

ad septentrionem declinat: accipiantur à punctis
A C ex parte D circunferentiæ A I, C K ipsi p β
æquales: perq; I & centrum E ducatur linea occul
ta I E L, & per K & E alia ducatur K E M. postre
mo

Fig. 4 Federico Commandino, "Liber de Horologiorum descriptione" (Romae, 1562) necessary for the construction of a sundial with the orthographic projection on a plane of the points of the celestial sphere) seems to have been introduced by Hipparchus (190–120 B.C.), but Ptolemy can be attributed with its practical development. The study of the determinations made by Ptolemy in "De Analemmate" formed a theoretical and practical basis so that Federico Commandino could publish with the reform of the calendar a work entitled "de Horologiorum descriptione", aimed at describing the various types of existing clocks (horizontal, equinoctial, horizontal angle, vertical angle, etc.).

In 1565, ordered by Cardinal Ranuccio Farnese, Federico Commandino published in Bologna with Alex Benacii "De iis quae in aqua vehuntur" by Archimedes of Syracuse (287–212 B.C.) (Fig. 5). Despite Commandino owning a damaged Latin edition, translated by the Flemish-born Dominican Willem van Moerbeke (1215–1286), given to him by Cardinal Marcello Cervini, the work turns out to be an excellent edition and far superior to the previous versions by Niccolo Tartaglia (1499–1557) of 1543 and 1565. The version with commentary by Federico Commandino integrated the numerous gaps in the text, but, above all, popularized

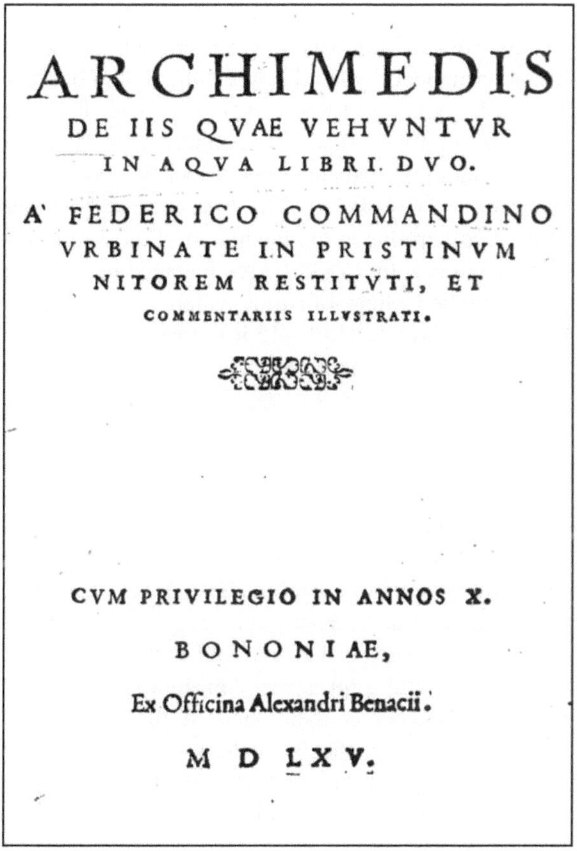
ARCHIMEDIS
DE IIS QVAE VEHVNTVR
IN AQVA LIBRI. DVO.
A' FEDERICO COMMANDINO
VRBINATE IN PRISTINVM
NITOREM RESTITVTI, ET
COMMENTARIIS ILLVSTRATI.

CVM PRIVILEGIO IN ANNOS X.
BONONIAE,
Ex Officina Alexandri Benacii.
M D LXV.

Fig. 5 Federico Commandino, "Archimedis De iis quae vehuntur in aqua libri duo" (Bononiae, 1565)

in the scientific community a fundamental work in which Archimedes anticipated some of the fundamentals of mathematical physics. In this new and important study, Archimedes described the buoyancy principle (Archimedes' principle), introducing the concept of relative specific gravity, dealing with problems related to the determination of the centre of gravity of solids which were subsequently relevant implications in the fields of architecture and civil engineering. In general, he taught a theory of floating bodies in which—according to Joseph-Louis Lagrange (1736–1813)—those who succeeded him changed nothing and added very little.

Following the study of the Treaty of Archimedes of Syracuse and in order to apply the methods used for plane figures to solids, Commandino published in 1565 his own contribution entitled "Liber de centro gravitatis solidorum", believed to date to be one of the most significant scientific contributions in the context of solid mechanics and, therefore, natural integration to the investigations of Archimedes. In this "Liber", he elaborated the concept of gravity as "Punctus aequaliurn momentorum", determining the centre of gravity of some solids and their trunks (especially, pyramids and cones, but not hyperboloids and ellipsoids of rotation) and demonstrated the determination of the centre of severity of the paraboloid of rotation, the result of which had been mentioned without any demonstration by Archimedes in the second book of "De iis quae in aqua vehuntur" (Fig. 6).

This contribution was long considered the most authoritative on the subject, he was reprimanded by his disciple Guidobaldo del Monte (1545–1607) and criticized by Galileo Galilei (1564–1642). Similarly, these determinations are contained in some of the notebooks—then unknown—of Leonardo da Vinci (1542–1519).

Commandino dedicated his "Liber de centre gravitatis solidorum" to Francesco Maurolico (1494–1575). The first contact between Commandino and Francesco Maurolico was due to the mediation of the Jesuit mathematician Baldassarre Torres (XVI century). In 1553, Torres entered the College of Palermo and shortly after moved to the College of Rome, where he was a doctor of the founder of the Jesuits, Ignatius of Loyola (1491–1556), and professor in the Faculty of Arts of "natural philosophy and mathematics". In Rome, Baldassarre Torres met Federico Commandino, who was in the papal city as a relative of Cardinal Ranuccio Farnese (1530–1565).

Their shared interest in mathematical studies allowed Torres to inform Commandino on the similar research carried out by Francesco Messina Maurolico, leading to the two mathematicians contacting each other. Further contact between Commandino and Maurolico occurred through an epistolary correspondence with lively discussions on Archimedean issues and the problem of conics, which are documented in an undated draft letter from Commandino to Maurolico (kept at the University Library of Urbino) and two letters from Maurolico to Commandino of which, the first is kept in the Vatican apostolic Library and dated October 8, 1557, while the second is mentioned in the biography of Maurolico written by his nephew. However, the few letters on the interpretation of the Archimedean text attest to a deep and mutual respect that existed between the two mathematicians, in addition to the recognition of Maurolician's work by Commandino on revalorising the Greek classics.

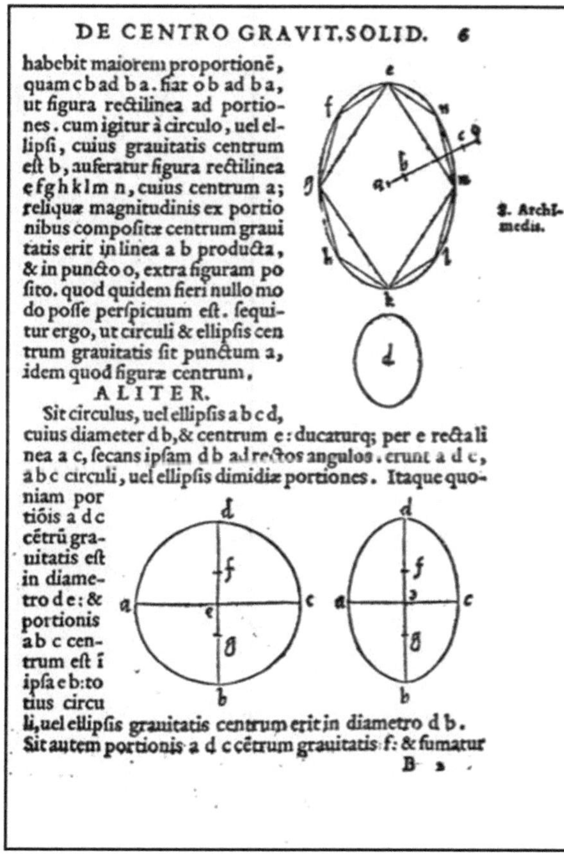
DE CENTRO GRAVIT. SOLID. 6

habebit maiorem proportionē, quam c b ad b a. fiat o b ad b a, ut figura rectilinea ad portiones. cum igitur à circulo, uel ellipsi, cuius grauitatis centrum est b, auferatur figura rectilinea e f g h k l m n, cuius centrum a; reliquæ magnitudinis ex portionibus compositæ centrum grauitatis erit in linea a b producta, & in puncto o, extra figuram posito. quod quidem fieri nullo modo posse perspicuum est. sequitur ergo, ut circuli & ellipsis centrum grauitatis sit punctum a, idem quod figuræ centrum.

8. Archimedis.

ALITER.

Sit circulus, uel ellipsis a b c d, cuius diameter d b, & centrum e: ducaturq; per e recta linea a c, secans ipsam d b ad rectos angulos. erunt a d c, a b c circuli, uel ellipsis dimidiæ portiones. Itaque quoniam portiōis a d c cētrū grauitatis est in diametro d e: & portionis a b c centrum est ī ipsa e b: totius circuli, uel ellipsis grauitatis centrum erit in diametro d b. Sit autem portionis a d c cētrum grauitatis f: & sumatur

B 2

Fig. 6 Federico Commandino, "Liber de centro gravitatis solidorum" (Bononiae, 1565)

In 1566, upon returning to Urbino and always with the publisher Alex Benatii of Bologna, Federico Commandino produced the translation from the Greek of "Apollonii Pergaei Conicorum libri quattuor. Una cum Pappi Alexandrini Lemmatibus, et commentariis Eutocii Ascalonitae. Sereni Antinsensis philosophi Libri duo nunc primum in lucem editi. Quae omnia nuper Federicus Commandinus Urbinas mendis quamplurimis expurgata è Graeco convertit, & commentariis illustravit", i.e. the first four books of the "Coniche" by Apollonius of Perga (262–190 B.C.). To Apollonius, the great mathematician of the classical period, can be attributed important work dedicated in eight books to the conic sections, one of the most sublime works of literature of all times and nations, and therefore almost unique theoretical foundation for many subsequent studies, including those taken up and expanded by Johannes Kepler (1571–1630) and Isaac Newton (1642–1727). In particular, Kepler would not have had the theoretical foundations needed to compose his theory on the elliptical orbits of the planets without the aforementioned first translation of the work of Apollonius by Federico Commandino (Fig. 7).

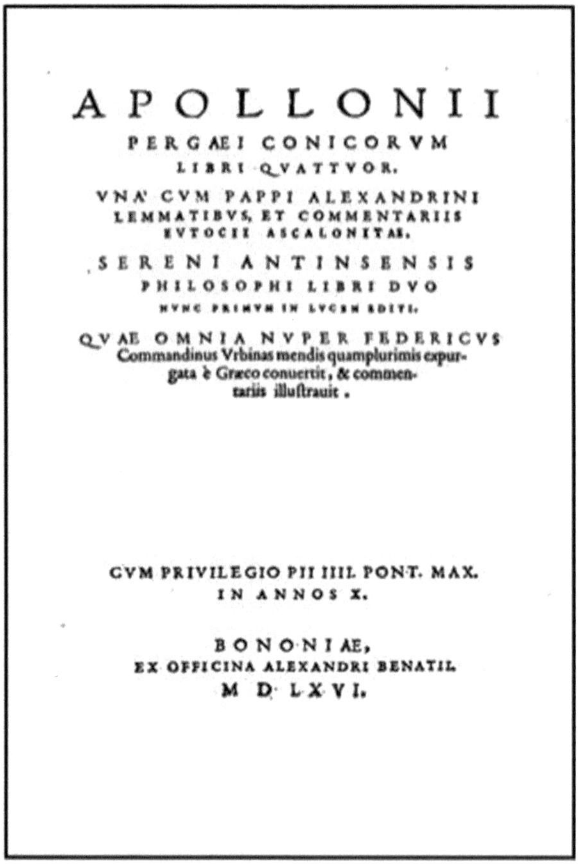
APOLLONII
PERGAEI CONICORVM
LIBRI QVATTVOR.
VNA CVM PAPPI ALEXANDRINI
LEMMATIBVS, ET COMMENTARIIS
EVTOCII ASCALONITAE.
SERENI ANTINSENSIS
PHILOSOPHI LIBRI DVO
NVNC PRIMVM IN LVCEM EDITI.
QVAE OMNIA NVPER FEDERICVS
Commandinus Vrbinas mendis quamplurimis expurgata è Græco conuertit, & commentariis illuſtrauit.
CVM PRIVILEGIO PII IIII. PONT. MAX.
IN ANNOS X.
BONONIAE,
EX OFFICINA ALEXANDRI BENATII.
M D LXVI.

Fig. 7 Federico Commandino, "Apollonii Pergaei Conicorum libri quattuor" (Bononiae, 1566)

The text of Federico Commandino constituted the first printed edition of the first four books of Apollonius, whose translation and comments were taken from the original manuscripts. Apollonius was not the first to introduce the theory of conic sections to scientific thought; Aristaeus, Euclid and Archimedes had probably already dealt with this type of curves. Nevertheless, Apollonius is accredited with a different conceptual approach. In fact, he conceived these curves as sections generated by any planes in a cone with a circular base, contrary to the previous position that these curves were generated exclusively by planes perpendicular to the generatrices. Moreover, Apollonius introduced the term "opposite sections" when he found himself having to describe the hyperbole as a complete curve consisting of two branches (as previously thought).

According to the structure of the four books of Apollonius, Book I contains: definitions and theorems concerning circular cones; the essential properties of conic sections (today expressed with a second-order Cartesian equation); definitions of the three conic sections (ellipse, parabola, hyperbola) as well as "right side" and "transverse side"; a series of important propositions (of parallel ordinates, cords and tangents). Asymptotes of the hyperbola and relationships with points and tangents to the conic are dealt with in Book II and the unusual definition of a conical as a "place of a point" which allowed attribution to Apollonius the possible resolution of the fundamental problem of "describing a conic determined by five points", even if the Greek mathematician never actually described the assumption in detail. Metric proposition, categories of theorems on conics as envelopes of straight lines, focuses of the conical centre (except the focus of the parabola), the genesis of conics using projective bundles of straight lines are the main topics covered in Book III. Book IV, however, has a greater unity and is dedicated to finding the number of points common to two conics or to a conic and a circle in the various positions of interchange where the conics could be.

The Latin translation by Federico Commandino was dedicated to Guidobaldo II. The volume was structured by preceding the first three books of Apollonius with a corresponding book by Pappus of Alexandria (third century A.D.), one of the most important Greek mathematicians of the late Hellenistic period, which dealt with several problems on the conics of Apollonius by commenting on them through a large number of lemmas. In particular, the first of the three books by Pappus was in turn commented on by Commandino. Whereas, the four books by Apollonius were presented both with a commentary by Federico Commandino as well as with that by Eutocius of Ascalon (V–VI century A.D.), a commentator, who added nothing of relevance to the scientific thought of the great Greek mathematician but contributed through his work to returning the epochal context of knowledge of ancient mathematics and geometry.

Federico Commandino followed the Latin translation of the conical section of Apollonius with "Sereni Antinsensis philosophi Libri duo nunc primum in lucem editi", i.e. the Latin translation from the original Greek of the two contributions of Sereno of Antinoe (fourth century A.D.) entitled "De Sectione Cylindri" and "De Sectione Cones", also published and illustrated for the first time as well as commented by Commandino (Fig. 8). The first book of this Greek mathematician from ancient Middle Egypt can be considered a complement to the conic sections of Apollonius in relation to the flat sections of rotation cylinders, i.e., ellipses, and the resolution of some interesting issues with cones and cylinders passing through a given ellipse. In the second book, Sereno of Antinoe deals with an area that seems to anticipate the geometric theory of maxima and minima. Specifically, he deals for the first time with the determination of the relative magnitude of triangles generated in a right circular cone (limited by the base and the vertex) by planes passing through the vertex, resulting in different behaviour of the acute-angle and obtuse-angle cones.

In 1570, Commandino published in Pesaro with Hieronymus Concordiam, "De superficierum divisionibus liber Machometo Bagdedino ascriptus, nunc primum Ioannis Dee londinensis & Federici Commendini urbinatis opera in lucem editus". This work had been given to him in 1563 in Urbino by the English mathematician John Dee (1527–1608), who considered it to be by Euclid (III–II century B.C.) and had copied it from an Arabic version of the manuscript by Muhammed ibn Muhammed al. Bagdadi (?–1141) found at the Cotton Library in London.

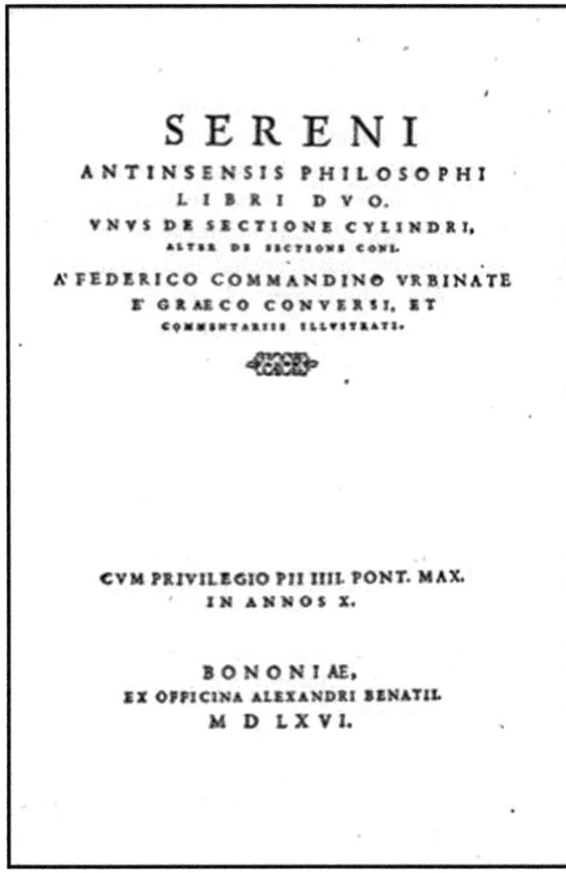
SERENI
ANTINSENSIS PHILOSOPHI
LIBRI DVO.
VNVS DE SECTIONE CYLINDRI,
ALTER DE SECTIONE CONI.
A' FEDERICO COMMANDINO VRBINATE
E' GRAECO CONVERSI, ET
COMMENTARIIS ILLVSTRATI.
CVM PRIVILEGIO PII IIII. PONT. MAX.
IN ANNOS X.
BONONIAE,
EX OFFICINA ALEXANDRI BENATII.
M D LXVI.

Fig. 8 Federico Commandino, "Sereni Antinsensis philosophi Libri duo" (Bononiae, 1566)

There was no original Greek version of this important work by Euclid on the division of the figures, but there was an Arabic copy that allowed Commandino to reconstruct it. This work by Euclid discusses a significant class of problems, dealing with the division of flat areas in parts linked by relationships established by straight lines subject to special conditions. Specifically, the treaty refers to the geometric shapes of triangles and quadrangles. The problem is the division of such figures in two parts by means of a transverse line passing through a given point or parallel to one of the sides, so that these two parts are either equal, or such that one of them may be in a given part of the total figure or in a given relationship with this (Fig. 9). To solve these problems, Euclid starts from elementary knowledge (taken from the "Elements") and then proves the most difficult propositions. In this sense, the work is organised according to a scale of increasing difficulty.

The translation by Muhammad al-Baġdādī, however, is organised differently regarding the growing number of sides of the figure considered, with the latter dealing with the pentagon.

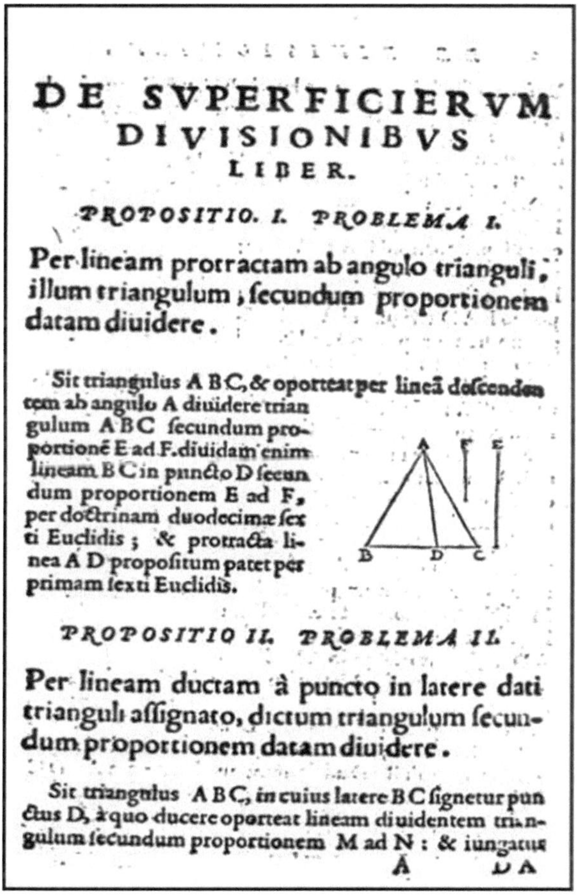

DE SVPERFICIERVM DIVISIONIBVS LIBER.

PROPOSITIO. I. PROBLEMA I.

Per lineam protractam ab angulo trianguli, illum triangulum, ſecundum proportionem datam diuidere.

Sit triangulus A B C, & oporteat per lineã deſcendentem ab angulo A diuidere triangulum A B C ſecundum proportionẽ E ad F. diuidam enim lineam B C in puncto D ſecundum proportionem E ad F, per doctrinam duodecimæ ſexti Euclidis; & protracta linea A D propoſitum patet per primam ſexti Euclidis.

PROPOSITIO II. PROBLEMA II.

Per lineam ductam à puncto in latere dati trianguli aſſignato, dictum triangulum ſecundum proportionem datam diuidere.

Sit triangulus A B C, in cuius latere B C ſignetur punctus D, à quo ducere oporteat lineam diuidentem triangulum ſecundum proportionem M ad N: & iungatur D A

Fig. 9 Federico Commandino, "De superficierum divisionibus liber Machometo Bagdedino ascriptus, nunc primum Ioannis Dee londinensis & Federici Commendini urbinatis opera in lucem editus" (Pesaro, 1570)

The translation by Commandino has parts in Latin, while most of the text is in Italian. Furthermore, Commandino followed the Treaty of Euclid (revisited according to the Arabic version by Muhammad al-Baġdādī) with his own work entitled "Short treatise translated around the same subject by the same", which presents the division of hexagons and heptagons, and concludes that "similarly will divide the other rectilinear figures of how many sides you want according to a given proportion with a line equidistant to a given straight line".

In 1572, Commandino published in Pisa with Camillum Francischinum the Latin translation of the Greek "Euclidis Elementorum books XV", which were followed in 1575 by the Italian version (Fig. 10). The Latin version by Commandino marks a turning point in the understanding of the principles and fundamentals of mathematics and geometry: the work is commented and accompanied by figures, while in the "Prolegomena" he traces a history of mathematics and clarifies the erroneous attribution of works to the eponymous philosopher of Megara. Moreover, the interest of Commandino for Euclid was precocious. In fact, there is a

DE GLI
ELEMENTI
D'EVCLIDE
LIBRI QVINDICI·
CON GLI SCHOLII ANTICHI.
TRADOTTI PRIMA
IN LINGVA LATINA DA M.
FEDERICO COMMANDINO DA VRBINO,
& con Commentarij illuſtrati,
ET HORA D'ORDINE DELL'ISTESSO
TRASPORTATI NELLA NOSTRA VVLGA-
re, & da lui riueduti.
CON PRIVILEGIO.
IN VRBINO, APPRESSO
DOMENICO FRISOLINO·
M. D. LXXV.
Con licentia de' Superiori.

Fig. 10 Federico Commandino, "Degli Elementi d'Euclide Libri Quindici" (Urbino, 1575)

correspondence dated 1537 between Commandino and the mathematician Thomas Leonardi da Fano (XVI century), whose argument concerns two versions of "Euclidis Elementorum books XV": the medieval one (1255–59) the work of Campano Novara (?–1296), then printed in Venice by Ratdolt in 1482, and the one published in 1505 by the mathematician and Venetian humanist Bartolomeo Zamberti (the second half of the XV century).

In the history of mathematical thinking, Euclid's "Elements" represents a work of extraordinary popularity. Printed for the first time in 1482, the work of Euclid competes for the number of editions and translations with the "Divine Comedy" by Dante Alighieri (1265–1321), printed for the first time just a decade earlier, and perhaps only exceeded by the "Holy Bible". "Elements" is made up of fifteen books, preceded by "Preliminaries" in which Euclid gives the reader the option of using only a ruler and compass to carry out geometric operations and constructions. The work contains all the human knowledge of the time, formulated and essential to the understanding of the geometric-mathematical fundamentals: theorems and problems up to the theorem of Pythagoras (570–495 B.C.); bases of the procedure commonly called geometric algebra to solve grade I and II problems; theorems regarding the circumference and regular polygons inscribed or circumscribed to it; theory of proportion to any size; similarity theory of plane figures; arithmetic of rational numbers; theory of irrational; fundamental relationships between lines and planes, parallelism, perpendicularity, etc.; measurement of the surface and the volume of solids; theory of regular convex polyhedras.

Both the Latin and Italian versions are set out the same way with a total of more than five hundred pages (nearly six hundred in the Italian edition): the demonstrations of Euclid are accompanied by line drawings and comments by the mathematician from Urbino (indicated with "Il Commandino") that, in order to stand out from Euclid's text, are in italics. In the last years of his life, Federico Commandino worked hard on the Italian version of the "Elements", aided by his students who took care of the illustrations, including Bernardino Baldi (1553–1617). For this ambitious project, Commandino had also obtained permission to set up a typographer on his property. He died in Urbino on September 3, 1575 and part of the programs remained unfinished. In fact, in 1575, the Italian translation of Euclid's "Elements" was published posthumously in Urbino at the "home of Federico Commandino", "first translated into Latin by Federico M. Commandino from Urbino, and illustrated with comments, and now translated by the same in our vernacular, and revised by him".

In 1572, he also published the only surviving work of the mathematician, physicist and astronomer Aristarchus of Samos (310–250 B.C.) translating into Latin the treatise entitled "Aristarchi De Magnitudinibus, et Distantiis Solis, et Lunae, Liber, cum Pappi Alexandrini explicationibus quibusdam" (Fig. 11). The work by Aristarchus is an important contribution to mathematics of Greek astronomers and geodesists and, above all, to the general theory of the heliocentric motion of planets. Aristarchus, also known as the "Copernicus of antiquity" (Niccolò Copernicus, 1473–1543), is accredited with work, never discovered, in which he tried to instil a revolution in the astronomical theories of his time that the Sun and the fixed stars were motionless, while the Earth revolved around the Sun along a circular orbit. The only surviving work by Aristarchus is a booklet on the dimensions and distances of the Sun and the Moon. In it, he estimates the size of the Sun and Moon by calculating the relative distances from the Earth. The ratio between the distances was computed using geometric reasoning, i.e. the trigonometric tangent of an angle. Thus, this work by Aristarchus has the merit of being considered as one of the first works of trigonometry in spite of the incorrect measurement of the distances of the Sun and Moon.

Commandino's Latin version proposes the customary model, presenting the author's (Aristarchus) text in Latin, the comments of Pappus of Alexandria (320?–? B.C.) and the mathematician from Urbino. Specifically, the small treatise of about eighty pages contain many full page line drawings.

In the last years of his life, Federico Commandino worked hard on Italian versions of Heron of Alexandria (II–I century B.C.) and Pappus of Alexandria (third century A.D.). At the same time, according to the biography by Baldi,

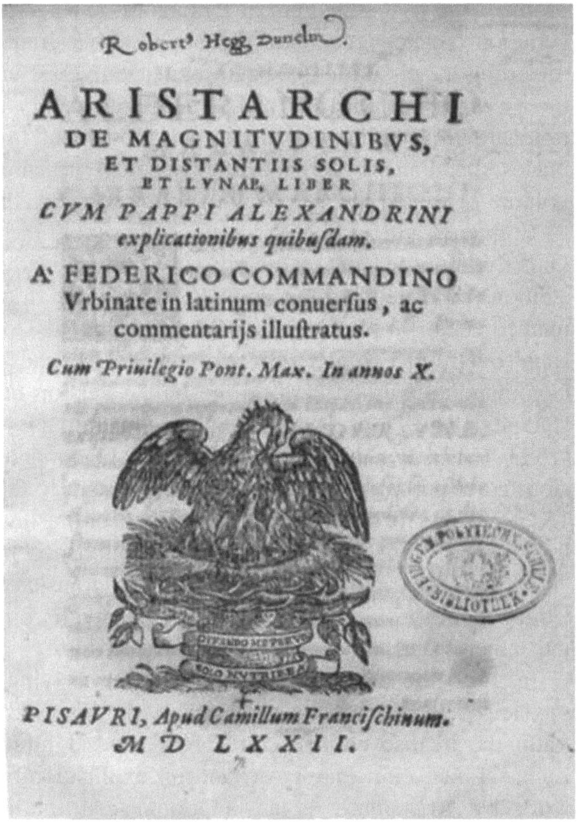

ARISTARCHI
DE MAGNITVDINIBVS,
ET DISTANTIIS SOLIS,
ET LVNAE, LIBER
CVM PAPPI ALEXANDRINI
explicationibus quibuſdam.
A' FEDERICO COMMANDINO
Vrbinate in latinum conuerſus, ac
commentarijs illuſtratus.
Cum Priuilegio Pont. Max. In annos X.
PISAVRI, Apud Camillum Franciſchinum.
M D L X X I I.

Fig. 11 Federico Commandino, "Aristarchi De Magnitudinibus, et Distantiis Solis, et Lunae, Liber" (Pisauri, 1572)

Commandino was going to print the "Practica Geometriae" by Fibonacci (1170–1240) and the "Summa" of Luca Pacioli (1445–1517). Upon his death in 1575, the version from the Greek Heronis Alexandrini. Spiritalium Liber, a Federico Commandino urbinate, ex graeco, nuper in latinum conversus" by Heron of Alexandria (first century B.C. to the second century A.D.) was published, while in 1588 an edition of "Pappi Alexandrini Mathematicae collectiones à Federico Commandino Vrbinate in latinum conversae, et commentariis illustratae" of Pappus of Alexandria (third century A.D.) was edited and published by Guidobaldo del Monte.

The Greek version of "Spiritalia" by Hero of Alexandria, also known as "Pneumatica", was translated into Latin from the Greek by Commandino and edited by his son, Valerio Spaccioli (XVI century). This edition was published full of illustrations and instructions for building various objects such as siphons, vases, lamps, as well as hydraulic tools. This version was subsequently translated into Italian in 1589 by Giovanni Battista Aleotti (1546–1636) and in 1592 by Alessandro Giorgi (XVI–XVII century).

Hero of Alexandria, of which very little is known, became interested in physics and mathematics, with his most important work being translated by Commandino. This treaty was devoted to practical mechanics and the building of amazing machines working with pressure (air, water or steam) including a water organ called "Hydraulis" (Fig. 12).

Commandino's translation of Heron had considerable importance in the field of architectural design since it constituted the main source on which the Renaissance architects based the construction of the artifices and fountains that adorned some of the most admired gardens of Italy and Europe.

The Latin version of the work by Pappo, even if edited by Guidobaldo del Monte (1547–1607), has a similar structure of Commandino's previous works (Fig. 13): the translated version is followed by the commentaries and graphic illustrations accompanying the texts. The work is noteworthy (more than 650 pages) and consists of Books III, IV, V, VI, VII and VIII, as surviving from antiquity, with Books I and II being lost (there is only one fragment of the latter). Specifically, Book III

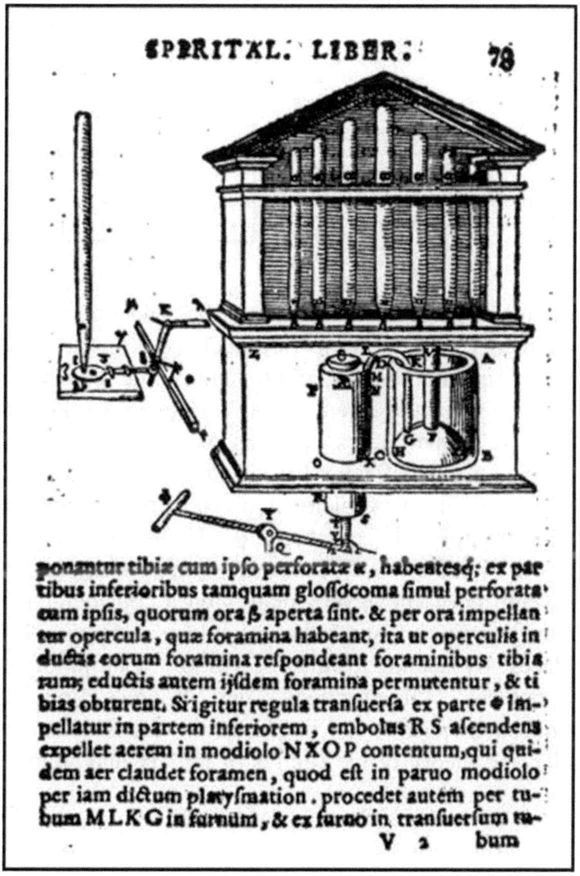
SPIRITAL. LIBER. 78

ponantur tibiæ cum ipso perforatæ α, habentesq; ex par
tibus inferioribus tamquam glossocoma simul perforata
cum ipsis, quorum ora β aperta sint. & per ora impellan
tur opercula, quæ foramina habeant, ita ut operculis in
ductis eorum foramina respondeant foraminibus tibia
rum; eductis autem ijsdem foramina permutentur, & ti
bias obturent. Si igitur regula transuersa ex parte θ im-
pellatur in partem inferiorem, embolus R S ascendens
expellet aerem in modiolo N X O P contentum, qui qui-
dem aer claudet foramen, quod est in paruo modiolo
per iam dictum platysmation. procedet autem per tu-
bum M L K G in furnum, & ex furno in transuersum tu-
V 2 bum

Fig. 12 Federico Commandino, "Heronis Alexandrini. Spiritalium Liber" (Pisa, 1575)

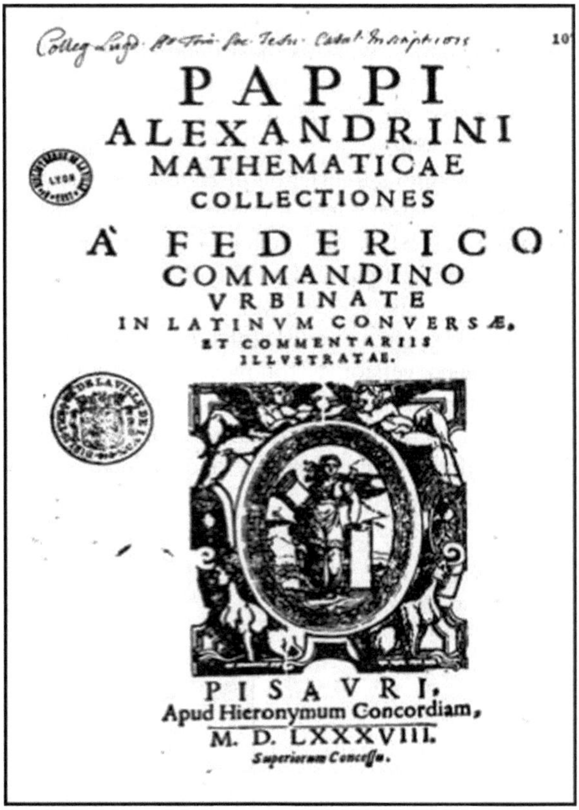

Fig. 13 Federico Commandino, "Pappi Alexandrini Mathematicae collections" (Urbino, 1588)

discusses the problem of Delos (whose resolution is used to calculate the cube root), the theory of proportions (arithmetic, geometric, harmonic, against harmonic, fifth, sixth, seventh, eighth, ninth, tenth), the theory of regular polyhedra with the inscription of these remarkable figures in a sphere. Book IV deals with the properties of some curves (such as Archimedes spiral and the squaring of Hippias), proving a generalization of the Pythagorean theorem applied to oblique triangles, while solving a problem of division of an angle in parts having predetermined relationships between them. Book V is devoted to isoperimetric theory, while Book VI superficially discusses several issues related to spherical astronomy. The seventh book, however, focuses on a considerable number of arguments put forward by important works, now lost, such as: "Data", "Porisms" and "Surface places" by Euclid; "Cutting of a ratio", "Cutting of an area", "Determinate Section", "Contacts", "Intersections" and "Conics" by Apollonius; "Proportions" by Eratosthenes (276–194 B.C.). Book VIII is dedicated to mechanics and, in particular, the theory of the centres of gravity and simple machines.

On the Circulation of Works

The work of Federico Commandino has had a significant influence on the process of formation of mathematical and experimental science. Commandino was one of the best known representatives of mathematical humanism of the Italian Renaissance. In fact, between the sixteenth and seventeenth century, Commandino was one of the leaders of the scientific school of Urbino, important in Europe, commissioned by Federico da Montefeltro (1422–1482), enlightened patron, who had called important figures to Urbino including Piero della Francesca (1412/17–1492), Luciano Laurana (1420–1479), Francesco di Giorgio Martini (1439–1501), making the Duchy of Urbino, one of the main European centres of humanistic and mathematical studies at the time.

In the historical context of the rebirth of scientific knowledge, the dissemination of Greek and eastern culture thanks to the recovery of the original texts represented a fertile ground for Federico Commandino. His humanist education, along with his knowledge of Greek, Latin and mathematics allowed the mathematician from Urbino to stand out in the cultural landscape of his time. In addition, following the erudite cardinals of the era introduced him to a lively environment of scholars, whose purpose was to bring to light a wealth of knowledge dispersed during the Middle Ages through the recovery and translation of ancient texts. Commandino played a central role in this resetting of the humanistic and scientific knowledge, serving both as a refined translator and commentator of ancient texts in mathematics and geometry as well as a promoter of the birth of an exact science.

The learned and innovative actions of Federico Commandino in this general operation of translating into Latin and later printing were such, so as to be not only to recover and translate the missing determinations, but also to comment and reconstruct them. The works of Archimedes, Pappus, Apollonius, Euclid and many other mathematicians and scientists of ancient Greece found in Commandino an insightful interpreter, capable of laying solid foundations for the emergence of the so-called scientific revolution, later executed by Galileo Galilei (1564–1642) and Johannes von Kepler (1571–1630). Suffice it to say that the versions of the works of Euclid, Apollonius and Pappus translated by Commandino had a remarkable diffusion during the sixteenth and seventeenth centuries, and that on the basis of the version of "Conics" by Apollonius translated and commented by him, Kepler stated the importance of knowledge of mathematics to trace the elliptical orbits of the planets: "Itali somniant, praeter unum F. Comandinum, Clavius enim germanum est". Similarly, Commandino played a central role in the formation of his disciples from Urbino including Guidobaldo Dal Monte (1545–1607).

In this sense, the most important intervention of Federico Commandino is the recovery of ancient science. Thanks to his translating into Latin and the vernacular from the Greek or Arabic, the works of the great Greek mathematicians such as Apollonius, Archimedes, Pappo, Eutocius of Ascalon, Ptolemy and many others in addition to the already known Euclid were brought back to light and, therefore, to the knowledge of the community of scholars. In this work of revision, Commandino not only translated but enriched his own scientific contribution with significant prefaces and additional comments to the original texts, such as the works of Archimedes, with it being widely assumed that it was the mathematician from Urbino who introduced the term "centre of gravity".

As anticipated, his scientific contribution was significant, where in the work "Ptolomaei Planisphaerium Commentarius", Commandino enriched the determinations of Ptolemy with unusual geometric and mathematical demonstrations concerning the practice of building perspective. With Federico Commandino, the discussion of perspective (always a practice exclusive of painters, artists and architects) is elevated to scientific speculation about the formalization of the rules of representation, marking a clear separation between art and science. Thus, artists continued to use perspective as a simple drawing technique, while thanks to the studies of Commandino and his initial determinations, the study of perspective in mathematics initiated a fruitful process, with Girad Desargues (1591–1661), approximately 80 years after the publication of "Ptolomaei Planisphaerium Commentarius", formalising the foundations of projective geometry.

Although still in connection with the science of painters, but more and more separated from this due to the tools and objectives, Commandino's work on perspective participates—albeit in a completely intuitive way—to the birth of projective geometry as an autonomous discipline.

In fact, Commandino, thanks to his studies and before Blaise Pascal (1623–1662) and Guidobaldo Del Monte (1545–1607), says that in certain positions the perspective of a circle is a conic. In the vast work "Ptolomaei Planisphoerum", the theme of linear perspective is addressed for the first time by a pure scientist and not a painter. In this contribution, Commandino exposes two new geometric constructions: the first, the description of tipping the main visual plane with the determination of the "point of view" (observer) as "point of distance"; the second, the determination of the distance from the context of the aforementioned point and its height from a horizontal plane. The projective procedure is founded on the basis of a spatial reference system consisting of two mutually perpendicular planes (in a vertical and horizontal arrangement) and on the tilting of the horizontal plane onto the vertical plane where the figure is assigned in true form. The original result that Commandino applied to this process was the construction of the prospect of a circle, which he acknowledges to be an ellipse. Furthermore, he states that for particular positions of the circle, the corresponding image can be transformed into a parabola or a hyperbola in the case in which the observer has the station point on the circumference (parabola) or inside of the circle (hyperbole).

However, the new and original determinations proposed by Commandino on linear perspective long remained unexplored by painters, since the work in question was totally in Latin and was deliberately not aimed at painters, but rather an educated public, learned in the sciences such as mathematicians whose formation was such as to fully understand these determinations. The consequence of this was the subsequent discussion of this issue in Italian by the humanist Daniele Barbaro (1513–1570), who in 1568 published in Venice "The Practice of Perspective". In this paper, Barbaro took most of the work previously printed by Piero della

Francesca (1412/17–1492), without mentioning the source, while in relation to the work of Commandino, recognising its value, he commented on the difficulty of understanding and the lack of clarity of the demonstrations, justifying the fact that the purpose of the work was obviously aimed at a doctorate in mathematical-geometric sciences and not only in the pictorial arts.

However, Commandino's determinations on perspective were the subject of subsequent considerations, such as in the work of François D'Aguillon, better known as Anguilonius (1546–1617), who in 1613 published in Antwerp the work entitled "Opticorum libri sex", without doubt based on the study of the Latin translations by Commandino. In the sixth book of this work, dedicated to perspective and entitled "De projetionibus" Anguilonius incorporates the concept of projection and correcting Commandino, introduces the term "proiectio" as a projection of a solid object on a plane.

Girad Désargues (1591–1661) also based some of his studies on Commandino's scientific determinations. In particular, for his studies on gnomonics, Désargues moved away from the classic procedures contained in the works "Analemma" by Claudius Ptolemy (100?–175) and Vitruvius (80–15 B.C.) to accept, at least in part, the innovation introduced by Commandino in his book entitled "De horologiorum descriptione" of 1581, which provided for the use of conicals.

Similarly, both Commandino's Latin and vernacular translations of Euclid's "Elements" by and, above all, the so-called "prolegomena" present considerations that are still of considerable relevance. Commandino states that the purpose of his work is twofold: it is definitely aimed at educated and erudite scholars, but his writing is such that it is easy for even "uncouth readers" to understand. The aim not to write just for the privileged was based on the opinion that mathematics must have a relationship with the real world, otherwise it would remain incomprehensible. Commandino, with a leading position, stated that only a theoretical and practical knowledge of mathematics could be of help to the rebirth of his time. Similarly, the illustrations of the buildings and geometric figures that accompany the text (the work of his pupil and biographer Bernardino Baldi) have the same function: to satisfy an audience of educated readers as well as that of a more technical training, while the drawings that decorate the initials in the opening the "Books" of which the treaty is made are attributed to the painter Federico Barocci (1535–1612) and his school. Finally, we must not neglect to reaffirm the merit attributed to Commandino, since he was the first to distinguish between the philosopher Euclid of Megara and the surveyor Euclid of Alexandria, separated by about a century but until then often confused with each other.

In conclusion, Federico Commandino was one of the greatest exponents of mathematical-geometric Renaissance thought as well as founder of the contemporary school of Urbino, whose students gradually integrated the determinations. In particular, these included: Guidobaldo Del Monte (1545–1607) specialised in studies on mechanics, great help to the work of Galileo Galilei (1564–1642); Bernardino Baldi (1553–1617) was a humanist, architect and first historian of scientific-mathematical thought as well as Commandino's biographer; Muzio Oddi

(1569–1639), specialised in the study of military and civil architecture as well as the design of sundials and scientific instruments.

Modern Interpretation of Main Contributions

The personality and work of Federico Commandino have been extensively analysed in the modern age, both from a biographical profile as well as on his contribution to the development of scientific thought in the fields of mathematics and mechanics.

Among the most recent biographies, Concetta Bianca (1951), student of late-medieval and humanistic philology as well as humanistic and Renaissance literature, in the “Biographical Dictionary of Italians” (1982) gives a detailed biography of Federico Commandino, going over in detail the stages of his life and placing the scientific and cultural importance of his works within a contemporary historical context. In presenting the stages of his life based on the sources examined, Concetta Bianca feels that it is somewhat simplistic to attribute the notoriety of Federico Commandino merely to the activity of expert philologist and accurate translator. In fact, she points out that Commandino’s choice to restore and bring to light the texts of ancient mathematicians underlies a significant turning point in the way of understanding and disseminating mathematical disciplines. According to Bianca, Commandino was, in fact, fully aware of the cultural significance of his task of translating ancient texts, as can be seen clearly in both the dedications he put in his books as well as the comments with which he always accompanied translations, that were appreciated by his contemporaries and disciples.

More specifically, and around the 1930s, the Italian mathematician Gino Benedetto Loria (May, 19 1882–January, 30 1954), primarily known for his significant and substantial contribution to the development of studies in Italy in the field of history of mathematics, in his work in three books, “Storia delle matematiche dall’alba della civiltà al tramonto del secolo XIX” (Milan, 1929–33), presented the work of Federico Commandino as “excellent” when compared with that of his humanist contemporaries. In fact, Gino Loria believes that the work of translating ancient texts carried out by Federico Commandino not only reactivated the circulation of forgotten ideas and methods but also, with ease of understanding and development through intelligent comments and new applications, gave new life to the resumption of studies in the field of geometry. Furthermore, Loria considers that the work of Commandino was also of fundamental help to the world of artistic culture of the sixteenth century, with his scientific studies on linear and systematized perspective exceeding those of the painter Piero della Francesca (1416?–October, 12 1492), author of a mathematical treatise on perspective entitled “De prospectiva pingendi” (1460–1480?), and laid a solid foundation for his student Guidobaldo Del Monte (January, 11 1545–January, 6 1607) scientifically defining the “punto di concorso” in the linear perspective construction. The position of Gino Loria has recently been taken up by the researcher Anna Sgrosso (The geometry of the image, II): in fact, she believes the process formulated by Commandino 'remarkable and certainly original'.

In 2009 (September, 18–19), an international conference on Federico Commandino, entitled "Federico Commandino. Mathematics and humanism in Renaissance Urbino was held in Urbino. Sponsored by the International Centre for Studies "Urbino e la Propsettiva" to celebrate the fifth centenary of the birth of Commandino as well as remember him and the so-called "school of mathematics of Urbino" (of which Federico Commandino was founder), the conference was attended by numerous Italian and foreign scholars and was sponsored by the University of Urbino and the Accademia Raffaello. The Italian research institutions present included: Alessandra Fiocca (University of Ferrara), Veronica Gavagna (University of Salerno), Pierluigi Graziani (University of Urbino), Pier Daniele Napolitani and Martin Frank (University of Pisa), Vico Montebelli (International Centre of Studies "Urbino e la Prospettiva). The foreign institutions included: Elio Nenci and Antonio Becchi (Max Planck Institute for the History of Science, Berlin), Sabine Rommevaux (Centre d'Etude de la Renaissance Supérieures, CNRS), Ken Saito (Osaka Prefecture University). The chairmen were Ken Saito and Paul Freguglia (University of L'Aquila) and Gian Italo Bischi (University of Urbino) closed the proceedings.

Aimed at the study of the history of science and the development of scientific thought in the field of mathematical and mechanical studies, the conference proceedings of Urbino have produced new acquisitions and critical considerations on the figure and role of Federico Commandino. In particular, the issues raised have focused on the knowledge of mathematical sciences at the University of Ferrara and Padua at the time of Commandino (A. Fiocca), the dialectic ability in terms of logic and interpretation at the school of mathematics of Urbino (P. Graziani), on the drafting of the sixteenth-century editions of Euclid's "Elements". On this theme, some contributions compared versions by Commandino, Giovanni Campano (1220–1296), Bartholomeo Zamberti (XVI, second half), Christophorus Clavius (March, 25 1538–February, 12 1612), others analysed both the handling of certain propositions (V. Gavagna, S. Rommevaux) as well as the iconographic perspective, recognising Commandino's use of the prospective technique as a recovery of the "scaenographia" (V. Gavagna). On this last point, a request for further critical examination was proposed in order to verify the influence of Commandino on the evolution of the construction of iconographic illustrations in scientific texts (P. Graziani).

Other topics discussed included a critical analysis of the studies and works by Commandino of Archimedes, Euclid and Vitruvius, as well as the different interpretations between Commandino and Giudobaldo Del Monte on Proposition 6 of the quadrature of the parabola and the question of balance. In particular, in relation to the analysis of the correspondence between Maurolico and Commandino, there is a deification of Archimedes by the mathematician from Urbino, with the clear intention to use the comments as a further explanation but not to amplify the scientific scope of the Archimedean text (D. Napolitani). The critical analysis of Proposition 6 of the quadrature of the parabola and the issue of the balance has however highlighted a different interpretation by Giudobaldo Del Monte to that of his maestro, confirming the discovery of the indifferent equilibrium to the isostatic balance as one of the most important and original contributions of guidobaldiana mechanics, since it allows one to provide an explanation of the working of simple machines, with it being connected to a full understanding of the concept of "centre of gravity" and has drastic consequences for the guidobaldiana "Theory of balance" (M. Frank).

The considerations made about the contribution that Federico Commandino gave algebra were particularly original (V. Montebelli). In his commentary on "Book X" of "Elements", Commandino systematically uses algebra and, in particular, the radical calculation; however, he was not the first to use it, since others had already. The difference is that this numerical comment does not seem to be episodic and marginal but rather systematic and characterised by a cultural amplitude that relates the geometric quantities to their measurements as if evoking the Arabic and abacus tradition. In both Greek and Euclid mathematics, in particular, there is a clear separation between arithmetic and geometry, as well as number and size, which is found in neither Arabic nor abacus mathematics where arithmetic and algebra are preferred, problems are solved with numerical data and in geometry there is a process of identification between greatness and its measurement. In this sense, when faced with the theoretical difficulty of the subject, Federico Commandino adopted "practical" methods and fully demonstrates his cultural conception of mathematics.

In "Prolegomena" to the edition of Euclid, Commandino places mathematical knowledge in an intermediate position between natural and speculative philosophy. Mathematics is not, therefore, pure speculation but a form of knowledge that has significant links with reality. For Commandino, mathematics is knowledge that can be useful to society and allows large applications in mechanical engineering, astronomy, optics, geodesy, music and the art of "counting". Even the minor arts need arithmetic and geometry: painting needs perspective; military art, geometry for sieges and the construction of war machines.

The "digitizing" of Euclid's propositions carried out by Commandino in his translation of Euclid's "Elements" allows for the work to be read not only by a learned public but also by the vast group of engineers interested in the numerical translation of Euclidean geometry. By adopting abacus proceedings, Commandino created an integration between theory and practice, a decisive ideological position for the new science that will be born with Galileo Galilei (February, 15 1564–January, 8 1642), as well as for the birth of the analytic geometry of Descartes that is based on the identification of algebraic and geometric entities, including size and number.

Commandino was the creator of all that can be found in the fact that his translation of "Elements" had a huge publishing success and numerous reprints in Latin throughout Europe and eventually English from the XVIII century up to the London of 1867. Johan Ludvig Heiberg (1854–1928), author of the ponderous critical edition of "Elements", writes that the edition by Federico Commandino constituted for more than two centuries an essential reference point; therefore, it is possible to think that it consolidated this culture of identifying geometric and algebraic magnitudes. Similarly, it can be stated that, in the scientific and cultural development of mathematical thinking, Federico Commandino was one of the greatest European mathematicians-humanists of the sixteenth century, significantly influencing the history of science.

Acknowledgments The author wishes to thank Sacha Berardo for the English translation.

Bibliography

Baldi B (1707) Cronica de matematici, Urbino, pp 137–138
Baldi B (1714) Vita Federici Commandini, Giornale de' letterati d'Italia, XIX, 140–185
Bertoloni Meli D (1992) Guidobaldo dal Monte and the Archimedean revival. Nuncius Ann Storia Sci 7(1):3–34
Biagioli M (1989) The social status of Italian mathematicians 1450–1600. Hist Sci 27, 75, 1:41–95
Bianca C (1982) Commandino Federico, Dizionario Biografico degli Italiani. Istituto dell'Enciclopedia Italiana, Roma 27
Brams J (2003) La riscoperta di Aristotele in Occidente. Jaca Book, Milano, pp 105–130
Castellani C (1896–1897) Il prestito del codici manoscritti della Biblioteca di S. Marco, Atti dell'Istituto veneto, LV, 350–351
Ceccarelli M, Cigola M (1995) On the evolution of mechanisms drawing. In: Proceedings of IXth IFToMM world congress, Politecnico di Milano, vol 4, pp 3191–3195
Ceccarelli M, Cigola M (2001) Trends in the drawing of mechanisms since the early middle ages. J Mech Eng Sci 215:269–289
Centro Internazionale di Studi 'Urbino e la prospettiva' (2009) Convegno Internazionale Federico Commandino (1509–1575) Umanesimo e Matematica nel Rinascimento Urbinate, Urbino. http://urbinoelaprospettiva.uniurb.it/commandino.asp
Cigola M (2012) In praise of parallel theories: descriptive geometry and applied mechanics. In: Carlevaris L, Filippa M (eds) In praise of theory. The fundamentals of the disciplines of representation and survey, Roma, pp 39–46
Clagett M (1964) Archimedes in the middle ages I, 13. Wisc, Madison
Crozet P (2002) Geometria: la tradizione euclidea rivisitata. www.treccani.it
de Nolhac P (1887) La bibliothèque de Fulvio Orsini, Paris, 9
De Rosa A, Sgrosso A, Giordano A (2001) La Geometria nell'immagine. Storia dei metodi di rappresentazione, Torino, II, 58, 114, 156–160, 202, 226, 228, 230, 239, 264, 343
Drake S, Drabkin I (1969) Mechanics in sixteenth-century Italy, Madison, pp 41–44
Field JV (1997) The invention of infinity: mathematics and art in the Renaissance, Oxford
Gamba E (1994) Documents of Muzio Oddi for the history of the proportional compass. Physis Rivista Internazionale Storia Scienza 31(3):799–815
Gilbert NW (1963) Renaissance concepts of method, New York-London, 34, 82, 89
Greco A (1961) Lettere familiari, Firenze, III, 81
Grossi C (1819) Degli uomini ill. di Urbino Commentario, Urbino, pp 53–57
Kemp M (1992) The science of art, New Haven
Loria G (1929a) Lo sviluppo delle matematiche pure durante il secolo XIX. La Geometria: dalla geometria descrittiva alla geometria numerativa, Scientia: rivista internazionale di sintesi scientifica 45:225–234
Loria G (1929) Storia delle matematiche. Antichità, Medio Evo, Rinascimento, Torino, I
Loria G (1931) Storia delle matematiche. I secoli XVI e XVII, Torino, II, pp 131–141
Loria G (1933) Per la storia della prospettiva nei secoli XV e XV, Bologna, pp 11–12

Moscheo R (1998) I Gesuiti e le matematiche nel secolo XVI. Maurolico, Clavio e l'esperienza siciliana, Società Messinese di Storia Patria, Biblioteca dell'Archivio Storico Messinese, XXV, Messina, pp 1–461
Moscheo R (2008) Maurolico Francesco. Dizionario Biografico degli Italiani 72
Napolitani PD (1985) Maurolico e Commandino, Il Meridione e le scienze, secoli XVI-XIX, Palermo, pp 281–316
Napolitani PD (1995) Commandino and Maurolico: publishing the classics, in Torquato Tasso and the University, Ferrara, pp 119–141
Napolitani PD (1997) Le edizioni dei classici: Commandino e Maurolico. Torquato Tasso e l'Università, Firenze, pp 119–141
Napolitani PD, Saito K (2004) Royal road or labyrinth? Luca Valerio's De centro gravitatis solidorum and the beginnings of modern mathematics. Bollettino di storia delle scienze matematiche, XXIV(2)
Neville P (1986) The printer's copy of Commandino's translation of Archimedes, 1558. Nuncius Annali di Storia Scienza 1(2):7–12
Polidori L, Ugolino F (1859), Versi e prose scelte di Bernardino Baldi, pp 513–537
Riccardi P (1870) Biblioteca matematica italiana. Modena, pp 644–648
Rose PL (1971) Plusieurs manuscrits autographes de Federico Commandino à la Bibliothèque Nationale de Paris. Revue d'Histoire des Sciences XXIV(4):299–307
Rose PL (1972a) Commandino, John Dee, and the De superficierum divisionibus of Machometus Bagdedinus. Isis, 63, 216, 88–93
Rose PL (1972b) John Dee and the De Superficierum Divisionibus of Machometus Badgedinus, Isis, LXIII, pp 88–93
Rose PL (1973) Letters illustrating the career of Federico Commandino. Physis - Rivista Internazionale Storia Scienza 15:401–410
Rose P L (1975) The Italian renaissance of mathematics, Genève, pp 185–221
Rosen E (1968) The invention of the reduction compass. Physis 10:306–308
Rosen E (1970–1990) Biography, dictionary of scientific biography, New York
Rosen E (1970) John Dee and Commandino. Scripta mathematica, XXVIII:321–326
Rosen E (1981) Commandino Federico, dictionary of scientific biography, Scribner's, New York, II
Rosen E (2008) Commandino Federico, www.encyclopedia.com
Russo L (1997) La rivoluzione dimenticata. Il pensiero scientifico greco e la scienza moderna, Milano
Sinisgalli R (1983) La prospettiva di Federico Commandino, Firenze
Sinisgalli R, Vastola S (1994) La rappresentazione degli orologi solari di Federico Commandino, Domus perspectivae, p 245
Swetz FJ, Katz VJ (2011) Mathematical treasures. Billingsley Euclid, Loci
Timpanaro Cardini M (1978) Commento al I libro degli 'Elementi' di Euclide, Pisa
Treweek AP (1957) Pappus of Alexandria. The Manuscript Tradition of the Collectic Mathematica, Scriptorium, XI, 228–233
Zerlenga O (1997) La forma 'ovata' in architettura. Rappresentazione geometrica, Napoli, 11, 13, 89, 113, 261

Egnazio Danti (1536–1586)

Mario Centofanti

Abstract Egnazio Danti, mathematician and cosmographer, is deep down a Renaissance man. A complex personality, characterized by great cultural and multiple interests in the relationship between Art, Science and Technology. Professor of mathematics in Florence and then at the University of Bologna, Cosmographer at the court of the Grand Duke of Tuscany, he was a skilled designer and manufacturer of scientific instruments. But also geographer, 'descriptor corographicus' (chorograph), expert measurer, iconographer creator of allegories and iconographic programs, engineer, painter. His contribution to the science of perspective is significant. In fact, he published two important works in the European panorama of scientific studies and production of the sixteenth century: "La prospettiva di Euclide (The perspective of Euclid)" in 1573 and "Le due regole della prospettiva pratica di J.B. da Vignola (The two rules of the practical perspective of JB da Vignola)" in 1583. Remarkable and innovative was also his contribution to the design and construction of instruments for the realization of perspective from real, and in the invention and development of a particular type of vertical anemoscope. Equally important was his contribution in the sixteenth century to the dissemination of knowledge for the construction and use of the astrolabe, of the armillary sphere, and of the Latin radium, a widespread measuring instrument.

Biographical Notes

Egnazio Danti, son of Giulio (member of the Perugia aristocracy) and Biancofiore degli Alberti, was born in Perugia and was baptized in S. Domenico, April 29, 1536, with the name of Carlo Pellegrino.

M. Centofanti (✉)
Department of Civil, Construction-Architectural and Environmental Engineering,
University of L'Aquila, via G. Gronchi 18, 67100 L'Aquila, Italy
e-mail: mario.centofanti@univaq.it

© Springer International Publishing Switzerland 2016

M. Cigola (ed.), *Distinguished Figures in Descriptive Geometry and Its Applications for Mechanism Science*, History of Mechanism and Machine Science 30, DOI 10.1007/978-3-319-20197-9_6

Danti's family context is particularly significant for the development of his cultural and artistic education.

He was first trained by his father in goldsmith's art, but also in literature and mathematics.

His grandfather, Pier Vincenzo (fine XV sec—1572), son of the notary Bartolomeo Rinaldi, member of the Guild of Goldsmiths of Perugia, was a mathematician and an architect; he called himself "Dante dei Rinaldi" (in honor of the poet), to permanently replace his patronymic with Danti.

The father Giulio (Perugia 1500–1575) was an architect, goldsmith and smelter. The older brother Vincenzo (Perugia 1530–1576) was goldsmith and sculptor. The younger brother Girolamo (Perugia 1547–1580) was goldsmith and painter. Another culturally influential character in the family was Teodora Danti (Perugia 1498–1573), the sister of Giulio, a scholar of astronomy, mathematics and geometry, as well as a pupil of the painter Perugino.

At the age of sixteen he was sent with his brother to Rome to continue his apprenticeship as a goldsmith and gemstone cutter (Fig. 1).

In 1554 he entered the Dominican Order, and in 1555, after the novitiate, became a monk with the name of Egnazio, and followed, up to the age of 26, the

Fig. 1 Bartolomeo Passarotti (1529–1592), ritratto di Egnazio Danti, XVI sec

canonical 7 years of training at the convent of San Domenico in Perugia: 3 years in philosophy (dialectic, philosophy and metaphysics) and four in theology.

After completing his studies, he moved to Florence in 1562, first at San Marco, and then at the convent of Santa Maria Novella.

In Florence he entered the sphere of the Grand Duke Cosimo Medici where he initially carried out the duties of a teacher of mathematics for children of the aristocracy. This activity turned public in 1571, as he became reader of Studio Fiorentino. Egnatio was "geographer at the Palace". He called himself "Master and Cosmographer of the Serene Grand Duke of Tuscany".

He was then instructed to attend a typically engineering activity in order to study a waterway connection between Livorno and the Adriatic, with the aim of giving shape to an ambitious idea conceived by Cosimo himself.

But the most important commitment for Danti was certainly the task of beautifying his official residence, Palazzo Vecchio, and specifically the 'Wardrobe', destined to guard the most precious objects of the Duke. The exhibition included a hall of geography: a terrestrial globe in the center and the maps of the territories of the Old and the New World on the doors of the cabinets placed along the walls.

Over a period of 12 years, he executed the drawings, then colored under his direction, of thirty of the fifty-three realized maps (Europe, fourteen; Africa, eleven; Asia, fourteen; America, fourteen).

In Florence Danti started his activities as a manufacturer of scientific instruments, beginning with the terrestrial Globe for the 'Wardrobe' of the Grand Duke. In 1572 he built a marble astronomical dial on the facade of Santa Maria Novella, to which he added in 1574 an equinoctial armilla. The Florentine period also saw the realization of the first vertical anemoscope for Giambattista Francesco Cini, at the Villa delle Rose in Val di Greve.

After the death of Cosimo, Danti was moved away from Florence and in 1575 he moved to Bologna at the Convent of San Domenico. He then assumed the chair of mathematics for the afternoon classes.

During his stay in Bologna he built several vertical anemoscopes that were installed in the courtyard of the archbishop's palace for Cardinal Gabriele Paleotti (1575), in the villa of Lorenzo Costa, in Palazzo Bianchetti in Ozzano, and in the cloister of the convent of San Domenico.

His extensive activities as measurer and chorographer are also to be ascribed to this period, when he executed the land survey of the countryside and the city of Perugia. He was then instructed to execute the survey of the Papal States that he carried out for sure in Romagna and in parts of Umbria, Sabina and northern Lazio.

The outcomes of this work were the "Perusini Agri" map, printed in Rome in 1580 by Cartaro and added by Ortelio in the 1601 Antwerp edition of "Theatrum Orbis terrarum", and the Chorographic map of the "Territory of Orvieto" (1583), added in 1593 to the Italian edition of the "Theatrum" by Ortelio.

In 1580, on the invitation of Pope Gregory XIII, he was called to Rome as papal cosmographer and mathematician. He participated in the activities of the Congregation for the calendar reform promulgated by the Pope in the February of 1582. Danti also played the role of 'iconographer' for the realization of the decoration of the so-called gallery of the Maps in the western arm of the Vatican Belvedere. In fact he was responsible for drawing up the program and the inscriptions and operating the synthesis of geographical information.

For the gallery, nowadays included in the Vatican Museums, he drew the cardboards of the 40 geographic maps representing regions, cities and Italian possessions, that were then frescoed by various painters.

In conjunction with the gallery, Gregory XIII ordered the construction of a tower called 'dei venti (Wind Tower)', on the architectural design of Ottavio Mascherino (1536–1606). Danti was commissioned to provide the iconography and painting of the tower. He realized a sundial and a complex vertical anemoscope mechanism, whose indications were related to wall and ceiling paintings of the Meridian Hall.

In 1583 he was accepted among the members of the Academy of San Luca in Rome and, on November 11 of that year, Pope Gregory XIII appointed him Bishop of Alatri, where he was active until his death occurred on October 19, 1586.

List of (Main) Works

Works on Perspective

LA PROSPETTIVA DI EUCLIDE

Nella quale si tratta di quelle cose, che per raggi diritti si veggono: et di quelle, che con raggi reflessi nelli specchi appariscono.
Tradotta dal R.P.M. Egnatio Danti Cosmografo del Serenissimo Gran Duca di Toscana. Con alcune sue Annotationi de' luoghi piu importanti insieme con la Prospettiva di Eliodoro Larisseo cavata della Libreria Vaticano, e tradotta dal medesimo, nuovamente data in luce.
In Fiorenza, 1573, nella Stamperia dei Giunti.

Danti translated from Greek to Italian the Euclid opera (late fourth century—beginning of the third century BC) originally titled "Optika kai katotoptrica" working on a copy printed and edited in 1557 by Joanne de la Penè at Andrea Wechelium in Paris.

He added the work of Heliodorus Larisseo (sixth century), printed for the first time, with the greek and latin text in comparison, and in its translation into Italian.

He also inserted in the Euclid text extensive comments with xylographic illustrations to the twelve synthetic 'guesses' [assumptions], and brief notes to the sixty 'theorems'. The comments of Danti are perfectly distinguishable as they are signed M. Egnat.

LE DUE REGOLE DELLA PROSPETTIVA PRATICA DI M. IACOMO BAROZZI DA VIGNOLA
Con i comentarii del R.P.M. Egnatio Danti dell'ordine dei Predicatori, per Francesco Zannetti, Roma, 1583.

Jacopo Barozzi da Vignola (1507–1573) had initiated around 1530 the compilation of his treatise, later repeatedly revisited and refined, but never published. Certainly, in the "Dedica ai lettori (Dedication to readers)" published in "Regola delli cinque ordini d'architettura (The Rule of the five orders of architecture)" (1562) he had anticipated the release of the treatise: "… poi in ciò si procede per certe belle Regole di Prospettiva la cui prattica è necessaria a questo et alla pittura insieme, in modo ch'io m'assicuro vi sarà grata spero anco di tosto donarvi (then we proceed for certain beautiful Rules of Perspective whose practice is required in this and painting together, and that I hope to give you soon in such a way that I am sure you will appreciate)".

The son of Jacopo, Giacinto (1535/40—post 1584), committed to Egnazio Danti the editing of the posthumous publication, as he says in the preface: "[Jacopo Barozzi] sempre accresceva di perfettione le Regole da lui scritte, di che può far fede la differenza che è in tra più esemplari, che egli cortesissimo della sua industria in diversi tempi dette a diversi e il presente testo che à me da Iacinto suo figliolo fu dato di poi che l'Autore l'hebbe l'ultima volta rivisto e riordinato…. ([Jacopo Barozzi] always enhanced in perfection the Rules he wrote, as it is proved by the difference between the various drafts, which he, most courteous of its industry, gave in different times to several people and this text given to me by his son Giacinto after the author had for the last time revised it and rearranged it.)"

Works on Astronomical Instruments

A group of works concern the Astrolabe and other astronomical instruments. Of particular interest is the 1569 edition of the

> TRATTATO DELL'USO E DELLA FABBRICA DELL'ASTROLABIO di Frate Egnatio Danti dell'Ordine di S. Domenico. Con l'aggiunta del planisfero del Roias. In Fiorenza appresso i Giunti.

Dedicated to Cardinal Ferdinando de' Medici, the treatise contains the first complete description of the astrolabe printed in Italy. Danti exposes the instrument's operations and how to build it, examining some of the most important samples of the time, invented by Gemma Frisius (1508–1555) and Juan de Rojas (XVI century).

A second edition, greatly expanded and representing the compendium of all the Danti treatises about instruments, dates back to 1578:

> PRIMO VOLUME DELL'USO ET FABBRICA DELL'ASTROLABIO, ET DEL PLANISFERIO
> di Maestro Egnatio Danti publico Lettore delle Mathematiche nello Studio di Bologna. Nuovamente ristampato, et accresciuto in molti luoghi, con l'aggiunta dell'Uso, et Fabbrica di nove altri istromenti Astronomici: Sfera armillare, Torquetto Astronomico; Di Cl. Tolomeo: Astrolabio Armillare, Gran Regola Astronomica, Quadrante Astronomico, Armilla Equinoziale; Diottra d'Hipparco, Gnomone Astronomico, & Geometrico, Anemoscopio verticale, in Firenze, Appresso i Giunti.

In 1583 we have the publication of the

> TRATTATO DEL RADIO LATINO
> istrumento giustissimo e facile piu d'ogn'altro per prendere qual si voglia misura, e positione di luogo, tanto in Cielo, come in Terra: il quale oltre alle operationi proprie sue, fà anco tutte quelle della gran Regola di C. Tolomeo, e dell'antico Radio astronomico, inventato dal illustrissimo et eccellentissimo Signor Latino Orsini. In Roma, appresso Vincenzo Accolti.

Danti published the treatise of Latino Orsini, inventor of the instrument, with his usual large commentaries. The competence of Danti is linked to his ability as a measurer having long used the radium in his known campaigns of territorial survey.

Review of Main Works

La prospettiva di Euclide (The Perspective of Euclid)

It is really significant that Danti began to address his attention to the issues of perspective by moving from the translation of the greek text on Optics by Euclid.

Even the inclusion of the unpublished work by Heliodoros Larisseo would seem to refer to a scholarly approach. But in the preface Danti instead provides a very precise motivation of his interest in the perspective, anticipating his curatorship for the treatise of Vignola, that had to come 10 years later (1583).

Danti is keen to stress the importance of perspective for all the liberal arts, and its being a necessary precondition for the Geography "… poiche ella sola ci mostra il modo di ridurre in piano, in forma o ovata, ò circulare, e in diverse altre maniere proporzionevolmente il sito di tutta la terra insieme, e delle provincie particulari e ci fa avere perfetta cognitione delle ditanze de' luoghi facendoci conoscere chiaramente non solo la proportione, e convenienza di sito, che ha una regione con l'altra…".

It is evident the reference to his activities as cosmographer, cartographer and chorographer, through the use of projections: "the way to reduce to flat".

And then he turned to the 'source' for excellence, the Optiké, widespread with the advent of printing and its numerous editions published between the sixteenth and seventeenth centuries.

Euclid (330–275 BC) of Alexandria, Egypt, is in fact considered to be the putative father of the 'perspectiva naturalis (natural perspective)' as the author of the first organic scientific treatise on geometrical optics that reached us and concerning, in large part, the geometric problems of vision: geometric relationships that exist between the actual size of the objects, their position in space and the apparent measures deductible from visual images.

Its sixty-one propositions or theorems concern partly (twenty) the optical phenomena and for the rest the perspective phenomena.

The tenth theorem is of particular interest: "The more distant parts of the flat surfaces, which are below the eye, appear taller."

Many scholars intended it as a sort of anticipation of the Renaissance 'legitimate construction' and of the construction of the 'visual pyramid' intersected by the projection plane.

The illustration on page 29 shows how the visual rectilinear rays form the image intersecting the projection plane seen in profile (Fig. 2).

Among the notable things: the second supposition that introduces, for the first time in history, the concept of 'cone of vision', and the sixth theorem that refers to the idea of the apparent convergence of parallel straight lines towards a point at infinite distance (Fig. 3).

Le due Regole della Prospettiva pratica (The two rules of the practical perspective).

The treatise is divided into 'definitions' by Vignola, in great typographic body, accompanied by the thirty engraved copper plates he drew, and 'Commentaries' by Danti, in small italic typographic body, illustrated by one hundred and twenty xylographies.

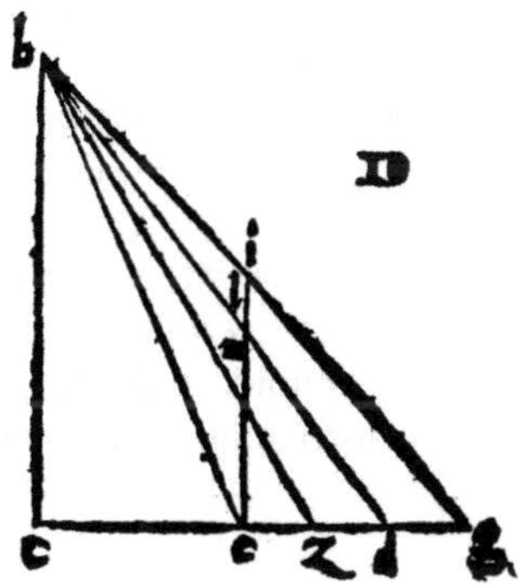

Fig. 2 Tenth theorem, p. 29. Intersection of the visual rays with the projection plane

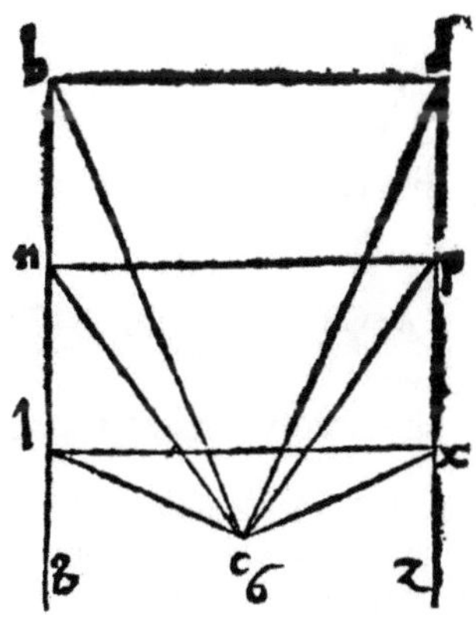

Fig. 3 Sixth theorem, p. 24. Convergence of *parallel straight lines*

The text of Vignola is about forty pages, while that of Danti is more than one hundred pages, and therefore (Camerota 2006) represents a treatise in the treatise as Barbaro against Vitruvius.

Danti opens his work with an interesting biography of Barozzi "Vita di M Jacomo Barrozzi da Vignola architetto e prospettivo eccellentissimo (Life of M Jacomo Barrozzi da Vignola, most excellent architect and prospective)", with much unpublished information.

The preface is followed by a broad introductory part where Danti operates a prior theoretical systematization of the matter, with twenty-seven definitions of the art of perspective, twelve assumptions, and fifty-three theorems, problems and propositions.

Such an extensive development of arguments and insights "…avevano lo scopo di legittimare geometricamente i procedimenti che molti pittori conoscevano solo per pratica (…were intended to geometrically legitimize the proceedings that many painters knew only through practice)" (Camerota 2006).

The discussion of the first rule is opened by an exemplary table by Vignola (Fig. 4) which shows the mechanism of formation of the perspective image of a solid with octagonal base. The image is obtained by the intersection of the projection plane with the visual rays (projecting straight lines) joining the point of view with the significant points of the object.

Vignola defines it as "il principal fondamento della prospettiva (the main foundation of perspective)" (Fig. 5).

Danti resumes and substantiates from the theoretical and methodological point of view.

The foundation of the first rule, 'l'essentia della Prospettiva (the essence of Perspective)', is nothing more than the figure that is formed in the section of the 'visual pyramid' with the projection plane (Fig. 6), in conformity with the legitimate construction by Leon Battista Alberti (1404–1472) (Fig. 7).

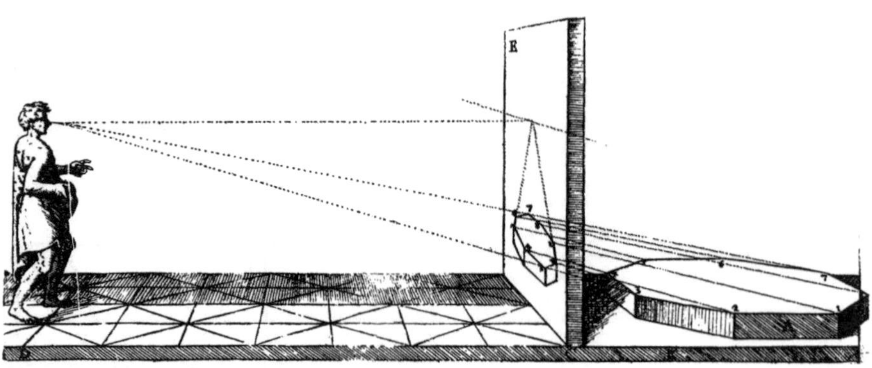

Fig. 4 Jacopo Barozzi da Vignola, perspective image as intersection of the visual rays on the projection plane, "in che consiste il fondamento della prospettiva (in which consists the foundation of perspective)" chap. II, p. 55

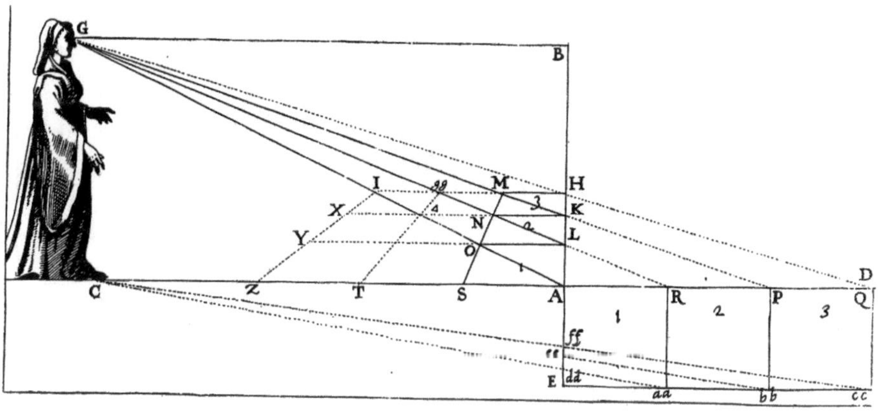

Fig. 5 Jacopo Barozzi da Vignola, perspective of plane figures: "Della pratica dei cinque termini per digradare le superfici piane (About the practice of the five terms for sloping flat surfaces)", chap. VI, p. 69. The foreshortening or the sloping of the measures is obtained by intersection of visual rays on the projection plane

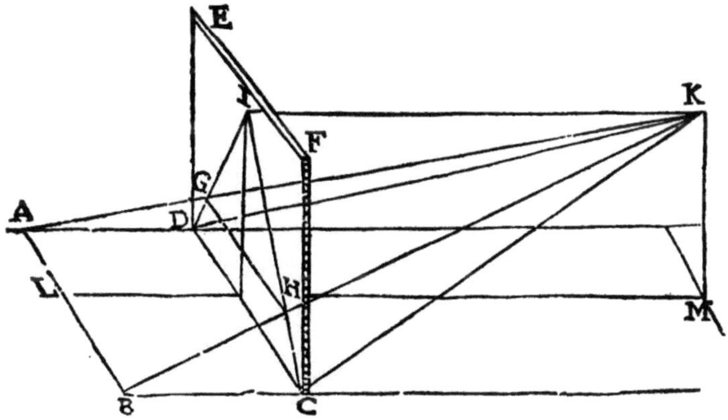

Fig. 6 Egnatio Danti, the perspective of a line parallel to the horizon, Theorem XXVII, p. 38

Closing the presentation of the first rule, the treatise also addresses the representation from the bottom to the top, the problems of the theatrical scenography, the anamorphic principles and even some mechanical optical game.

The first rule is canonical, in the sense that it explicitly uses the intersection of the visual pyramid on the projection plane.

The second rule, that however Vignola himself defines conforms to the first, is to use the main point together with the distance points, called by Danti 'helpers', without the intersection of the visual pyramid on the projection plane.

Although, as noted by Danti, the process of section is still virtually underlying.

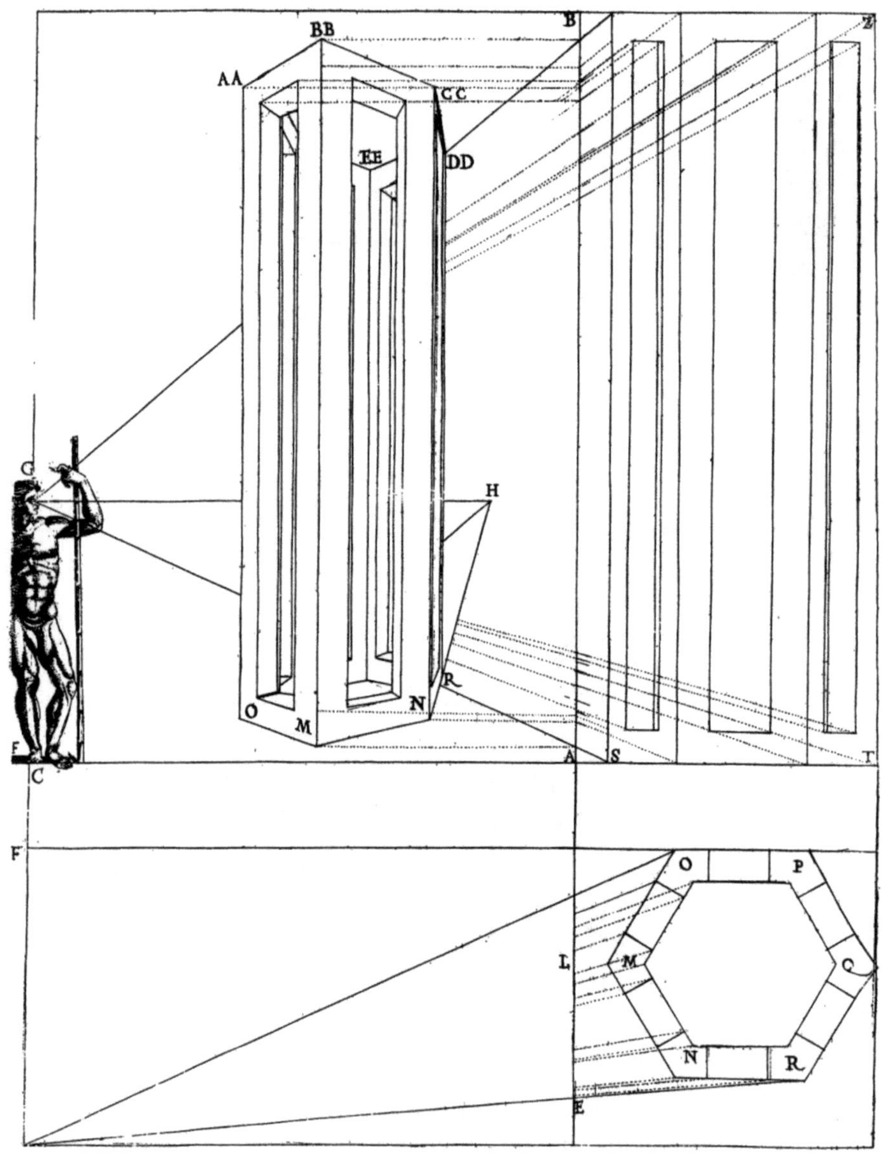

Fig. 7 Jacopo Barozzi da Vignola, the perspective of a solid with hexagonal base, p. 80

An exemplary case is the perspective of just one distance point at Chap. VI, p. 108 (Fig. 8). Given the main point and the right distance point, only vertical straight lines and straight lines inclined at 45° and passing through the singular points of the octagon are used to determine the tracks. The perspective image of the plane figure is constructed by intersecting the 'linee erette (erected lines)' (vertical)

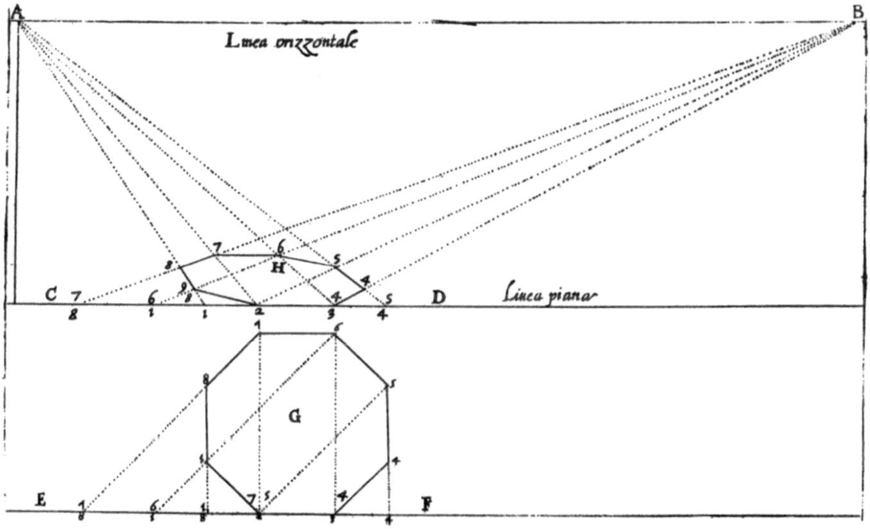

Fig. 8 Jacopo Barozzi da Vignola, the perspective of an octagon, chap. VI, p. 108

converging in the main point, with 'diagonal lines' that converge at the distance point.

Perspective with two lateral distance points on the horizon, right and left of the main point (Fig. 9).

And finally the perspective with four distance points, two on the horizon, and the other two one above and the other below the main point. During the tracking phase, as noted by Danti, it is even possible to operate without the main point to put the cube in perspective (Chap. VI, p. 107), using only the diagonal lines of the faces of the solid converging to the distance points (Fig. 10).

This is precisely the main innovative contribution that Vignola proposes on the basis of his considerable practical experience.

With the second rule he thus provides a way to resolve the perspective problem, which simplifies, in terms of convenience and immediacy, the common yet complex regular procedure, often cited by Danti: to draw the object in plan; to determine the views ('digradazione (degradation)') through an intersection operation of the projection plane with the visual rays passing through the points of the object; to delineate the elevation in perspective using the profile of the solid.

"Il risultato della integrazione [di Egnazio Danti] è stato ottimo e particolarmente pregevole apparve subito la chiarissima esposizione della seconda regola, quella relativa all'impiego dei punti della distanza, indubbiamente già nota ed utilizzata dai prospettici pratici e da alcuni teorici precedenti, ma mai esposta tanto lucidamente e convincentemente come in questo testo. (The result of the integration [by Egnazio Danti] was great, and the very clear exposition of the second rule, that relating to the use of the distance points, undoubtedly already known and used by

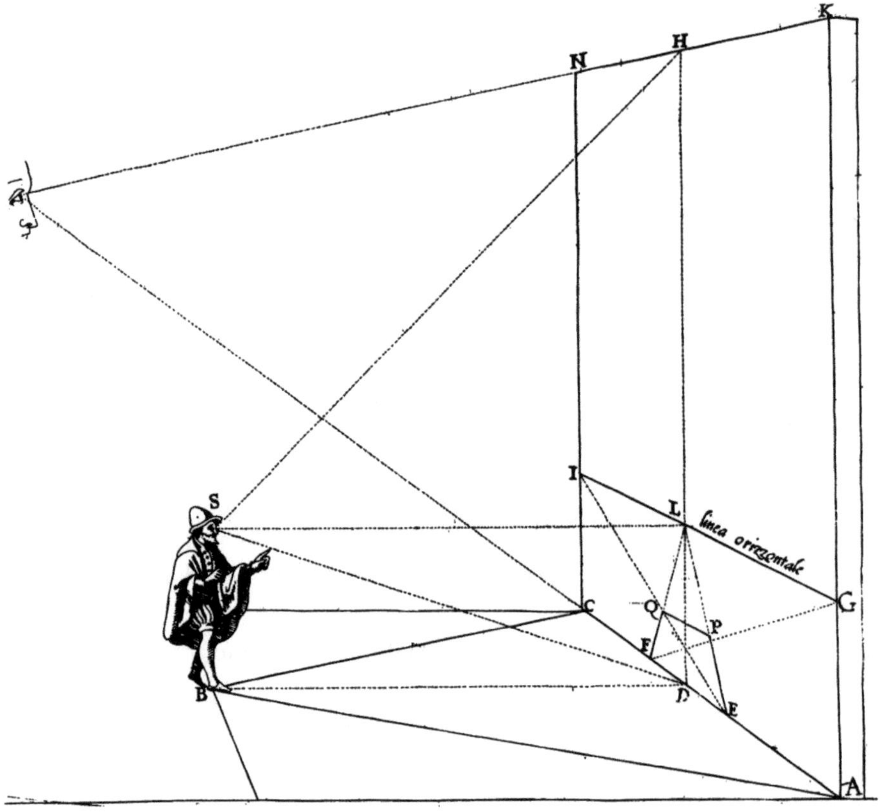

Fig. 9 Jacopo Barozzi da Vignola, perspective of a flat shape with two distance points: "Quanto si deve star lontano a vedere le Prospettiva, da che si regola il punto della distanza (How far should you be to see the perspective, from which you adjust the distance point)", V, p. 105

perspective technicians and by some prior theoretical, but never exposed so lucidly and convincingly as in this text, immediately looked very fine.)" (Vagnetti 1979).

The Perspectographs

One of the topics of major interest in the Danti-Vignola treatise "Le due regole (The two rules)", is the illustration of the tools for the experimental and mechanical implementation of perspective (Fig. 11).

Danti presents a series of perspectographs, as many as seven, conceived on basic mechanical principles.

The tool presented by Danti at page 56 (Fig. 12) is nothing more than the replica of the 'door' by Albrecht Dürer (1471–1528) (Fig. 13). Danti however, for the

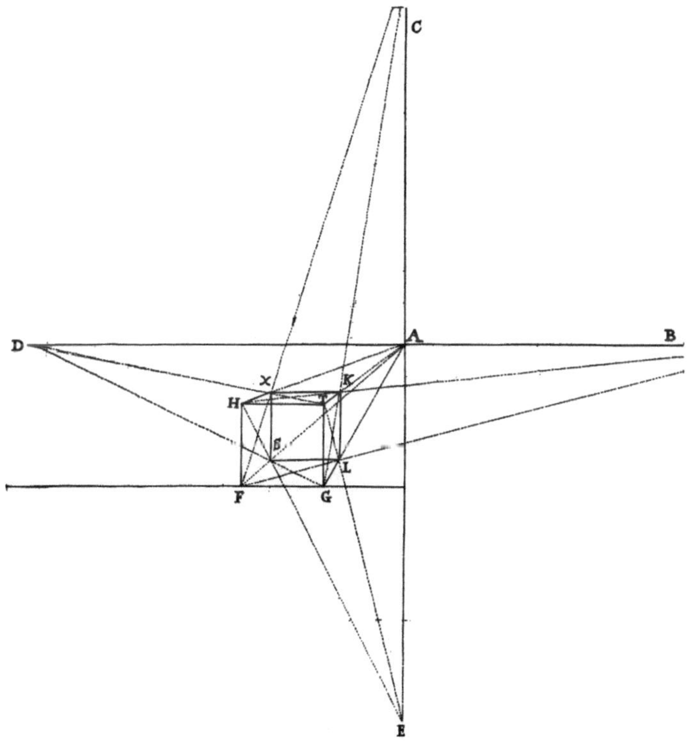

Fig. 10 Jacopo Barozzi da Vignola, perspective of a cube with the use of four distance points, p. 107

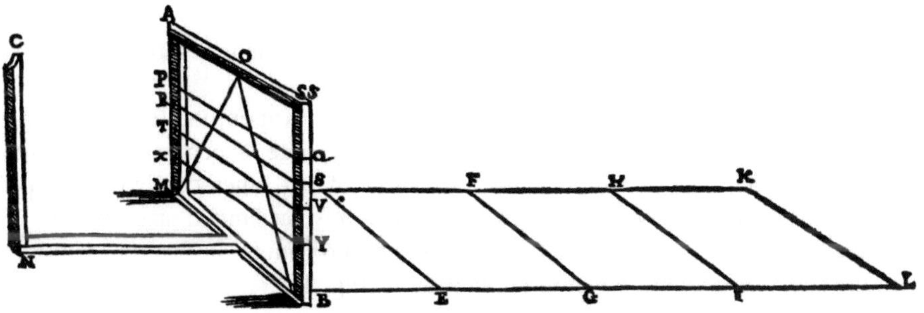

Fig. 11 Egnatio Danti, Perspectograph of Tommaso Laureti (1530–1602), for the demonstration of the operating principles of perspective, p. 39

purpose of identifying the position of the intersection point of the visual ray with the projection plane, replaces the orthogonal grating frame of Dürer with a pair of mobile wires. An innovative shrewdness that allows a substantial increase in the accuracy obtained in the operational venue.

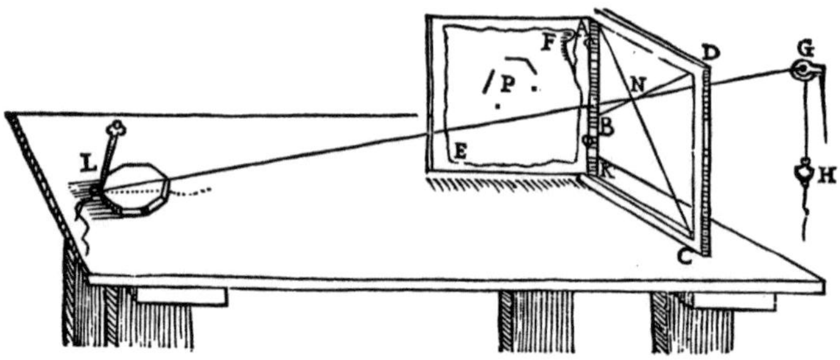

Fig. 12 Egnazio Danti. Perspectograph by Dürer, p. 56

Fig. 13 Dürer, Albrecht. (1525), Underweysung der Messung, Nürnberg, lib. IV, the "door"

In the sample at p. 57 (Fig. 14), Danti uses, instead of the wires, two sized rulers hinged to the base of the door.

In the sample at page 57 Danti proposes, as an innovation, the use of a diopter (Fig. 15) to perform remote collimation and put very large objects into perspective. He thus exceeded the conceptual intrinsic limit of the other contemporary perspectographs, given by the possibility to operate only on limited size objects.

And finally the perspectograph of Lanci. The projection surface is cylindrical and the sheet of drawing paper is applied on it (Fig. 16).

The other table prepared by Vignola without comments (Fig. 17) represents instead the mechanical instrument to perform in a practical and operational way the perspective drawing, as an imitation of reality, that is through direct observation of the figure to be represented.

From the description of Danti:

- AB horizontal base of the instrument, grooved and sized
- CD sized vertical ruler
- AB and CD represent the 'door' and the projection plane

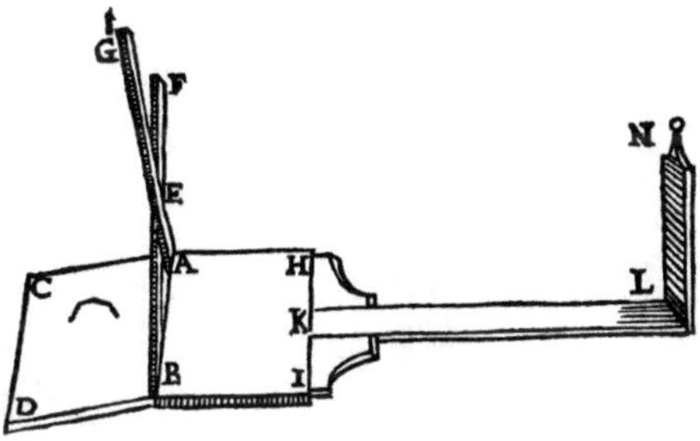

Fig. 14 Egnazio Danti, perspectograph with sized rulers of Girolamo da Perugia (about 1580), chap. III, p. 57

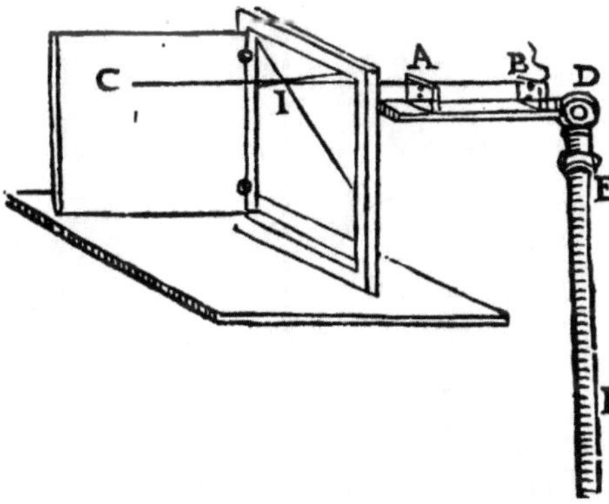

Fig. 15 Egnazio Danti, perspectograph with diopter and mobile wires, chap. III, p. 57

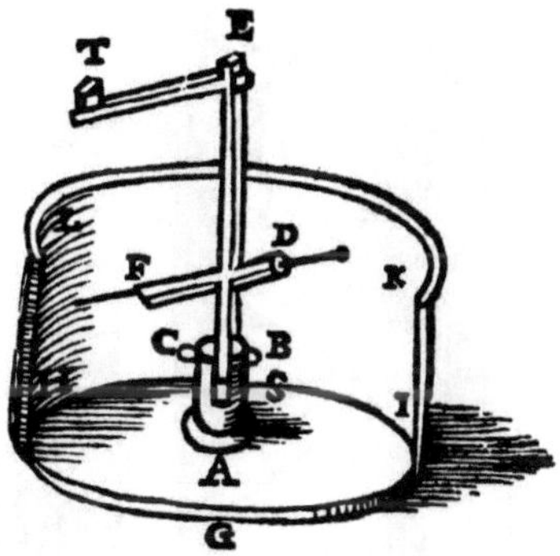

Fig. 16 Egnazio Danti, Perspectograph of Lanci, chap. III, p. 61

- MN vertical fixed rod, brings the target N which can slide along the rod upward and downward, and can rotate around the axis of the rod
- The grooved base AB is equipped with feet that allow it to be moved forward or backward, thereby varying the distance of the point of view from the projection plane

Fig. 17 Jacopo Barozzi da Vignola, perspectograph, p. 60

- The ruler CD can move laterally on the line AB to the right or to the left, maneuvered manually by a rope and a winding cylinder said 'subbio (beam)'
- You perform the collimation between the point of view (target) and the point of the figure to be detected, by sighting the edge of the vertical ruler
- You read the positions (coordinates) on the vertical sized ruler and on the sized base on the point C of intersection between the two
- You report the positions of the points on the palette where an orthogonal grid is designed, corresponding to the horizontal and vertical graduations of the ruler and of the base.

The instrument assumes, as shown, the simultaneous presence of two operators. The first, presumably the artist, who selects the significant points of the figure to be detected and operates the target of the same, and the second, who reports on the sheet with the grid the position of the detected points.

The Vertical Anemoscope

The anemoscope is a tool that indicates the direction of the wind with a weathervane moving around a vertical axis, free to orientate according to the wind direction.

The anemoscopes were known since the classic age, however Egnazio Danti extended the metallic axis revolving with the weathervane, so that this could pass through the roof and go down into a room, where, equipped with an index, it would be able to mark the direction on a wind rose.

Danti introduced substantial improvements, so much so that the invention of this kind of vertical anemoscope is attributed to him.

The anemoscope designed and built by Danti (Fig. 18) is described in the Latin treatise "Anemographia" published in 1578. The treatise was simultaneously published in Italian in the "Primo volume dell'uso e della fabbrica dell'Astrolabio… (First volume of the use and factory of the Astrolabe …)".

The vertical anemoscope is constituted by a metal rod AB (called 'stile (style)' by Danti) which carries at the top the weathervane A; some rings F, E, D, on which the attacks needed to fasten the rod to the walls are pivoted; a base with a central hole where the bottom tip of the rod is inserted; a horizontal toothed wheel united with the rod C, and with a square section in that point; a vertical toothed wheel G with pivoted in the middle two indicator arrows, one facing the inside and the other facing the outside; the movement of the component rod-weathervane-horizontal wheel around the vertical axis is transmitted to the vertical wheel, and then to the arrows which mark, on the painted wind rose, the direction detected. The two toothed wheels are equal for the diameter of an arm (about 600 mm) and the number of teeth equal to 48. About the teeth Danti says they must be "non solo aguzzi in punta, ma che tondeggino anche in traverso (not only pointed at the tip, but they also have to circle in transverse)" (p. 177).

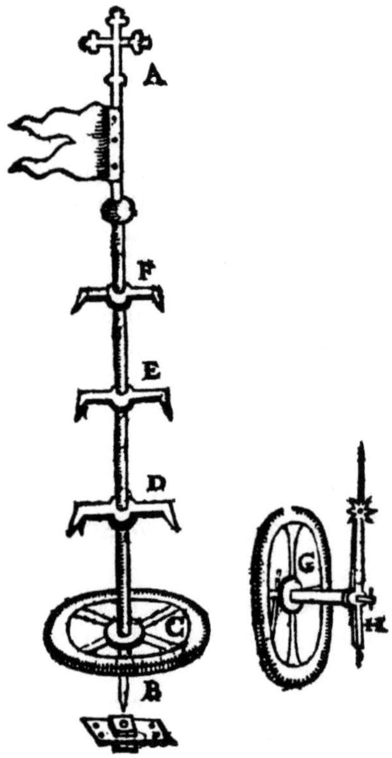

Fig. 18 Egnazio Danti, vertical Anemoscope, "First volume of the use…" 1578, Part Six, p. 20

This device allows you to read the wind direction even within an inner room. Danti designed and built anemoscopes in Florence for Giovanni Battista Cini, and during his stay in Bologna he built other anemoscopes in the courtyard of the archbishop's palace for Cardinal Gabriele Paleotti (1575), in the villa of Lorenzo Costa, at Palazzo Bianchetti in Ozzano and in the cloister of his convent of San Domenico, moreover the only one partially preserved.

Particularly complex is the mechanism of the anemoscope of the Wind Tower in the Vatican, where the gears are three, with five toothed wheels. The reading is performed on the ceiling of the Meridian Hall, where the wind quadrant is depicted (Pinelli 1996, pp. 174–182). The Vatican Anemoscope is represented in Anemographia, Rome 1581, BAV, Cod.Vat.Lat 5647, F.14 verso.

On the Circulation of Works

There is an obvious matter concerning the classification of "Le due Regole (The two rules)" within the complex and articulated cultural frame of the period between the XV and the XVI century. The matter regards in particular the analysis of the references that Danti uses for its theoretical and practical discussion of development and deepening of the application methods designed by Vignola.

First of all Piero della Francesca (1418–1492) with his "De Prospectiva pingenti", manuscript, first published in 1841. On this work of Piero, Danti notes: "il primo in tempo e che con miglior metodo e forma ne abbia scritto, del quale abbiamo oggi tre libri scritti a mano eccellentissimamente disegnati (the first in time who wrote about it with the best method and form, of which we have today three excellently drawn handwritten books)".

Daniele Barbaro (1513–1570) "La pratica della Prospettiva… (The practice of perspective…)", published in Venice in 1556, where he served mainly Piero della Francesca's work; so that Danti himself says "ne ha trascritto una gran parte nel suo libro sulla Prospettiva (he transcribed a large part of it in his book on Perspective)".

Sebastiano Serlio (1475–1554) "Trattato di Architettura, Libro secondo, Di Prospettiva (Treatise of Architecture, Book II, Perspective)" published in Paris in 1545 and allegedly inspired to the teaching of Baldassare Peruzzi (1481–1536). To these he adds: Leon Battista Alberti (1404–1472) "De Pictura", published in Latin in 1540 in Basel, while the vulgar version of 1435 was not published until 1847; Leonardo da Vinci (1452–1519), "Trattato della pittura (Treatise on Painting)", posthumous around 1550, published only since 1651; Albrecht Dürer (1471–1528), "Underweysung der Messung, Nuremberg", Book IV, published in Nuremberg in 1525.

And more, Jehan Cousin the Elder (1490–1560) "Livre de perspective", Paris 1560; Jean Pelerin, Viator (1435/40–1524) "De artificiali perspectiva", published in Toul in 1505; Federico Commandino (1509–1575) "Ptolomaei Planispherium commentarius", Venice 1558; Jacques Androuet du Cerçeau (active between 1549 and 1584), called Iacomo Andreotti by Cerchio, author of "Leçons de perspective positive", Paris 1576.

The resulting picture is particularly significant especially with respect to the difficulties that at the time were encountered for the perspective construction in the operational venue. All of this in the absence of the availability of a reliable text with which to solve the geometrical-constructive problems that occurred, especially in the artistic field.

Moreover, it is hardly necessary to emphasize that the text of Piero della Francesca was manuscript, little known and not published until 1841; that the treatise of Dürer had been published only in 1525 in German and in 1532 in Latin; that the one by Viator Pelerin had poor diffusion in Italy in the early decades of the sixteenth century and was written in French and Latin; that the text of Commandino had a scientific and mathematic approach, and therefore it was only affordable to a few.

The widespread use of the Vignola-Danti treatise was also conveyed by the great international success of the "Regola delli Cinque Ordini (Rule of the Five Orders)" by Vignola, published for the first time in 1562. The publishing history of "Le due regole (The two rules)" comprises many as six editions during the seventeenth century: 1602 Rome, 1611 Rome, 1635 Siena, 1644 Rome, 1682/83 Bologna, 1684 Rome.

The Vignola-Danti treatise, with everything induced by its exemplary clarity, marks a significant moment in the fulfillment of the transition from the artistic matrix experimentalism to a rigorous scientific approach that had already been initiated by Commandino.

In this sense the notation by Vagnetti appears appropriate: "L'opera del Vignola e del Danti è indubbiamente molto pregevole ed ha svolto una funzione fondamentale nella storia della teoria prospettica, tanto per la diffusione della conoscenza dei procedimenti esatti elaborati per la nuova scienza, quanto perché ha in qualche modo anticipato le conclusioni alle quali giunse più tardi Guidobaldo del Monte alla fine del secolo XVI sulla vera natura dei punti di concorso; essa costituisce momento di passaggio obbligato tra i numerosi tentativi dei 'pratici' e le prime dotte formulazioni degli 'scienziati'. (The work of Vignola and Danti is undoubtedly very valuable and has played a vital role in the history of the perspective theory, both for the dissemination of knowledge of the exact procedures developed for the new science, and for the fact that it has in some ways anticipated the conclusions later expressed by Guidobaldo del Monte at the end of the sixteenth century about the true nature of the concourse points; it represents an obliged time of passage from the many attempts of the 'practical' to the first scholarly formulations of 'scientists'.)" (Vagnetti 1979).

Kemp himself expresses an extremely positive opinion about the role that the Vignola-Danti treatise played in the sixteenth and seventeenth century: "Considerato nel suo insieme "Le due regole" era il libro più intelligente, utile ed esaurientemente informativo sulla costruzione prospettica fino ad allora pubblicato, e può ben reggere il confronto con qualsiasi testo successivo che abbia tentato di bilanciare i procedimenti pratici e l'analisi matematica. (Taken as a whole, "The two rules" was the most intelligent, useful and exhaustively informative book about perspective construction hitherto published, and may well stand comparison with any subsequent text that has attempted to balance the practical procedures and mathematical analysis)" (Kemp 1990).

The treatises of Danti as "Anemographia" and "Uso e fabbrica dell'Astrolabio (Use and factory of the Astrolabe)" had a wide spread in Europe, even in manuscript copies (Dubourg 2011, p. 134).

The perspectographs were perfected and used not only in painting, sculpture and architecture, but also in the military art and in topography. The perspectographs, despite having reached a high degree of mechanical sophistication, began after the middle of the seventeenth century to be less and less used, to be gradually supplanted by the camera obscura (Kemp 1990).

Modern Interpretation of Main Contributions

The "Prospettiva di Euclide (Perspective of Euclid)" by Egnatio Danti gets topical again when, in the early twentieth century, the publishing house Hoepli, founded in Milan in 1870 by Ulrico Hoepli (1847–1935), decides to print a collection of manuals in the technical and scientific field. Among the others, the Optics of Euclid in the Italian translation of Giuseppe Ovio (1863–1957), Senator of the Kingdom, a scholar of physiological optics and professor of clinical ophthalmology: Giuseppe Ovio, "L'Ottica di Euclide (The optics of Euclid)", Ulrico Hoepli editore, Milan 1918.

The Optics is discussed and explained in its practical uses because it was still considered the foundation for the teaching of the science of vision, and in particular the concept related to the rectilinear propagation of the light beam.

Giuseppe Ovio chose to publish the text in the version of Theon of Alexandria (335–405 BC), also using the Italian translation by Danti (1573). Other references: "Euclidis Megarensis … elementorum libri XII", Venice 1505, edited by Bartolomeo Zamberti; the Latin translation of Joanne de la Penè, Paris 1557; "La perspective d'Euclide, traduit en François…", edited by Roland Fréart de Chanteloup; the edition of David Gregory (1659–1708), in greek and latin, Oxford 1703; and finally the "Euclidis Opera Omnia" in nine volumes (1883–1916), edited by Johan Ludvig Heiberg and Heinrich Menge.

The publishing success of the treatise "Le due regole (The two rules)" continues even in the eighteenth and nineteenth centuries, as already shown for the sixteenth century, with further six editions: 1743 Venice, 1744 Bologna, 1770 Rome, 1816 Milan, 1828 Rome, 1830 Milan. It appears however clear how, with the spread of the contributions of Guidobaldo Burbon del Monte (1545–1607), Pesaro 1604, and Gerard Desargues (1593–1661), Paris 1636, the attention to the Danti-Vignola treatise tends to move progressively towards interests mostly related to the history of perspective.

Leopoldo Cicognara (1767–1834), in his famous bibliographic summary, "Catalogo ragionato dei libri d'Arte e di antichità posseduti dal Conte L.C. (Reasoned catalogue of the art and antiquities books owned by Conte L.C.)", Pisa, 1821, vol. I, pp 149–163, defined "Le due regole (The two rules)" "the best book known to us for similar institutions", referring to the Bologna 1644 edition.

And we get to the twentieth century. Noteworthy is the new edition in facsimile, Vignola (MO) 1974, of the 1583 copy preserved in the Biblioteca Hertziana, with a preface by Maria Walcher Casotti. By the same author a paper on "Jacopo Barozzi da Vignola nella storia della Prospettiva (Jacopo Barozzi da Vignola in the history of Perspective)" (Casotti 1953). Another important critical contribution is the 1962 one by the Japanese scholar Timothy Kaori Kitao "Prejudice in perspective: a study of Vignola's perspective treatise", in The Art Bulletin, XLIV, pp. 173–94 (Kitao 1962). In 1985, Sala Bolognese (BO) reprinted in facsimile the 1743 Venetian edition. In 1988 in Alburgh, we still have an anastatic version of the original edition of 1583, edited by Martin Kemp, who in 1990 published the monograph "The science of Art, Optical themes in western art from Brunelleschi to Seurat", Yale University Press, New Haven and London, where in the chapter "the linear perspective from Dürer to Galileo" he critically analyzes the contribution of Vignola and Danti.

More recently, the critical edition with translation into French, edited by Pascal Dubourg Glatigny, Paris 2003; by the same author an interesting essay

(Dubourg 2004) that valorizes the role of Egnazio Danti as the founder of a personal theory on artistic perspective; the essay by Filippo Camerota "Giacomo Barozzi da Vignola and Egnazio Danti" within the monograph on "La prospettiva del Rinascimento. Arte, architettura, scienza (The perspective of the Renaissance. Art, architecture, science)", Mondadori Electa, Milan 2006; in 2007 the reprint of the 1828 Carlo Antonini edition including both the "Regola dei cinque ordini (Rule of the five orders)" by Vignola, and "Le due regole (The two rules)" by Vignola and Danti, edited by Diego Maestri and Giovanna Spadafora. Lastly, we have to report the extensive and documented biographical essay (Dubourg 2011) with the edition of the Correspondence.

Bibliography

Almagia' R (1952) Monumenta cartographica vaticana. Vol III: Le pitture murali della Galleria delle carte geografiche, Roma, BAV

Bartolini S (2006) I Fori gnomonici di Egnazio Danti in Santa Maria Novella, Firenze, Polistampa

Bartolini S (2008) Gli strumenti astronomici di Egnazio Danti e la misura del tempo in Santa Maria Novella, Firene, Polistampa

Bonelli G (1977) Il grande astrolabio del Museo di storia della scienza di Firenze, in "Annali dell'Ist. e Museo di storia della scienza di Firenze", II (1977), 2, pp 45–66

Brink S (1983) Fra Ignazio Danti: das Programm der Sala Vecchia degli Svizzeri in Vatikan und Ripas Iconologia, in "Mitteilungen des Kunsthistorischen Institutes" in Florenz, 2 XXVII, pp 223–254

Camerota F (2006) La prospettiva del Rinascimento. Arte, architettura, scienza, Mondadori Electa, Milano

Courtright N (2003) Il Papato e l'arte della Riforma nel XVI secolo Roma: Torre di Gregorio XIII dei Venti in Vaticano. Monumenti di Roma papale. Cambridge University, Cambridge/New York

Casanovas J (1983) The vatican tower of winds and the calendar reform. In: Gregorian reform of the calendar [...], Coyne GV, Ho-skin MA, Pedersen O (eds) Città del Vaticano, Pontificia Academia Scientiarum, pp 189–198

Casotti MW (1953) Jacopo Barozzi da Vignola nella storia della prospettiva, in Periodico di Matematica, vol 31.2, pp 73–103

Casotti MW (1974) Nota introduttiva, in Vignola - Danti "Le due regole della prospettiva pratica, ristampa a cura della Cassa di Risparmio di Vignola

Cecchi A, Pacetti P (eds) (2008) La sala delle carte geografiche in Palazzo Vecchio: capriccio et invenzione nata dal Duca Cosimo, Firenze, Pagliai

Cheney I (1989) The Galleria delle Carte Geografiche at the Vatican and the Roman Church's view of the history of Christianity. Renaissance Pap 17:21–37

Cigola M (2012) In praise of parallel theories: descriptive geometry and applied mechanics. In: Carlevaris L, Filippa M (eds) In praise of theory. The fundamentals of the disciplines of representation and survey. Roma Gangemi, pp 39–46

Cigola M, Ceccarelli M (1995) On the evolution of mechanisms drawing. In: Proceedings of IXth IFToMM World Congress, Politecnico di Milano 1995, vol 4, pp 3191–3195

Conticelli V (2007) Guardaroba di cose rare et pretiose. Lo Studiolo di Francesco I de' Medici. Arte, storia e significati, Lugano, Agorà Publishing

Courtright NM (1990) Gregory XIII's Tower of the Winds in the Vatican, Ph.D., New York, New York University, vol 2

Courtright NM (2003) The papacy and the art of reform in sixteenth-century Rome. Gregory XIIT's Tower of the Winds in the Vatican, Cambridge, Cambridge University Press
Daly Davjs M (1982) Beyond the Primo Libro of Vincenzo Danti's. Trattato delle perfette proporzioni, Mitteilungen des Kwzsthistorischen Institutes in Florenz, XXVI, 63–84
Dubourg Glatigny P (1999) La «merveilleuse fabrique de l'oeil»: illustration anatomique et théorie de la perspective à la fin du XVIe siècle, Roma moderna e contemporanea, 7, 3, 369–394
Dubourg Glatigny P (2002) Egnatio Danti O.P. (1536–1586). Itinéraire d'un mathématicien parmi les artistes, Mélanges de l'École Française de Rome - Italie et Mediterranée, 114, 2, pp 543–605
Dubourg Glatigny P (2003) Egnatio Danti et la perspective, in Egnatio Danti, Les deux règles de la perspective pratique de Vignole, 1583, Traduction et édition critique de Pascal P. Dubourg Glatigny, Paris, CNRS, pp 1–85
Dubourg Glatigny P (2004) Egnatio Danti as the founder of the authentic theory of artistic perspective as compared to late Renaissance ideas on the authenticity of texts. S Afr J Art Hist 19:48–68
Dubourg Glatigny P (2008) La place des arts mécaniques dans les "Scienze matematiche ridotte in tavole (Bologne, 1577) d'Egnatio Danti, in Dubourg Glatigny P, Vérin H (eds) "Réduire en art. La technologie de la Renaissance aux Lumières", Paris, Éditions de la Maison des sciences de l'homme, pp 199–212
Dubourg Glatigny P (2011) Il disegno naturale del mondo. Saggio sulla biografia di Egnatio Danti con l'edizione del carteggio. Milano, Aguaplano
Fidanza GB (1996) Vincenzo Danti, Firenze, Leo S. Olschki
Fidanza GB (1997) Vincenzo Danti architetto, "Mitteilungen des Kunsthistorischen Institut" in Florenz, 41, 392–405
Fiorani F (2003) Danti edits Vignola: the formation of a modern classic on perspective. In: Massey L (ed) The treatise on perspective: published and unpublished. New Haven-London, Yale University Press, pp 127–159
Fiore FP (1986) Danti, Egnazio, in Dizionario Biografico degli italiani, Roma, Istituto della Enciclopedia Italiana Treccani, vol 32, pp 659–663
Gambi L (1996) [1994] Egnazio Danti e la Galleria delle Carte geografiche, in "La Galleria delle Carte Geografiche in Vaticano", in Gambi e Pinelli (eds) Modena, Panini, pp 62–72
Goffart W (1998) Christian pessimism on the walls of the Vatican Galleria delle Carte Geografiche. Renaissance Q 51, 3:788–828
Kemp M (1990) The science of Art, Optical themes in western art from Brunelleschi to Seurat, Yale University Press, New Haven and London
Kitao TK (1962) Prejudice in perspective: a study of Vignola's perspective treatise. The Art Bull XLIV:173–94
Langedijk K (2009) The Medici, Egnazio Danti and Piazza Santa Maria Novella. Medicea 3:60–85
Levi Donati G (ed) (1995) Le tavole geografiche della Guardaroba Medicea di Palazzo Vecchio in Firenze ad opera di Padre Egnazio Danti e Don Stefano Buonsignori (sec. XVI), Perugia, Benucci
Levi Donati G (2002) Le trentacinque cartelle della guardaroba medicea di Palazzo Vecchio in Firenze, Perugia, Benucci
Maestri D, Spadafora G, a cura di (2007) Jacopo Barozzi. La regola dei cinque ordini. Le due regole della prospettiva pratica. Nella edizione del 1828 proposta da Carlo Antonini. Roma, Dedalo
Mancinelli F, Casanovas J (1980) La Torre dei Venti in Vaticano, Città del Vaticano, Archivio Segreto Vaticano. pp 7, 11, 16, 30s, 36–39, 42, 46s
Milanesi M (1996) [1994] Le ragioni del ciclo delle Carte geografiche, in Gambi e Pinelli (eds) "La Galleria delle Carte Geografiche in Vaticano", Modena, Panini, vol II, Testi, pp 97–123
Moscati A (2012) La prospettiva pratica. Gli strumenti per costruire la prospettiva, in Carlevaris L, De Carlo L, Migliari R, a cura di (eds) "Attualità della geometria descrittiva", Gangemi, Roma, pp 457–472

Pacetti P (2008) La sala delle Carte Geografiche o della Guardaroba nel Palazzo ducale fiorentino, da Cosimo I a Ferdinando I de' Medici, in Cecchi e Pacetti (eds) 2008, pp 13–40
Paltrinieri G (1994) Le meridiane e gli anemoscopi realizzati a Bologna da Egnazio Danti, "Strenna Storica Bolognese", pp 367–386
Pinelli A (1996) [1994] Il bellissimo spasseggio di papa Gregorio XIII Boncompagni, in " La Galleria delle Carte Geografiche in Vaticano", in Gambi e Pinelli (eds) Modena, Panini, II, Testi, pp 9–71
Pinelli A (1996b) [1994] Sopra la terra il cielo. Geografia, storia e teologia: il programma iconografico della volta, ibid., pp 99–128. In Gambi e Pinelli (eds), Modena, Panini, vol II, *Testi,* pp 125–154
Pinelli A (2004) La bellezza impura: arte e politica nell'Italia del Rinascimento, Roma-Bari, Laterza, pp 155–206
Righini Bonelli ML (1980) Gli antichi strumenti al Museo di Storia della scienza di Firenze, Firenze
Roccasecca P (2003) Danti e 'Le due regole', in Frommel C, Ricci M, Tuttle eRJ (eds) Vignola e i Farnese. Atti del convegno internazionale, Piacenza 18–20 aprile 2002, Milano, Mondadori Electa, pp 161–173
Rosen MS (2004) The Cosmos in the palace: the Palazzo Vecchio Guardaroba and the culture of Cartography in Early Modern Florence, 1563–1589, Doctoral dissertation, Berkeley, University of California
Schütte M (1993) Die Galleria delle Carte Geografiche im Vatikan, "Eine ikonologische Betrachtung des Gewölbeprogramms", Hildesheim-New York, Olms
Settle T (1990) Egnazio Danti and mathematical education in late Sixteenth-Century Florence, new perspectives on renaissance thought: essays in the history of science, education and philosophy. In: Schmitt B, Henry J, Hutton S (eds) Memory of Charles. London, Duckworth, pp 24–37
Settle T (2003) Egnazio Danti as a builder of gnomons: an introduction, "Musa Musaei", Beretta M (ed) Firenze, Olschki, pp 93–115
Stein JW (1938) La sala della Meridiana nella Torre dei venti in Vaticano, in "L'Illustrazione Vaticana", IX, 1938
Stein JW, S J (1950) The Meridian Room in the Vatican "Tower of the Winds", Miscellanea Astronomica vol III, art. 97, Specola Vaticana, Città del Vaticano
Vagnetti L (1979) De naturali et artificiali perspectiva, in "Studi e documenti di architettura", 1979, pp 9–10
Zucchini G (1936) Gli avanzi di un anemoscopio di Ignazio Danti, in "Coelum", vol VI, pp 1–4

Guidobaldo Del Monte (1545–1607)

Barbara Aterini

Abstract Guidobaldo, Marquis del Monte (1545–1607) developed mechanics theories by concentrating on the importance of Archimedes' teachings and by focusing on a rigorously geometrical approach to issues, never forgetting a consistent observation of experience. He investigated perspective thoroughly and was able to highlight some of its unique aspects as well as the topical value of others.

Biographical Notes

Guidobaldo Del Monte (Picture 1) was born in Pesaro on 11th January 1545. He studied at the court of Urbino, where he met grand-duke-to-be Francesco Maria II della Rovere (1549–1631) and poet Torquato Tasso (1544–1595). He became a close friend of the latter, as they both attended Padua University where, in 1564, he read philosophy, theology, law and mathematics.

He grew and studied within the Urbino cultural environment which promoted the best Renaissance traditions, strongly supported by Duke Guidobaldo II (1514–1574), a great patron of writers and artists. His teacher Federico Commandino (1509–1575) was a translator of ancient books: this background knowledge allowed precious insight into classical works such as "Spirali", "Equilibrio dei piani" and "Galleggianti" by Archimedes of Siracusa (circa 287 B.C.–212 B.C.), "Coniche" by Apollonius of Perga (262 B.C.–190 B.C.), and the "Mathematicae collections" by Pappus of Alexandria (end of 3rd century A.D.) including some contents from the works of Archimedes and Heron of Alexandria (1st century A.D.). Books on spherical geometry by Theodosius of Bithynia (circa 160 B.C.–100 B.C.) and the Analemma by Ptolemy (circa 100 A.D.–circa 175 A.D.) were part of Del Monte's formation. Commandino (1509–1575) had also

B. Aterini (✉)
Dipartimento di Architettura DIDA, Università degli Studi di Firenze,
via della Mattonaia 14, 50121 Florence, Italy
e-mail: barbara.aterini@unifi.it

© Springer International Publishing Switzerland 2016

M. Cigola (ed.), *Distinguished Figures in Descriptive Geometry and Its Applications for Mechanism Science*, History of Mechanism and Machine Science 30, DOI 10.1007/978-3-319-20197-9_7

Picture 1 Portrait of Guidobaldo Del Monte, by unknown, 17th century, oil on canvas. Milano, Veneranda Biblioteca Ambrosiana

translated the "Elementi" by Euclid (323 B.C.–286 B.C.), written between the 4th and the 3rd century B.C.

Euclid's (323 B.C.–286 B.C.) geometrical definition of space shapes Guidobaldo Del Monte's studies on one hand, whilst his approach on perspective draws mainly from the ideas and the research of two colossi: Filippo Brunelleschi (1377–1446) and Leon Battista Alberti (1404–1472). Brunelleschi (1377–1446) defined the form and aim of perspective through the basics of classical Euclidean and Ptolemaic optics. Guidobaldo learned and absorbed these concepts as well as their "forma mentis", which allowed him to think in terms of overturned planes and points which find their corresponding place on the drawing sheet. Such a modern notion of space enables a remarkable simplification leading to new observations—vanishing points to mention one- and connecting elements at a definite distance with other equally real ones at infinity.

During the Hungarian Wars, between 1566 and 1571, he served as a soldier as part of the expeditionary force sent by Duke Guidobaldo II (1514–1574) to help Emperor Maximilian II (1527–1576) against the Turks. He later joined the Holy League and fought with Francesco Maria II (1549–1631). Once the war ended he went back to his estate in Mombaroccio, where he focused his attention on mathematics as well as mechanics, astronomy and optics.

He met and became fond of another of Commandino's (1509–1575) pupils, Bernardino Baldi (1553–1617) whose education he took charge of at their master's death. He entertained ample correspondence with the major scholars and mathematicians of his time: Federico Bonaventura (1555–1602), Pier Matteo Giordani (1556–1636), Giacomo Contarini (1536–1595), Francesco Barozzi (1573–1604), his pupil Muzio Òddi (1569–1639) and Galileo Galilei (1564–1642).

Following in his father's footsteps of military architecture, in 1587 he was appointed chief surveyor of all the towns and fortresses in the Grand Duchy of Tuscany by Ferdinando I de' Medici (1549–1609).

During this time he began an assiduous correspondence with Galileo Galilei (1564–1642). Galileo was one of his son Orazio's (1570–1614) friends and had turned to Guidobaldo with questions on theorems about the centre of gravity in solids. Del Monte carried his appreciation for Galileo (1564–1642) into effect when he managed to secure him a place as a teacher at the University of Padua. Their friendship did not, however, prevent Guidobaldo from openly criticising some of Galileo's theories.

He was elected Gonfalonier of Pesaro, appointed Grandee of Spain and decorated with the most prestigious Order in France. He continued his research in Mombaroccio whilst keeping prolific contacts with his contemporary men of learning. Thanks to his scientific works he became well and widely known.

In 1602, following Francesco Maria II's (1549–1631) absence from Pesaro to rest in Casteldurante and fearing a conspiracy, Guidobaldo Del Monte was forced to withdraw definitely from public life. He spent the following years in his estate in Mombaroccio (Urbino), where he died on 6th January 1607.

List of Main Works

1. Mechanicorum Liber, Pesaro 1577
   (2ª ed., Mecanicorum Liber, Venezia 1615);
2. Planisphaeriorum Universalium Theorica, Pesaro 1579
   (2ª ed., Colonia, 1581);
3. De Ecclesiastici Calendarii Restitutione Opusculum, Pesaro 1580;
4. Meditatiunculae Guidi Ubaldi et Marchionibus Montis S. Mariae De Rebus Mathematicis, 1587, Ms. latin 10246, ex supp. latin 1058, Bibliotheque Nationale, Paris;
5. In Duos Archimedis aeque Ponderantium Libros Paraphrasis Scholijs Illustrata, Pesaro 1588;
6. Perspectivae Libri Sex, Pesaro 1600;
7. Problematum Astronomicorum Libri Septem, Venezia 1609;
8. De Cochlea Libri Quatuor Superiorum Permissu, et Privilegii, Venezia 1615;
9. De Proportione Composita Opusculum, Ms. 631, Biblioteca Oliveriana di Pesaro;
10. In Quintum Euclidis Elementiorum Librum Commentarius Opusculum, Ms. 630, Biblioteca Oliveriana di Pesaro.

Review of Main Works

With the Aristotelian philosophy constituting the pillar of knowledge at the time, Guidobaldo amply related to Archimedes (287 B.C. circa–212 B.C.) who had analysed issues formulated by Aristoteles (384 B.C.–322 B.C.): they all agreed on the defined mathematical and natural philosophy involvements within mechanics. Guidobaldo's works offered therefore tangible proof of the possibility to describe and confirm physical properties in strictly mathematical terms. The law of the lever or the position of the centre of gravity in solids are just examples of Guidobaldo's successful attempt to promise the development of a correct mathematical science of nature.

Guidobaldo del Monte formulated and explained his considerations with the support of geometry, making his approach to physics and mechanics issues geometrical as well as mathematical. By virtue of his knowledge of the Euclidean works, he tried to explain his intuitions through plane geometry and accomplished demonstration of its theorems. His theories were equally presented through one of his strongest topics, perspective. Mechanics is by far the subject he pondered on most: his first publication.

"Mechanicorum Liber" (Picture 2)- offers indeed the most relevant and extensive discussion on the subject. He describes the historical development of mechanics and he underlines the crucial role of Archimedean teachings punctuated by references to ancient authors.

Heron of Alexandria (1st Century A.D.) prompted his analysis of mechanical technologies whilst Pappus (end of 3rd Century A.D.), whose works were published posthumously in 1588, presents quotes on Heron's study of machines (1st Century A.D.).

Both Pappus and his translator Commandino (1509–1575) had categorised the simple machines in "cuneus" (lever), "vectis" (screw), "cochlea" (Archimedean screw), "polyspaston" (pulley), "axe" (axle). Guidobaldo revised this list with the addition of a sixth type: scales.

In the middle of the 16th Century, the issue around the equilibrium of the scales opened a proper debate which uncovered the heterogeneity of the basic concepts of mechanics. Defined "pre-classical" by some, the existing mechanics was the result of elements from Arabian science and from the Latin medieval science of weights (scientia de ponderibus). In line with this, it was commonly accepted that a weight varies its effect according to its position on the scales: the further it sits from the fulcrum, the bigger the effect. Scales will therefore be held even when the weights on both sides display the same "gravitas secundum situm". The idea of "positional gravity" has consequently a different yet parallel role to the Archimedean concept of "centre of gravity". The main problem was the reaction of equal-arm beam scales, being moved from a normal horizontal position to an oblique position. Three possible outcomes were outlined: the scales would revert to their original horizontal position, the scales would stay in an oblique position or they would move to a vertical position.

Picture 2 Cover of the second edition of the "Mecanicorum Liber", Venetiis, Apud Evangelistam Deuchinum 1615

Mathematicians like Niccolò Tartaglia (1499–1557) and Girolamo Cardano (1501–1576) opted for the first outcome of the scales reverting to a horizontal position, supporting their answer with daily experience as well as the science of weights.

Determined to somehow merge the Archimedean geometrical statics with the Aristotelean physics, Guidobaldo Del Monte demonstrated through geometrical reasoning that the scales equilibrium is stable in any given position, provided the scales are resting on their centre of gravity. This statement is a mere consequence of the definition itself of "centre of gravity" and it represents the logical evolution of

Guidobaldo's reasoning, who exhausted the subject by unveiling and proving the flaws of the medieval arguments. The discussion on the equilibrium of the scales required nonetheless more theoretical considerations on other mechanical instruments, like the pendulum and the inclined plane. The formulation of diverse concepts was equally necessary: they were based on practical elementary experiences involving weights and scales, focusing on non-equal arm beam scales.

Guidobaldo approached the explanation of how scales work through the law of the lever which states that non-equal arm beam scales are in equilibrium when the arms of the lever are inversely proportional to the weights positioned on the arms. In the "Mechanicorum" he gives a definition of "centre of gravity" following Commandino (1509–1575) literally, however he offers two different explanations. Besides referring to it as a point of indifferent equilibrium, he defines it as the point where any cut splits the whole in two parts of equal weight. Such definition, originally attributed to Posidonius (circa 135 b.C–50 B.C.) by Heron (1st Century A.D.) and mentioned in some works on mechanics by Archimedes (circa 287 B.C.–212 B.C.) -including "On the equilibrium of planes"- is however incorrect as it does not take into consideration the position of the centres of gravity of both sides. In order to solve the issue of balance, Guidobaldo investigates the centre of gravity in solids, in plane figures, both open and closed. He creates and builds tools along with experimental contrivances with the aim to verify and look into the physics laws involved: steelyards, systems of pulleys and hydrostatic scales are the first tools to be manufactured. The core idea of his explanation resides in resolving the motion of a body on scales in a component according to Nature - represented by a vertical segment- and in a component against Nature -represented by a horizontal segment. He demonstrates that the ratio according to Nature–against Nature grows with the use of larger scales; he transfers these results for scales and centre of gravity to other apparatuses. As a result, despite the absence of any substantial difference between non-equal arm beam scales (steelyard) and the lever, he discusses them in two separate chapters in his "Mechanicorum Liber". The opening pages are devoted to the "Libra", followed by a chapter dedicated to the lever, "De Vecte": the arguments are identical but the terminology differs. The centre of the scales is the "fulcimentum"; weights "pondus" and forces "potentia" impact on the lever.

Different groups of weights and forces are subsequently introduced: their effect on the lever, depending on their position relative to the "fulcimentum", is explained through geometrical demonstrations. Guidobaldo considers that the vertical projection of the centre of gravity on the lever can occupy either a horizontal or an oblique position. In the first instance the procedure provides correct findings, but with the second hypothesis the answer is controversial, since the different direction of the force is unaccounted for. Absorbed in this reasoning, Guidobaldo forgets to consider the instrument weight. When drawing his conclusions on the matter he will however acknowledge the issue and offer a solution to the simplest of cases when a load is suspended from a beam.

The tools he designs are manufactured in a highly specialised workshop founded by Simone Barocci (1530 ca.–1608) in Urbino: Guidobaldo is aware of the necessity to operate with real levers and pulleys in order to achieve reliable and accurate results, originating from the observation of the link between force and resistance. It is imperative to carry out the experiments on mechanical systems neutrally balanced first and then repeat the trials with specified weights. The simple existence of an independent balance becomes therefore crucial.

He allocated a whole chapter to the system of pulleys: he analyses all possible combinations of single, double and triple pulleys (Picture 3).

He commissions highly reliable, extra light and low friction pulleys to ascertain and define the connection between force and resistance in different modes.

The main objective is to reduce the functionality of the pulley mechanism to a lever mechanism.

Aristotele's (384 B.C.–322 B.C.) "Problemata mechanica" is the starting point to explain the core concept of a small force being able to move a large weight. The other machine involved, the axle of a wheel (De Axe in Peritrochio) (Picture 4) is described in detail in Chap. "Niccolò Tartaglia (1500c.–1557)": the force ("potentia") supporting the weight ("pondus") is proportional to the weight, like the pivot radius is proportional to the drum ("timpanum") together with its handles ("scytala"). Guidobaldo then solves the problem of how to choose a wheel able to move a given weight through a given force ("axe in peritrochio"). The Chap.

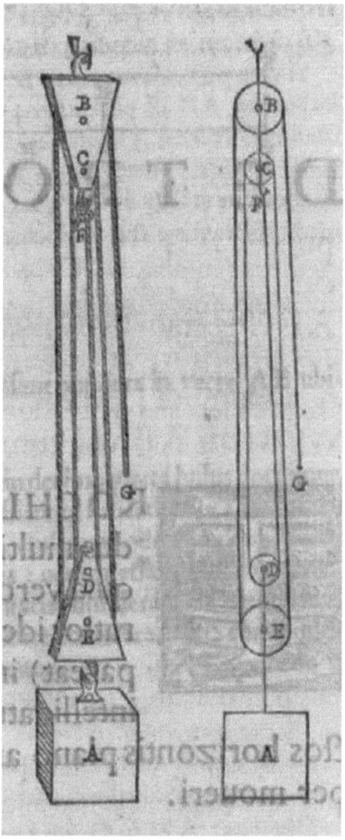

Picture 3 Drawing illustrating the pulley (or "trochlea") mechanism (see "Mechanicorum Liber", "De Trochlea", pp. 62–63)

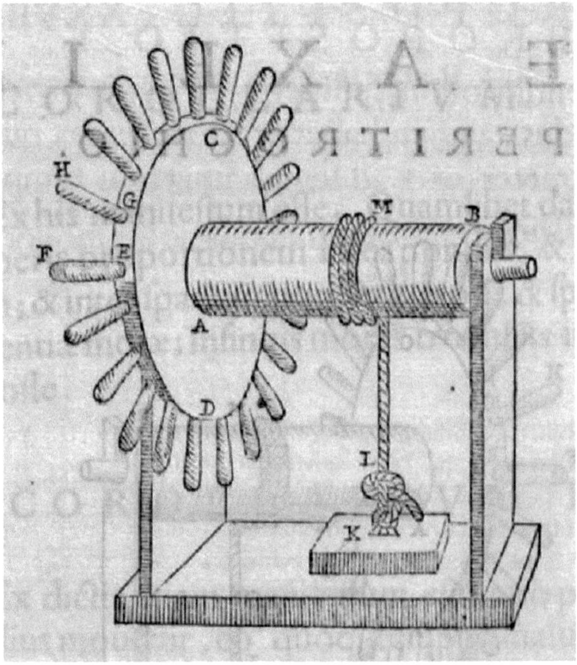

Picture 4 The mechanism of a wheel axle (see "Mechanicorum Liber", "De Axe in Peritrochio", p. 107)

"Federico Commandino (1509–1575)" deals with the wedge and the observations stem from the Aristotelean idea that two flanks of a wedge can split another material by acting like two levers. They can be considered as two inclined planes and their operating mode can be again reduced to the lever: the more acute the angle in the wedge, the easier the movement and the split of the substance in question. The force of the blows on the wedge is another factor, but at the time of analysing the impact of the strength of the strokes he cannot rely on any available theory allowing quantification of the relation between force and effect.

By comparing his studies of scales and the Archimedean hydrostatic tradition, Del Monte expounds a method to quantify the density of a given body, through the use of scales and immersion in water. He calculates the relative density, or specific weight, resulting from the ratio between the weight in air and the weight of the corresponding volume of water, measured through the decrease in weight of the immersed body. He is therefore able to achieve a density measurement through a ratio of lengths, without any weighing. He produces a tool to measure a physical quantity: the hydrostatic scales.

Del Monte also examines the Archimedean screw -also known as "cochlea"- a hydraulic machine used to move water or other liquids from a lower to a higher position (Picture 5). A huge screw sits within a cylindrical tube with its lower half immersed in water or in the substance to be lifted. By rotating the screw, each pitch collects some of the liquid which gradually climbs along the spiral until it exits at the top. Guidobaldo had already gathered several notes on the subject, mainly in the

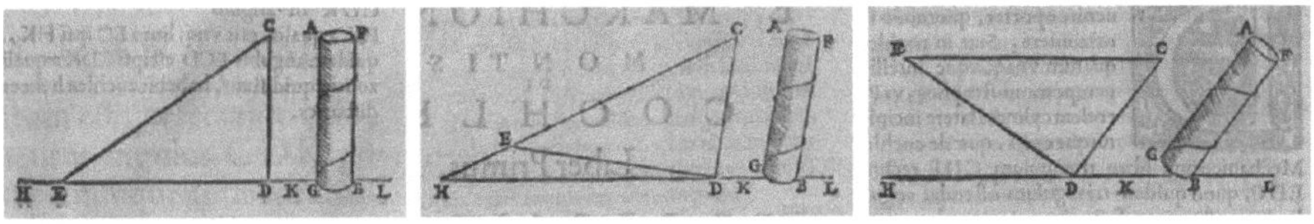

Picture 5 Drawings for the first propositions concerning the problem of the inclination of the "cochlea" relative to the horizon (see "De Cochlea Libri Quator", "Liber Primus", pp. 5–6)

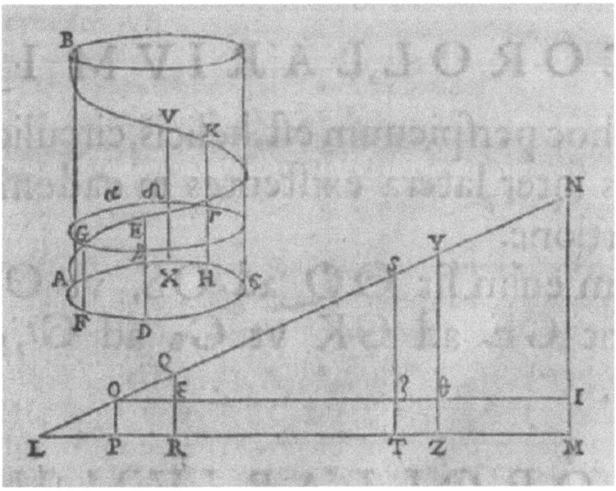

Picture 6 Illustration for the "Propositio XXX", study of the movement of the spiral (see "De Cochlea Libri Quator", "Liber Primus" p. 41)

"Meditatiuncolae": he observed that the weight of the body itself and its propensity to sink are exactly the same forces helping the ascent. The shape of the machine, the movement and the angle of the screw (Picture 6), the inclination of the "cochlea" in the river, aimed to allow an upward flow in the shortest possible time, were all significant factors (Picture 7).

Scientific tools have a remarkable role in Guidobaldo's work: newly emerging topographers, land-surveyors, civil and above all military architects increasingly demand accurate and reliable machines, aptly and swiftly supplied by the growing workshop founded by Simone Barocci (1530 ca.–1608).

Besides experimental contrivances such as the scales (1577), the system of pulleys (1577) and the hydrostatic scales (c.1587), he invents, improves and introduces various mechanical devices, recently classified according to their use and function. We have drawing tools such as compasses for long radius circumferences (1579), the elliptical compasses -or trammel-, a tool with rulers and cable designed to draw hyperbolas (c.1587). Amongst the surveyor tools we will mention the square (c.1589) and the astronomic theodolite (c.1579).

The proportional compasses (c.1570) and the mechanical multiplier of fractions of degree (1579) are two examples of calculation tools, whilst we remember the cup-shaped refraction sun-dial (1572) and the fountain refractive sun-dial (1587–

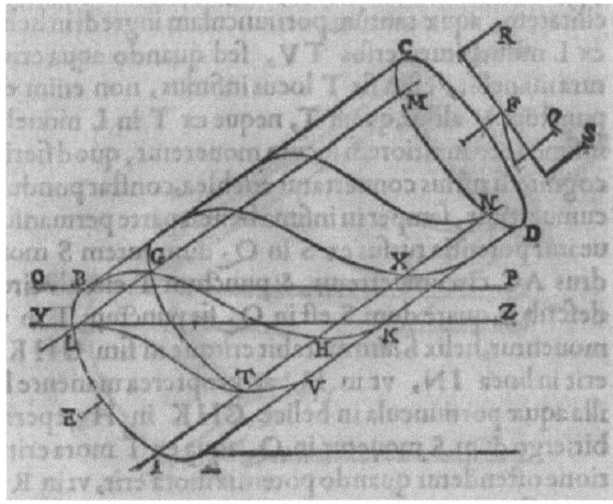

Picture 7 Illustration for the "Propositio III: "cochlea" with helixes (see "De Cochlea Libri Quator", "Liber Tertius", pp. 115–116)

1601) as tools to measure time. Bearing in mind the practical obstacles of drawing a planisphere, he suggests, in his "Planisphaeriorum Universalium", the manufacture of two new drawing tools aimed to facilitate the tracing of arcs in long radius circumferences and of ellipse arcs. Hence he devises and accurately describes compasses able to draw circumference arcs passing through three given points which are almost aligned (Picture 8).

This mechanical apparatus is composed of two prisms in the shape of wedges and a triangular hinged component made of two rulers hinged on pivot "B", which houses a stylus. Having rigorously and mathematically demonstrated for the first time in this same volume that the meridian circles are ellipse arcs in the equinoctial

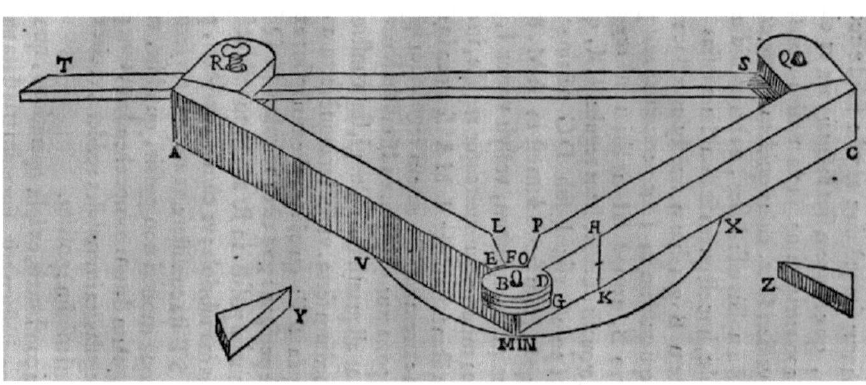

Picture 8 The compasses for long radius circumferences. For its manufacture Guidobaldo recommends the use of iron, bronze or hard wood (see "Planisphaeriorum Universalium Theorica", "Liber Primus", p. 54)

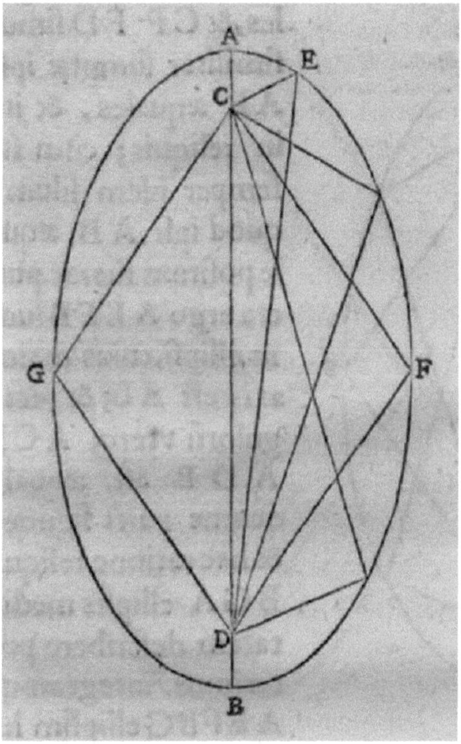

Picture 9 The method to draw ellipses used by builders (see "Planisphaeriorum Universalium Theorica", "Liber Secundus", p. 101)

orthographic projection, he examines two drawing approaches. The first method, derived from Apollonius' (262 B.C.–190 B.C.) third book of "Coniche" and usually adopted by builders, uses a stylus and a cable of uniform length (Picture 9) but, in Del Monte's opinion, it cannot be deemed safe and accurate as there is a risk of the cable not being constantly and equally taut. The second method derives from Guidobaldo's invention of a new tool: the elliptical compasses. The trammel is able to accurately draw quarters of ellipse in a continuous line starting from given semi-axes rather than the ellipse focuses. The tool is composed of a square and a ruler where a slot houses two lockable pegs or sliders. A stylus can be positioned at any end of the ruler in one of the two holes (the length of the ellipse major semi-axis corresponds to the distance between each hole and the furthest slider) depending on which quarter -upper or lower- needs to be drawn (Picture 10).

To start with, the pegs are spaced to reflect the difference between the two semi-axes. Once the square is resting on the ellipse semi-axes, the pegs are moved along the sides of the square and the quarter is drawn.

The way the elliptical compasses are presented and described is revolutionary, even more so in 1579: the tool is drawn both as a whole entity and as a detailed scanning of its parts (Pictures 11 and 12).

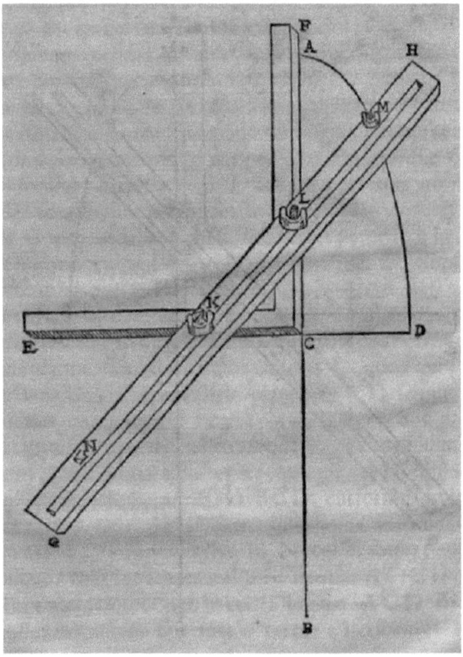

Picture 10 The trammel draws ellipse quarters in a continuous line (see "Planisphaeriorum Universalium Theorica", "Liber Secundus", p. 125)

Del Monte provides precise advice towards its manufacture, ultimately supplying the executive plan in (1560) by mathematician Nicolò Tartaglia (1499–1557).

Due to its mechanical configuration, it bears similarities with Nicomedes' (2nd Century B.C.) tool, conceived to draw conchoids and recognisable in a picture of the "General trattato de numeri et misure" (part V, Venezia).

In the manuscript "Meditatiunculae" we find descriptions of tools devised to draw conical sections and to advance the knowledge of projective trajectory. By suggesting a drawing approach through points, Del Monte describes a tool to draw hyperbolas, presenting both a ruler version and a cable version. He applies and puts in a concrete form the property stating that a hyperbola is the section of points where the difference between the distance of any point and the two focuses remains uniform (Picture 13).

Moreover, Guidobaldo deals with surveyor tools: the surveyor square (Picture 14), intended for the measurements of land surfaces and the survey of angles, adopts the system of splitting the relevant area into triangles as well as right-angle trapeziums, from one or more posts. He designs a cylindrical square which will remain in use for centuries to come as, unlike the wooden disc, provides planes instead of visual lines, thus allowing one to survey angles on uneven terrains. Another considerable technical feature is represented by the joint connecting the square to the support brace. This joint allows a horizontal position of the main

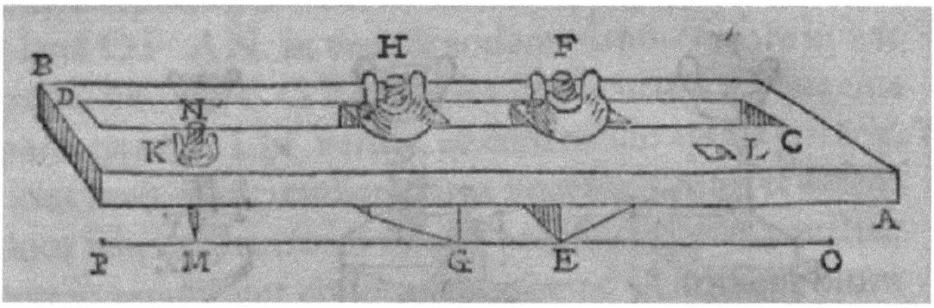

Picture 11 The elliptical compasses (see "Planisphaeriorum Universalium Theorica", "Liber Secundus", p. 109)

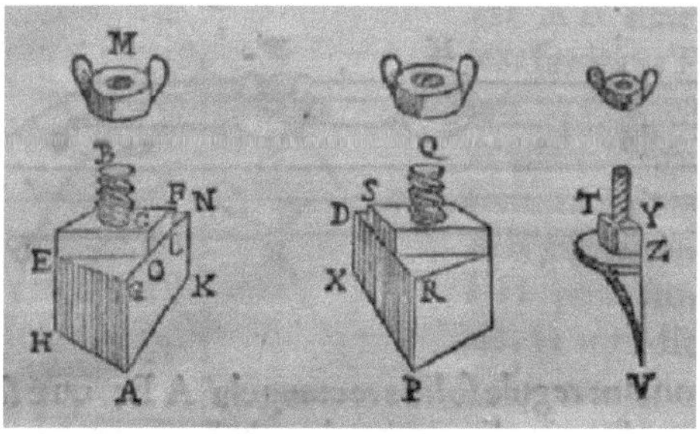

Picture 12 The elliptical compasses: detailed resolution of parts (see "Planisphaeriorum Universalium Theorica", "Liber Secundus", p. 108)

slit plane, which in turn offers a visual angle of 45° through the other slits, thus enabling the measurement of heights.

In the first book of the "Problematum Astronomicorum Libri Septem" (1609), Del Monte exposes his thoughts on scales and on fractions of degree, a highly relevant aspect of astronomic observations. From a general point of view, he analyses readings on a standard scale without accounting for the tool displaying the scale: this approach brings a recurring system able to track degree fractioning to infinity (thirds, quarters, fifths and so on) without being limited to the usual minutes and seconds. This could be preliminarily and geometrically achieved with standard compasses doing all the calculations on a common scale 0° to 60°, although such process would be time-consuming and difficult to translate in terms of accuracy for re-occurring measurements. Taking inspiration from the watch-making science and feeling confident about the feasibility of a procedure able to eliminate existing obstacles, Del Monte designs a mechanical tool which automatically multiplies by 60 and which is both fast and accurate. He calls it the "mechanical multiplier for the calculation of fractions of degree" (Picture 15).

The device is composed of a system of four cog-wheels where the rotation of an index on one dial corresponds to 60 rotations of another index on the opposite dial.

As a result the rotation of 1° of an arc coincides with the rotation of 60° on the other dial. This tool had already been conceived and sketched in 1579 as part of

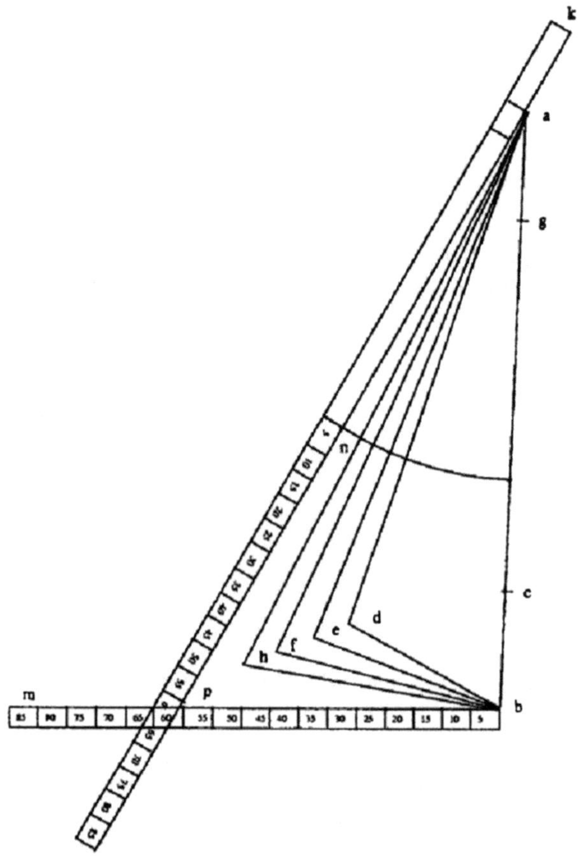

Picture 13 The tool to draw hyperbolas. It is composed of two rulers where the bottom one -bm- is entirely graduated and it is equipped with a fixed pin placed at end *b*. The other ruler -kl- is longer and graduated from point *n*, it also carries a slider fitted with a pin. The two pins are to be secured where the hyperbola focuses need to be, the two rulers can rotate around their respective pins. A branch of the hyperbola can now be traced and points *p*, *h*, *f*, *e* and *d* can be located through the intersection of the divisions marked by the same numbers on the two rulers. By doing so the difference between the distance of various points on the hyperbola from its two focuses remains uniform

general notes on the calculation of fractions of degree in the "Meditatiuncolae". However, the original idea to mechanically expand on this type of fractions belongs to Giacomo Contarini (1536–1595), who even invented a prototype of multiplier.

Muzio Oddi (1569–1639) points to Del Monte as the inventor of the proportional compasses, evolution of Commandino's (1509–1575) reduction compasses (see Gamba 1992).

The tool is composed of two rulers, connected by a hinge, with the scales engraved on them, enabling the user to divide segments and circumferences into a given number of equal parts. This system marks a major improvement over the traditional compasses and transforms the tool into the ancestor of the slide rule. It is, at the same time, a drawing tool as it allows division of a segment or a circumference in a given number of equal parts. The instrument will certainly influence Galileo himself (1564–1642).

In terms of astronomic tools, besides the scales he implements the astronomic theodolite, a device able to measure the height and the azimuth of celestial bodies (Picture 16).

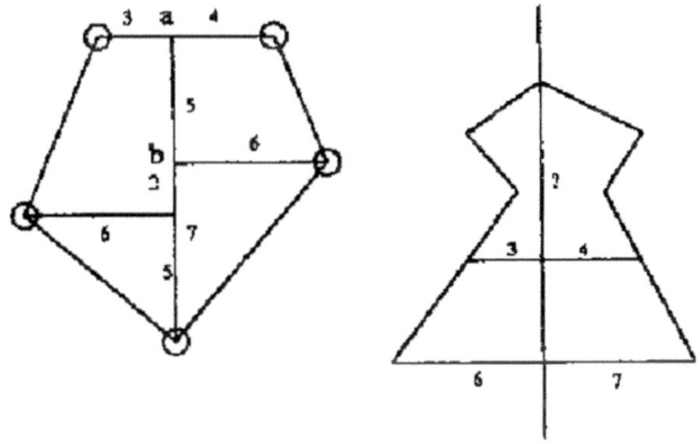

Picture 14 The earliest surveyor squares were wooden discs with two slits at a right angle; Guidobaldo's tool is made of a cylinder with longitudinal slits at angles of 90° and 45°, resulting in fixed final angles of 45°, 90°, 135°, 180°, 225° and 270°. At the Museo Galileo in Florence, the Medicean Collections include a square stamped "Urbino 1654," displaying a slit at 30°/60° besides the standard ones at 90° and 45° (see Meditatiunculae p. 12)

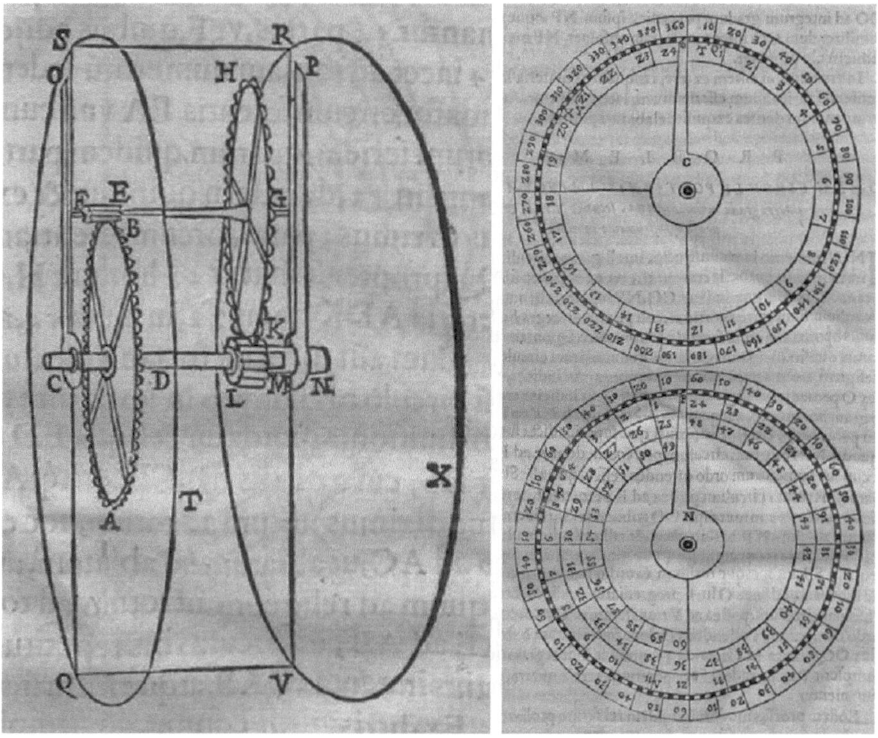

Picture 15 The system of cog-wheels and the dials of the "mechanical calculator of fractions of degree" (see "Problematum Astronomicorum Libri Septem", "Liber Primus" pp. 27–28)

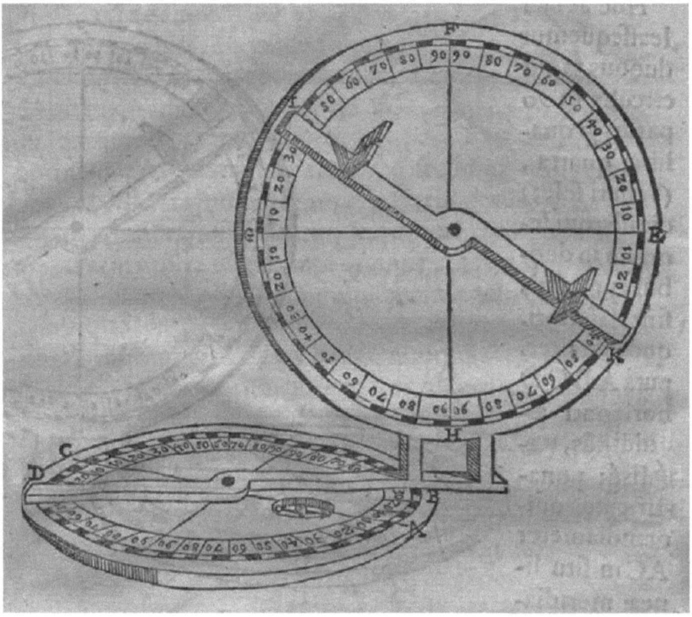

Picture 16 Guidobaldo's astronomic theodolite (see "Problematum Astronomicorum Libri Septem", "Liber Primus" p. 33)

The tool is composed of two graduated metal discs: one lying horizontal with a compass and a vertical one equipped with an "alidad", i.e. a rotating upper section for the measurement of angles. It presented several similarities with the more traditional astrolabes such as the scales engraved in the graduated discs and it allowed the measurement of any angle, both in plotting and in astronomy.

Within his astronomic studies, Guidobaldo tackled the manufacture of sun-dials. In his "Meditatiuncolae" he dedicated almost thirty pages -under the title of "Degl'horologgi"- to their manufacture on a horizontal surface, as described in the "De Analemmate" by Ptolemy and in the "De horologiorum descriptione" by Commandino (1509–1575). It is an application of the analemma in order to establish the length and the direction of the shadows projected by a stylus sitting perpendicular to the clock dial. He later analyses vertical clocks and describes the "Italic" clock, made of twenty-four equal hours where the last one corresponds to the time of sunset. In his work on sun-dials (1638), Oddi (1569–1639) maintains that Del Monte studied and developed special refraction sun-dials. It appears that these clocks, cup-shaped or positioned in a fountain bed, were already in use in the second half of the 15th Century, although the ones we know date to the first half of the 16th Century.

Commandino (1509–1575) engaged in the first observations on refraction in Urbino, it is then legitimate to assume that Del Monte himself became interested in the subject. Orazio del Monte (1570–1614) refers to some of his father's now lost works -the *De horologijs* and the *De radiis in aqua refractis* are clearly mentioned- in his letter to Galileo (1564–1642) dated 16th June 1610, thus confirming Guidobaldo's interest in the topic.

In 1572 Guidobaldo designed a portable refractive sun-dial, cup-shaped in a brass semi-sphere, which was manufactured by Simone Barocci (1530 ca.–1608). The internal surface was engraved with the hours and provided with a fixed stylus - or arm- where the inclination angle relative to the horizontal plane equalled the latitude angle of its allocated position. The cup was then filled to the brim with water so that the tip of the gnomon could reach the water surface. The time reading coincided with the projected shadow of the gnomon at the bottom of the cup.

In these astronomy studies the reference to descriptive geometry is equally crucial to solve issues arising from the representation of the celestial sphere on a plane. It is imperative to demonstrate a theorem with a three-dimensional view and to simultaneously represent it through plane praxis. All gnomonics demands, after all, this approach since the evaluation of the intersection between the cone of shadow, generated by the sun meeting the gnomon, and the plane of the clock leads to a plane construction, where the first step is the tracing of the analemma as well as the positioning of both the fixed elements and the ones depending on the latitude of the clock chosen location.

On the Circulation of Works

In 1596 Guidobaldo completes two works: the "Problematum Astronomicorum Libri Septem" and the "De Cochlea Libri Quatuor". Posthumously published by Orazio, the only son to follow in his scientific footsteps, they become public respectively in 1609 and 1615.

The first work concentrates on the search for a purely geometrical method to solve major astronomical issues: it had to be simpler but simultaneously more rigorous and accurate than its predecessors, which were calculation based. The treatise, written in 1580, reflects on astronomy and observational mathematics without however neglecting the improvement of some astronomic tools and the development of a number of mathematical tools such as the proportional compasses, the elliptic compass and a device to split the circle in degrees, minutes and seconds. The "De Cochlea Libri Quatuor" finalises his findings on the "cochlea", a hydraulic machine used to move water and other liquids from a lower to a higher position. In 1581 the "Mechanicorum Liber" (1577) had been translated into vernacular by Filippo Pigafetta (1533–1604): the book circulated from Venice under the title of "Le Mechaniche dell'illustriss. Sig. Guido Ubaldo de' Marchesi Del Monte. Tradotte in volgare dal Sig. Filippo Pigafetta".

In 1582 Oreste Biringucci (1558–1585) started from Del Monte's mechanics when incorporating an appendix about simple machines into his Italian translation of Piccolomini's (1508–1578) paraphrase of the Aristotelean mechanical matters:

"Parafrasi di Monsignor Alessandro Piccolomini Arcivescovo di Patras, Sopra le Mecaniche d'Aristotele" (Zanetti 1582, p. 127).

On Guidobaldo's death, Bernardino Baldi (1553–1617) celebrated his master with a monograph which culminated in both a eulogy ("Versi e prose scelte", Firenze 1859, pp. 208–209, 507) and the publication of some of Del Monte's unpublished works. Written around 1590 by Baldi himself under Del Monte's influence, "In Mechanica Aristotelis Problemata Exercitationes: Adiecta Succincta Narratione de Autoris Vita & Scriptis" gets published in 1621.

Modern Interpretation of Main Contributions

The study of perspective is, alongside his investigation of mechanics, a key topic for Guidobaldo: his concepts are strongly reflected and applied into today's figurative science. His drawings in perspective of mechanical tools show how perspective is his means of expression. Even the resolutions of details drawn to describe a tool are in perspective. His manuscript "Meditatiuncolae" already contained his first geometrical demonstration of perspective: he proved that the image of a line parallel to the picture plane remains parallel with it. This pivotal concept supports all his subsequent studies.

He provided practical rules to draw in perspective, including writing on a wall or a column using letters that, despite sitting at different heights, appear to be of the same size. Such matter of "anamorphosis", finding its roots in optics and already used in the sculpture and architecture of Ancient Greece, follows Euclid's (323 B.C.–286 B.C.) studies on the laws of vision and of visual perception. The "Perspectivae Libri Sex" (1600) is the compendium of concepts which will become the pillars of the science of perspective. It deals with central projection, i.e. the projection from a point at any given distance and it represents a crucial step forward where figurative science is concerned. In the first book Guidobaldo establishes the main elements of perspective: the datum-plane (datum-level), the eye, the object or the image of the object, the section, the section line, the apparent image, the vision line, the position of the eye relative to the object, its distance and the value of the angle where vision is at its best.

He analyses the projection of a line and he deduces the "theory of parallel lines", introducing at the same time the concept of "punctum concursus" intended as the point where the picture plane meets the parallel perceived by the eye to the given line. By doing so he scientifically proves, for the first time in the history of perspective science, the existence of the vanishing points, also called convergence points which are, by definition, the place of convergence of the perspective image of sheaves of parallel lines, the projection to infinity of their common point.

Today we express this concept by indicating the point at infinity as the point representing the direction of a line, therefore parallel lines share their point at infinity for having the same direction. Accordingly, the convergence point, i.e. our vanishing point, is the projection of this direction, from the centre of the picture plane.

This phenomenon had been previously noticed, but it had remained questionable and unexplained in its perspective essence, lacking the sheer concept of finite distance and the scientific knowledge of the vanishing points, now presented as essential to the theory of perspective.

Totally fresh and diverging from previous beliefs was the statement that the vanishing point is not necessarily situated at the same height as the eye on the ground line. Del Monte demonstrates the infinite number of vanishing points in a perspective plane, all located at the same height against the ground plane. This reflects the acquisition of the concept of vanishing line of a plane and emphasises Guidobaldo's determination to organise a solid theory supporting the practical side.

Remarkable and extremely modern is his approach through ideal planes projecting lines (today we would call them projecting planes) and their intersection. The importance of the projective element in the creation of the figurative image is highlighted by Del Monte also through the use of the word "umbra" preferred over "commensuratio". The choice recalls the ancient tradition in the origins of painting, relating to the wall sketching of figures achieved by tracing the outline of the shadow cast by a human body positioned by a source of light. Such projection assumes that the surface, the line and the point are visible; a body is therefore composed of all the points from where the visual rays depart to leap out from the body outline and reach the eye in the form of lines. As a result, any object placed in front of the observer will translate into a pyramidal or conical shape with the visual rays converging into the eye. The ancient concept of visual pyramid (Picture 17) and the more modern one of visual cone (Picture 18) are both described

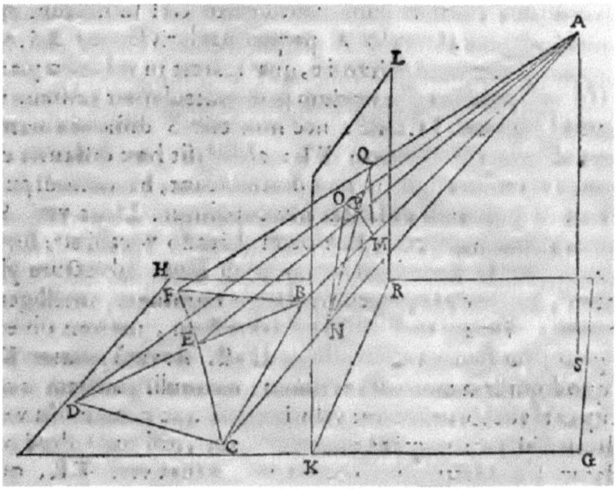

Picture 17 The drawing illustrates the concept of projection from a point at a given distance. The main elements of projection are present: point "A" which is the position of the eye, its orthogonal projection "S" on the horizontal datum-plane containing true-sized figure "BCDF" with "E" as its centre. When intersecting with the picture plane, the visual pyramid with vertex "A" returns the image in perspective "MNOQ", with "P" as its centre, of figure "BCDF" (see "Perspectivae Libri Sex", "Liber Primus", p. 5)

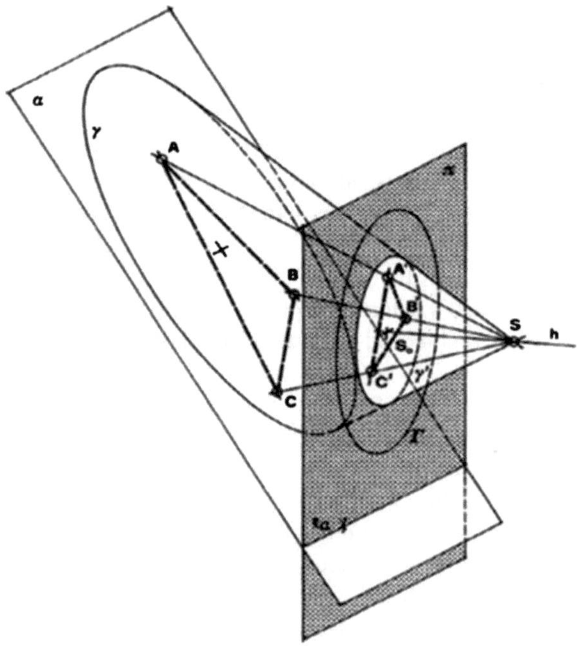

Picture 18 The drawing shows the modern but similar concept of central projection where the visual pyramid is shown as a visual cone

unmistakably and their recognition marks an important step forward in the science of representation.

When analysing how to represent the apparent body, Guidobaldo considers the section being overturned on the plane below; the geometrical artifice he suggests relies on the picture plane being overturned on a horizontal plane, i.e. the datum-plane, where the ground line functions as a hinge. Such analysis will spark one of the basic homological principles. In his various illustrations of the problem, he also considers the inverse issue to attain to the objective body starting from its perspective, all achieved through plane operations.

With the help of two given lines, one being the image in perspective of the other, he determines the position of the point of view (i.e. of the eye) both in height and depending on its distance from the plane (distance point). In other words he performs a "recovery of perspective" through a geometrical reconstruction of points for which the image in perspective is available. Nowadays this process is universally known as plane photogrammetry. Guidobaldo defines furthermore the so-called "method with the distance point" where the image in perspective of a given body originates, in contrast with his predecessors, from any two convergence points besides the main point and the distance point.

This approach makes the rule general.

Initially he considers the perspective of figures sitting on planes which are parallel to the horizon, he then moves onto figures resting on any plane and later he briefly outlines the perspective behaviour of figures on non-plane, curved surfaces -i.e. cylindrical-. Baroque artists will extensively deal with this topic in the following centuries.

It is therefore possible to trace the perspective of any figure once we know its orthogonal projection on the datum-plane and the height of each of its points. Interestingly and for the first time, Guidobaldo maintains the possibility to start from the perspective of the plan of the object and to subsequently move to the height of each of the points in the object, constantly relating to the horizontal datum-plane. We still use this procedure when we place the overturned picture plane, which is the true size of a building plan, so that the appropriate perspective view is achieved in elevation. At the forefront of Guidobaldo's studies is the fundamental role of perspective in the science of projection, a role highlighted by Jesuit Franciscus Aguilonius (1567–1617) in 1613 and clearly pointing to its 17th Century practical applications to celestial and terrestrial cartography as well as to astronomy and physics.

In his "Liber Quintus", keeping his focus on the concept of projection, Del Monte studies the behaviour of shadows generated by a lamp. The lamp is one point whilst the rectilinear rays of light coming from it can be considered lines; he then investigates the shadows cast from a point-wise source of light positioned at a given distance.

Given that, in order to see the shadow of an object sitting on the plane below, the lamp needs to rest higher than the object itself, he draws the lamp straight in the perspective section and derives the shadow without the help of ichnography (Pictures 19 and 20).

He instructs us to draw the dividers, i.e. the lines separating the opaque section from the one hit by the light; he also deals with the shadow of solids of rotation. The building and painting of stage scenes is where perspective comes alive and where it sets the pace for the development of Scenography (Pictures 21 and 22).

He shows how to work directly on the stage where he draws doors and windows to partition the inclined walls and, without the help of a plan, he represents

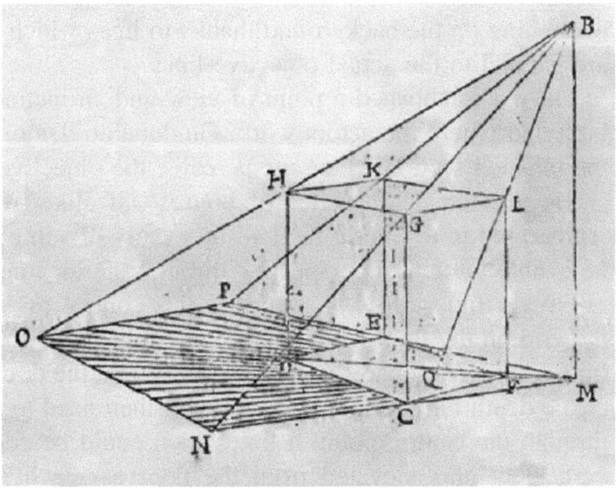

Picture 19 Drawing illustrating the still up-to-date concept of shadow intended as projection of a solid onto a plane through rays of light (see "Liber Quintus", p. 246)

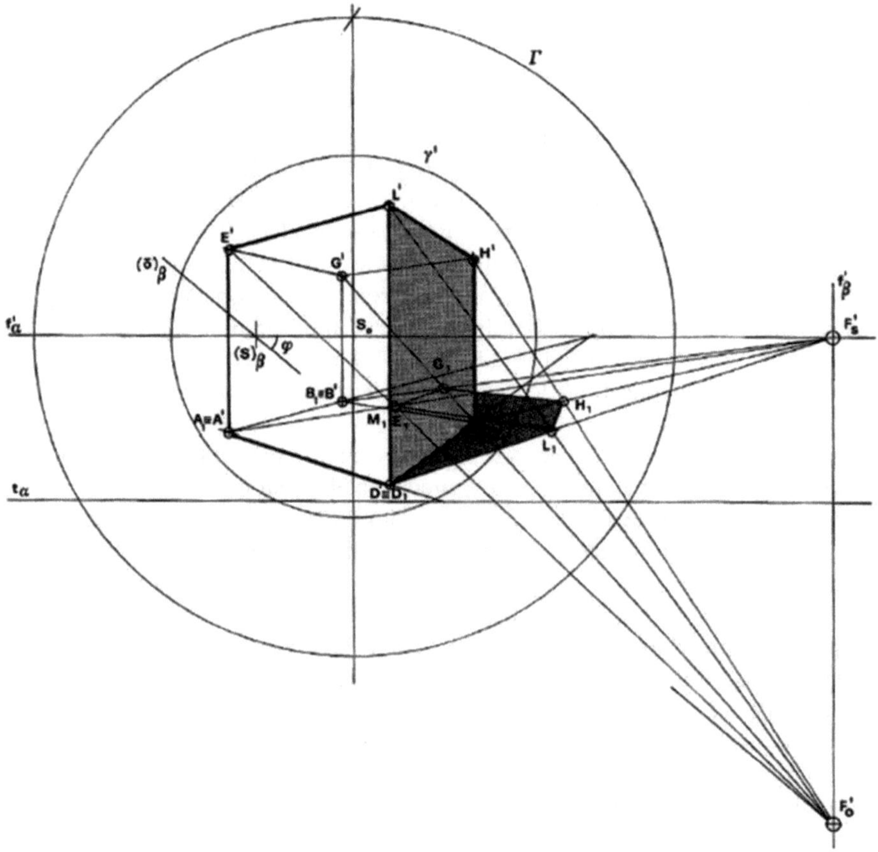

Picture 20 Drawing executed nowadays using the central projection method: the shadow is a projection through rays of light. The similarity is strikingly obvious

everything on the background thanks to lines which go through the eye and which are parallel to the actual objective lines.

Having established a point of view and an inclined plane where the scenes are built and where the actors work, Guidobaldo demonstrates that this same plane is the inclined section where we perceive the objective horizontal plane of the floor.

He also creates a series of houses, all alike, resting on inclined planes and represented in a generic section: by evenly dividing the inclined plane of the floor he is able to control perspective through the diagonals (Pictures 23 and 24).

Rocco Sinisgalli (1947) verifies this control by the diagonal and states that "if the grid on the floor is well executed, a diagonal going from angle to angle of the transferred grid squares will demonstrate that the drawing is correct; it [...] does not create depth but it controls it; you will then need to draw the so-called centric line through the centric point, a line which could be considered as the upper limit of vertical heights elevated from the floor as much as the eye". Thirty years ago

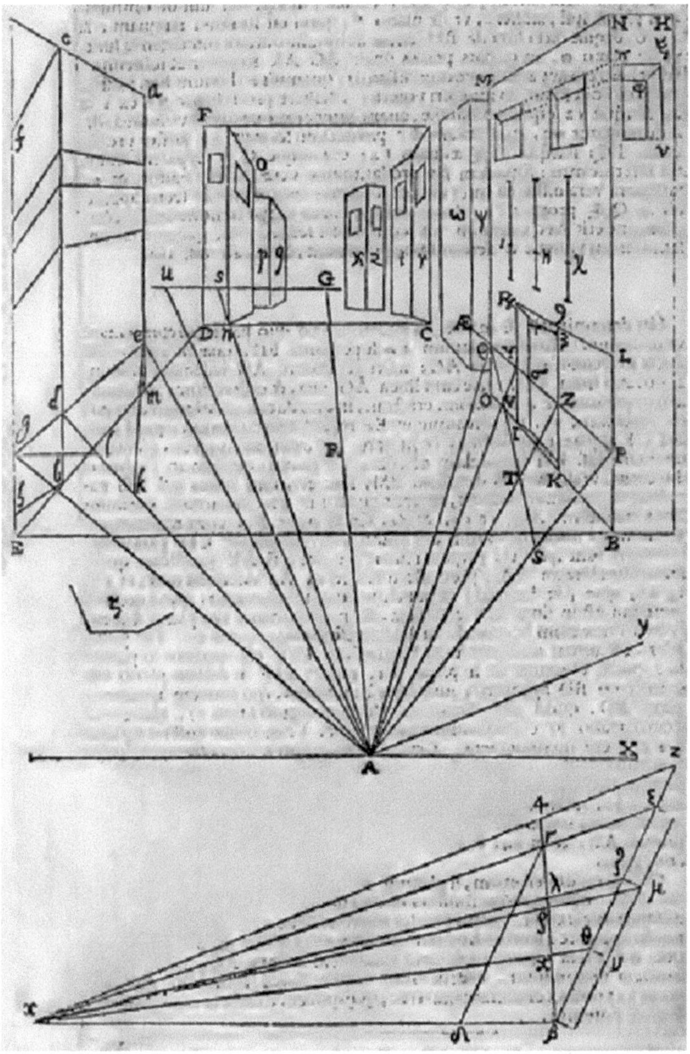

Picture 21 Drawing in perspective of a stage scene with indication of the buildings' heights; the inclined walls, or scenes, with the purpose of extending the view can be identified (see "Perspectivae Libri Sex", "Liber Sextus", p. 302)

Sinisgalli translated for the very first time Del Monte's "De perspectivae libri" making sure to do all the drawings again, in order to check and retrace the process. As pointed out by Gaspare de Fiore (1926–2011), Sinisgalli presents Guidobaldo as the initiator of the modern theories on perspective.

The supremacy of architecture and, contextually, of painting is unchallenged in the book; the rules of perspective are extensively applied to the requirements of the stage scenes and they will stem the development of the so-called "relief

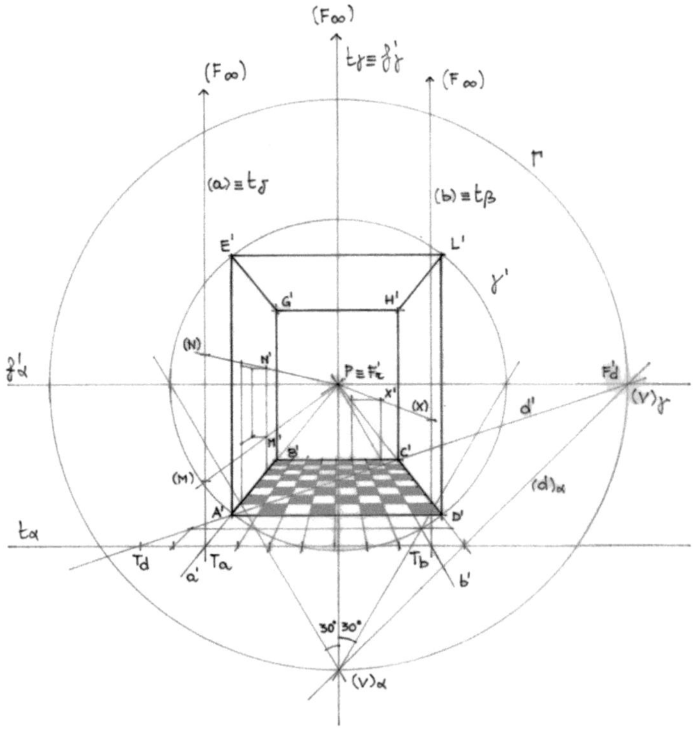

Picture 22 A modern drawing similarly represents the perspective of a room without the approach of central projection

Picture 23 Plan to correctly start a perspective. In the base grid the horizontal lines are parallel to the picture plane, whilst the orthogonal lines converge in a point outside the sheet (convergence point), the diagonal allows to control the process (see "Perspectivae Libri Sex", "Liber Sextus", p. 310)

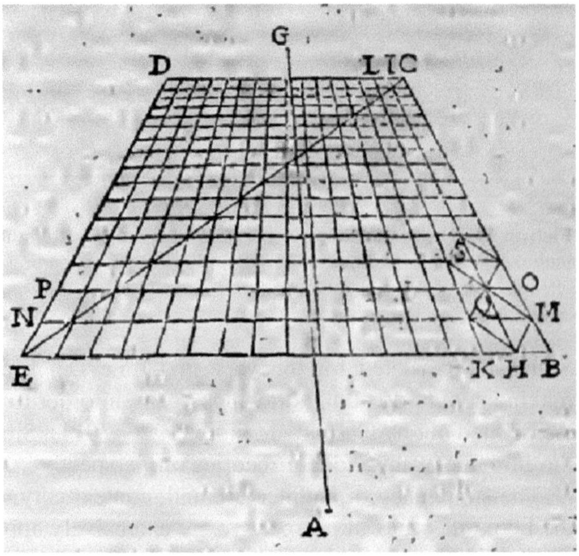

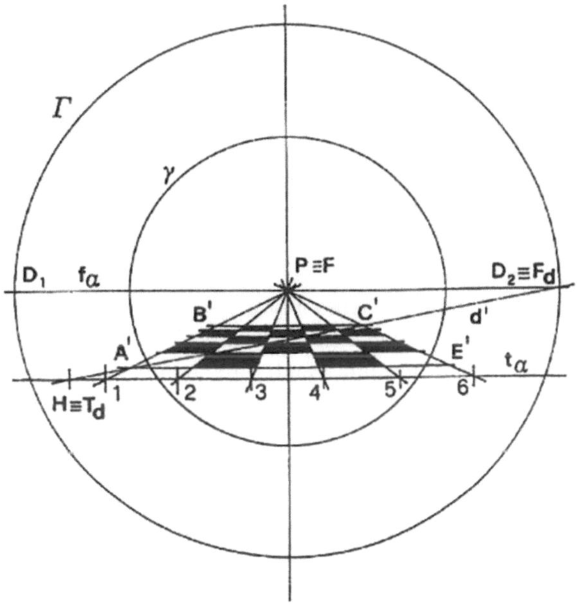

Picture 24 A similar drawing executed nowadays with the method of central projection. The control generated by the diagonal is equally obvious; convergence point "P" is the vanishing point of the lines orthogonal to the picture plane

perspective", an aspect of constant interest amongst scholars and artists. Guidobaldo's rigorously scientific approach did not, in fact, deal with the issues of "relief perspective" which Bramante (1444–1514) had experimented with in Santa Maria presso San Satiro in Milan and which had been applied by Scamozzi (1548–1616) in the stage scenes of the Teatro Olimpico in Vicenza.

Del Monte suggested the drawing of plane figures in perspective on the background and on the sides of the stage, thus opening the door to the most spectacular applications which materialised quite quickly. Prompted by his studies Breysig (1798) in Germany, Poncelet (1822) and Poudra (1862) in France, Staudigl (1868) in Austria all focused on "relief perspective". The German mathematician Ludwig Burmester (1840–1927) provided a simplified explanation which would have been accessible also to painters with a limited mathematical knowledge. Leading all those who experimented with "relief perspective", the "De scenis" remains unique in its genre and it allegedly influenced Francesco Borromini (1599–1667) when he created the perspective for Palazzo Spada.

We can certainly conclude that Guidobaldo dealt with perspective thoroughly and was able to highlight some of its key aspects as well as the topical value of others. He observed that every segment is projected into a segment, that the corresponding lines meet on the section line once the section, and the image in perspective with it, has been rotated around that same section line, now acting as the hinge of the overturned image. In other words we can safely state that corresponding lines (the real one and its image) meet on the overturning hinge which will be later named "axis" of the one-to-one correspondence. Some years later (1605), Simon Stevin (1548–1620) proved that lines joining corresponding points meet in the point of overturning of the eye, when the vertical line passing through the eye rotates, on its foot, parallel to the picture plane. He therefore realised that corresponding points are aligned with the centre (eye). This marked the start of homology which is a plane transformation allowing one to generate two corresponding figures, one in true size and the other in projection.

A few decades later the French mathematician Girard Desargues (1591–1661) started from this theory to formulate the theorem of homological triangles. His "Brouillon Project" analyses the conical sections which are at the heart of projective geometry.

Mathematician Gino Loria (1862–1954) mentions Del Monte's treatise in his "Storia della Geometria Descrittiva": "everybody who studied that book stood in admiration of its author and, if the number of admirers does not amount to an army, it is only because mathematicians assumed the work was aimed at artists and artists mostly found its pure Euclidean style difficult and obscure". Luigi Vagnetti (1915–1980) will express a similar opinion in 1979: he will praise both the historical and scientific value of this paper which systematically organises the subject of perspective 190 years after Brunelleschi's genius intuitions and less than two centuries before Gaspard Monge's (1746–1818) codification.

Guidobaldo's merit lies essentially in his complete and methodical approach to the whole discipline. His modern theories and explanations of the principle of parallel lines, the inverse problems of perspective, the discovery of the distance point, the overturned planes, homology, the representation of shadows and the creation of stage scenes make him the 'father' of the science of perspective. The "De Perspectivae Libri Sex" marks the end of the long Italian supremacy on the subject and prompts the development of theatre scenography by clarifying the connection between this relatively new topic and the science of perspective.

Bibliography

Andersen K, Gamba E (2008) Guidobaldo del Monte. In: Gale T (ed) New dictionary of scientific biography, Detroit 2008, vol 5, pp 174–178

Arrighi G (1965) Un grande scienziato italiano: Guidobaldo dal Monte in alcune carte inedite della biblioteca Oliveriana di Pesaro. Atti dell'Accademia Lucchese di Scienze, lettere ed arte XII 2:181–199

Arrighi G (1968) Un grande scienziato italiano Guidobaldo del Monte in alcune carte inedite della Biblioteca Oliveriana di Pesaro. Atti dell'Accademia Lucchese di Scienze, lettere ed arti 12:183–199

Baldi B (1621) In Mechanica Aristotelis Problemata Exercitationes: Adiecta Succincta Narratione de Autoris Vita & Scriptis, Fabricius Scharloncinus, Moguntiæ. Albinus, Moguntiae, p 194

Becchi A, Bertoloni Meli D, Gamba E (2013) Guidobaldo del Monte (1545–1607), Theory and practice of the mathematical disciplines from Urbino to Europe. In: PD Napolitani (eds) Editorial coordination: Lindy Divarci and Pierluigi Graziani, p 396 (Edition Open Access 2013)

Boffito G (1929) Gli strumenti della scienza e la scienza degli strumenti, pp 81–83. Boffito, Firenze

Camerota F (2003) Two new attributions: a refractive dial of Guidobaldo del Monte and the "Roverino Compass" of Fabrizio Mordente, Nuncius. Annalidi storia della scienza XVIII 1:25–37

Ceccarelli M, Cigola M (2001) Trends in the drawing of mechanisms since the early middle ages, in Journal of Mechanical Engineering Science, vol 215, pp 269–289. Professional Engineering Publishing Limited, Suffolk

Cigola M, Ceccarelli M (1995) On the evolution of mechanisms drawing. In: Proceedings of IXth IFToMM World Congress, Politecnico di Milano 1995, vol 4, pp 3191–3195

Cigola M (2012) In praise of parallel theories: descriptive geometry and applied mechanics. In: L Carlevaris, M Filippa (eds) In praise of theory. The fundamentals of the disciplines of representation and survey, pp 39–46. Gangemi editore, Roma

Commandino F (1558) Ptolomaei Planisphaerium. Jordani Planisphaerium. Federici Commandini Urbinatis in Ptolemaei Planisphaerium Commentarius. In quo universa Scenographices ratio quam brevissime traditur, ac demonstrationibus confirmatur. Venetiis 1558, Aldus (Paolo Manuzio)

Commandino F (a cura di) (1562) Claudii Ptolemaeus Liber de Analemmate, A Federico Commandino Urbinate instauratus, & commentarii illustratus, Qui nunc primum eius opera e tenebris in lucem prodit. Eiusdem Federici Commandini liber de Horologiorum descriptione, Apud Paulum Manutium Aldi F., Roma 1562, in 8° (203x139 mm.), (4), 93, (3) carte

Favaro A (1899–1900) Due lettere inedite di Guidobaldo del Monte a Giacomo Contarini. In: Atti del Reale Istituto Veneto di scienze, lettere et arti, LIX, II, pp 303–312. Venezia

Favaro A (1905) Per la storia del compasso di proporzione. In: Atti del Reale Istituto Veneto di scienze, lettere ed arti, LXVII, vol 2, pp 723–739

Favaro A (1992) Galileo e Guidobaldo del Monte, Serie ventesima di scampoli galileiani raccolti da Antonio Favaro. In: Atti della Reale Accademia di scienze, lettere ed arti di Padova, 30, 1914, pp 54–61, Anast. Reprint 1992, LInt, vol. II, pp716-723

Frank M (2007) Das erste Buchder "In duos Archimedis aequeponderantium libros Paraphrasis" von Guidobaldo dal Monte. Master thesis, Supervisors M Folkerts, J Teichmann, Ludwig-Maximilians-Universität, München 2007

Frank M (2011–2012) Guidobaldo dal Monte's mechanics in context. research on the connections between his mechanical work and his biography and environment. Ph. D. thesis, Supervisors PD. Napolitani, C Maccagni, J Renn, Università di Pisa and Max–Planck–Institut für Wissen Schafts Geschichte, 700 pp

Frank M (2012) A proposal for a new dating of Guidobaldo dal Monte's Meditatiunculae. Edizioni Dipartimento di Matematica, University of Pisa, Pisa

Galilei G (1890–1907) Le Opere, (ed. naz.). In: A Favaro (ed) Firenze 1890–1907, Barbera, correspondences, vol 10, pp 21–100, 166–167, 371–372

Gamba E (1992) La mano ministra dell'intelletto. Orologi e matematica a Pesaro nel secondo Cinquecento, Pesaro città e contà. Rivista della Società pesarese di studi storici, II, pp 81–86

Gamba E (1998) Guidobaldo dal Monte matematico e ingegnere. In: Giambattista Aleotti e gli ingegneri del Rinascimento, Atti del convegno, Ferrara 1996. In: A Fiocca (ed) pp 341–351. Olschki, Firenze

Gamba E, Morini M (2000) I quattrocento anni della 'Prospettiva' di Guidobaldo Dal Monte, Pesaro città e contà. Rivista della Società pesarese di studi storici XI: 73–78

Gambioli D (1916–1917) La controversia sull'esilio di Guidobaldo Del Monte, l'illustre matematico marchigiano. Atti e memorie della Deputazione di storia patria per le Marche, III 2:266–270

Gambioli D, Loria G (1932) Guidobaldo del Monte. Signorelli, Roma

Gatto R (2002) Tra la scienza dei pesi e la statica. Le meccaniche di Galileo Galilei. In: G Galilei (ed) Le mecaniche. Edizione critica e saggio introduttivo di R. Gatto, pp IX-CXLIV. Olschki, Firenze

Grossi G (1893) Cenno biografico sul marchese Guidubaldo Del Monte. Monografie storiche e scientifiche-R, pp CXIX-CXXCVI. Istituto tecnico 'Bramante', Pesaro

Guasti C (ed) (1852) Lettere di Torquato Tasso, Firenze 1852, Felice Le Monnier, vol 1, pp 250–254. Two letters of Tasso to Guidobaldo

Guipaud C (1995) De la représentation de la sphère céleste à la perspective dans l'oeuvre de Guidobaldo del Monte, pp 223–232. In: R. Sinisgalli (ed) La prospettiva:fondamenti teorici ed esperienze figurative dall'antichità al mondo moderno (Atti del Convegno Internazionale di Studi, Istituto Svizzero di Roma, 11–14 settembre 1995). Cadmo, Firenze

Libri G (1841) Histoire des sciences mathématiques en Italie, IV, pp 79–84, 369–398. Jules Renouard, Paris

Loria G (1921) Storia della geometria descrittiva dalle origini sino ai giorni nostri. Hoepli, Milano

Mamiani GC (1828) Elogio storico di Guido Ubaldo Del Monte letto all'Accademia pesarese. In: Elogi storici di Federico Commandino, G.Ubaldo Del Monte, Giulio Carlo Fagnani letti all'Accademia pesarese dal conte Giuseppe Mamiani, pp 43–87. Nobili, Pesaro

Marchi P (1998) L'invenzione del punto di fuga nell'opera prospettica di Guidobaldo dal Monte. Master–thesis, Supervisor PD Napolitani, Università di Pisa, Pisa

Micheli G (1995) Guidobaldo del Monte e la meccanica. In: L Conti, E Porziuncola (ed) La matematizzazione dell'universo. Momenti della cultura matematica fra 500 e 600, pp 87–104. Assisi (Reprinted in G Micheli, Le origini del concetto di macchina, pp 153–167. Olschki, Firenze)

Rosen E (1968) The invention of the reduction compass, ibid, vol X, pp 306–308

Rose PL (1971) Materials for a scientific biography of Guidobaldo del Monte, Actes du XIIème Congrès International d'Histoire des Sciences, vol 12, pp 69–72. Paris

Sinisgalli R (1728–1777) Guidobaldo dei Marchesi del Monte et Monge. In: R Laurent (ed) La place de JH Lambert dans l'histoire de la perspective. Cedic/Nathan, Paris

Sinisgalli R (1978) Per la storia della prospettiva 1405–1605. Il contributo di Simon Stevin allo sviluppo scientifico della prospettiva artificiale ed i suoi precedenti storici, "L'Erma" di Bretschneider, pp 103–110. Roma

Sinisgalli R (1982) La geometria della scena in Guidobaldo. In: Atti del I Convegno dell'Unione Italiana del Disegno U.I.D., Università di Roma "La Sapienza," Dipartimento RADAAR, Roma

Sinisgalli R (a cura di) (1984) I sei libri della prospettiva di Guidobaldo dei marchesi Del Monte dal latino tradotti interpretati e commentati da Rocco Sinisgalli, Università degli Studi di Roma «La Sapienza»-Facoltà di Architettura, Dipartimento di rappresentazione e Rilievo, Presentazione di Gaspare De Fiore, "L'Erma" di E & G Bretschneider Editrice, Roma 1984, pp 336

Sinisgalli R (2001) Verso una storia organica della prospettiva, pp. 101–104, 111–115, 126–128, 242–249, 279–292. Kappa, Roma

Sinisgalli R, Vastola S (a cura di) (1992) L'analemma di Tolomeo, p 157, Cadmo, Firenze

Sinisgalli R, Vastola S (a cura di) (1994) La teoria sui planisferi universali di Guidobaldo Del Monte. Firenze

Vagnetti L (1979) De naturali et artificiali perspectiva, Libreria Editrice Fiorentina

Giovan Battista Aleotti (1546–1636)

Fabrizio I. Apollonio

Abstract Giovan Battista Aleotti was a polyhedral man of science of the XVI century, who worked in the field of architecture, engineering, scenography, but also a scientist who studied hydraulics and mechanics. He contributed to dissemination of the 'Pneumatica' of Heron of Alexandria, translating it into vulgar and enriching it with new additional notes and 'theorems', and, by his treatise the 'Hydrologia', to the dissemination of topographic and chorographic surveying with the use of 'archimetro'.

Biographical Notes

Giovan Battista Aleotti (1546–1636) (Fig. 1) was born in Argenta (Ferrara). In 1560 he moved to Ferrara to study Mathematics, Civil and Military Architecture. His first work in 1566 was the territorial survey of Polesine di San Giorgio, near Ferrara. In 1575 he was appointed architect to the service of the Duke Alfonso II that, in the meantime, to prove his sympathy and his estimate, confidentially nicknamed him the 'Argenta', after his birthplace.

During his long career he worked on civil and military architecture, building many edifices including the restoration and extension of the Fortress of Ferrara and the construction of some bastions of defense walls, some bell towers, churches and theaters in Ferrara, the Palazzo Bentivoglio in Gualtieri (Reggio Emilia), the square and the clock tower in Faenza and the Farnese theater in Parma, perhaps his best known and most important architectural work.

F.I. Apollonio (✉)
Dipartimento di Architettura, Università di Bologna, Viale Risorgimento, 2, Bologna, Italy
e-mail: fabrizio.apollonio@unibo.it

© Springer International Publishing Switzerland 2016

M. Cigola (ed.), *Distinguished Figures in Descriptive Geometry and Its Applications for Mechanism Science*, History of Mechanism and Machine Science 30, DOI 10.1007/978-3-319-20197-9_8

Fig. 1 Giovan Battista Aleotti (Engravure of XVII century, Ferrara, Biblioteca Ariostea)

But his multifaceted profile led him to undertake works in surveying and cartography, as well as drainage of wetlands, regimentation and management of riverbeds and gardens and, in particular, construction of stage machinery for tournaments and theatrical productions.

Aleotti was, in fact, known during his life as a scenographer and builder of equipment and "machines" that he realized during special celebrations, anniversaries and weddings. Commissioned by Marco Pio di Savoia and Clelia Farnese in 1587 in Sassuolo, he built a theater with a scenography shaped like a temple, modeled on the Pantheon in Rome. The stage functioned as a kind of "magic box" with images, lights, sounds and colors, and was dynamically rendered by mechanical motion. One example of his most famous theatrical scenography, however, was a construction made in 1592 in Mantua for the 'Pastor Fido' by Giovan Battista Guarini, who considered him one of the best scenographers of the era.

Giovan Battista Aleotti pursued a long and intense professional activity, firstly at the court of the Este and then at the service of the Marquis Cornelio Bentivoglio, whose family had sheltered in Ferrara after their expulsion from Bologna in 1506. With the passage of the territories of Ferrara under Papal domination, in 1598 Aleotti was appointed to "Architect of the Royal Apostolic Chamber," in 1600 "Public Architect of the City" and again in 1602 "Architect of the Fortress" of Ferrara. In these fields Aleotti applied his technical knowledge to such roles as engineer-architect, intellectual, writer, musician, and scientist at the same time pursuing the professional career of a scholar who, "being wise and morally rich, discourses on the forces of nature, the range of a bow, the speed of a flood, the trajectory of a body launched" (Raimondi 2004). Throughout his life he wrote, and in many cases published, several memoirs, reports, and treatises ranging from rural surveying to the use of the 'Archimetro', from works of hydraulic defense to those on fortifications, to the "Hydrologia" (a work posthumously published in 2000), to the translation from Latin to the vernacular of the "Pneumatica" of Heron of Alexandria (second century BC) or to "the dialogues of Sebondio" (Spanish-Catalan philosopher who lived in the first half of the fifteenth century). In his studies and in his writings Aleotti adduced an encyclopedic knowledge, learned in the "Studio ferrarese" (university), at the Academy or at the Court of Ferrara, which had its basis in "natural philosophy", that is the study of the phenomena of nature through the tools offered by mathematics, geometry, physics and geography. He interpreted, as Adorni (1974) proposed, the figure of "an eclectic, though certainly no more the Renaissance universal man, but a specialist in many fields: hydrology, military engineering, mathematics, construction of theaters, stagecraft, and so on". Man of Science, no doubt, but he excelled in deepening and disseminating his knowledge and experiences, remaining in the mid-stream between conservative and innovative intellectuals. Respectful of traditional knowledge and of doctrinal "autorictates", of knowledge acquired from classical and medieval culture, he declined to adopt the experimental methods that would soon flourish, lavishing in reproductions of texts that related to the sciences of the Greek and Latin writers in a dogmatic fixation on the truth of faith, aware—as a man of his time—of the need to use the tools offered by knowledge in order to extend the power of man over the forces of nature. Faithful to this belief, in the preface of his Hydrology book (British Library, Ms., C.112.f14) he declared, within this direction, the qualities and skills that an important modern architect should have, "that he is a scholar and expert in Mechanics of Nature, which comprise the base of Geometry, Arithmetic and Natural Philosophy…".

List of Works

The following list contains the main work of Giovan Battista Aleotti published in Italy, during his life or posthumously.

1. Della scienza et dell'arte del ben regolare le acque di Gio. Battista Aleotti detto l'Argenta architetto del papa, et del publico ne la città di Ferrara /a cura di Massimo Rossi, Modena, Panini, 2000.
2. Descrizione del magnifico apparato fatto nella città di Ferrara nelle felicissime Nozze degli III.mi signori Don Cesare d'Este e Donna Virginia de'Medici, 1585 (attribuited to Aleotti).
3. Tavola delle cose notabili contenute nel Discorso di G. Battista Aleotti d'Argenta, 1600.
4. Difesa per riparare alla sommersione del Polesine di San Giorgio e alle rovine dello Stato di Ferrara e per confutar, con ragione, il Discorso del S. Cesare Mengoli da Ravenna …in materia della Navigazione del Po di Primaro et dell'essicatione delle paludi, che le sono a destra in Romagna, 1601.
5. Relazione sui confini e sul taglio (di Porto Viro) fatto dai sig. ri Veneziani al Po, per cacciarlo nella Sacca di Goro, 1735.
6. Relazione intorno alla bonificazione Bentivoglio 12 aprile 1612, 1612.
7. Relazione nella quale riferisce che senza danno della navigazione di Cento, e senza pregiudicio dello Scolo delle campagne di quel territorio, l'E.mo Card. Lodovisio può far edificare un Molino nelle sue terre di Casumaro mediante l'acqua del canale di detta navigatìone, 1628.
8. Relazione …al Card Ludovisio per l'edificio di un Molino a Casumaro, 1629.
9. Relazione nella quale mostra che i Molini fatti e da farsi dalli ministri del Sig. Card. Ludovisio nel territorio di Cento, non apportano, né sono per apportare pregiudizio al li Molini della R. Camera Bologna, 1629.
10. Hero, Alexandrinus, Gli artifitiosi et curiosi moti spiritali di Herrone. Tradotti da M.Gio. Battista Aleotti d'Argenta. Aggiontoui dal medesimo Quattro Theoremi … Et il modo con che si fà artificiosamte salir vn canale d'acqua viua, ò morta, in cima d'ogn'alta torre, Ferrara, Baldini, 1589.
11. Gli artificiosi e curiosi moti spiritali di Herone. Tradotti da M. Gio: Battista Aleotti d'Argenta. Aggiuntovi dal medesimo quattro theoremi non men belli & curiosi de gli altri, Bologna, Zenero, 1647.
12. Heronis Alexandrini Spiritalium liber a Federico Comandino Urbinate ex Greco in Latinum conversus. Huic editioni accesserum Jo. Bapt. Aleotti. Quatuor Theoremata Spiritalia, ex Italico in Latinum conversa. Amsterdam, Janssonio-Waesberg, 1680
13. Heronis Alexandrini Buch von Lufft- und Wasser-Künsten, welche von Fridrich Commandino von Urbin aus dem Griegischen in das Lateinische übersetzt. Deme beygefügt Joannis Baptiste Aleotti vier … mit einem Anhang von allerhand Mühl-, Wasser- und Grotten-Wercken aus Salomon De Cous … auch anderen berühmt- und erfahrnen Autoribus zusammen getragen, Franckfurt, Ammon,1688

14. Dall'interrimento del Po di Ferrara e divergenza delle sue acque nel ramo di Ficarolo. Discorso inedito di Giambattista Aleotti argentano, premesse le memorie per servire alla biografia dell'autore, cura di Luigi Napoleone Cittadella, Ferrara, Domenico Taddei, 1847.
15. Discorso dell'Argenta intorno alla difficoltà di metter il Reno nel Po grande. A. D. 1602, Ferrara, Tip. Argentana, 1901.
16. Relatione di Gio. Battista Aleotti detto l'Argenta nella quale riferisce, che senza danno della nauigatione di Cento, & senza pregiudicio dello scolo delle campagne di quel territorio l'illustriss. & reuerendiss. sig. card. Lodouisio può far edificare un molino nelle sue terre di casumaro, mediante l'acqua del canale di detta nauigatione, Bologna, Clemente Ferroni, 1628
17. Relatione di Gio. Battista Aleotti detto l'Argenta, nella quale mostra, che li molini fatti, & da farsi dalli ministri dell'illustriss. & reuer.mo sig. card. Ludouisio nel territorio di Cento, non apportano, ne sono per apportare pregiuditio alli molini, della reuer. Camera, ne impedimento alla nauigatione, ne danno al territorio, ma commodo, & utile; con la risposta alle relationi delli periti della communità di Cento, Bologna, Euangelista Dozza, 1629
18. Difesa di Gio. Battista Aleotti d'Argenta, architetto, per riparare alla sommersione del Polesine di S. Giorgio, & alla rouina dello Stato di Ferrara, e per confutar, con ragione, il discorso del S. Cesare Mengoli da Rauenna, scritto all'illustriss. ... cardinale Visconte, in materia della nauigazione del Po di Primaro, & dell'essicatione delle paludi, che le sono a destra in Romagna, Ferrara, Vittorio Baldini, 1601
19. Giovan Battista Aleotti architetto: i disegni dell'album Borromeo /a cura di Costanza Cavicchi, Argenta, Trentini, 1997

Review of Main Works

The most important work that Aleotti published during his life is the translation of 'Pneumatica' of Heron of Alexandria, a Mathematician, Physicist, Engineer geometer and worker in mechanics who lived, probably, in the first century AD. Heron's origin is uncertain although we can deduce that he taught at the Museum in Alexandria, he wrote in Greek on the measurement of geometric figures and invented many contrivances operated by water, steam, or compressed air, and he can be considered the inventor of numerous items including a fire engine, a Dioptra, a surveying device using triangulation long before English mathematician Leonard Digges' 16th-century telescopic theodolite, an odometer, a device used for indicating distance traveled by a vehicle, described by Vitruvius around 27 and 23 BC, although its invention may have been by Archimedes (287–212 BC) during the First Punic War, and a steam-powered engine called an 'aeolipile'. A large number of works by Heron survived, although the authorship of some of them is disputed. The works can be divided into several categories, technical works, mechanical works and mathematical works. Heron wrote on ‘The automaton theatre’ describing a puppet theatre worked by strings, drums and weights, on ‘Mechanica’ in three books written for architects, and ‘Metrica’ which gives methods of measurement, ‘Definitiones’ contains 133 definitions of geometrical terms, ‘Stereometrica’ measures three-dimensional objects and ‘Mensurae’ measuring a whole variety of different objects and is connected with parts of Stereometrica and Metrica although it must be mainly the work of a later author. He also examined the theory of motion, certain statics problems, and the theory of the balance. In ‘Mechanica’ there is a discussion on centres of gravity of plane figures and he examines methods of transporting objects by such means as sledges, the use of cranes, and looks at wine presses (Fig. 2).

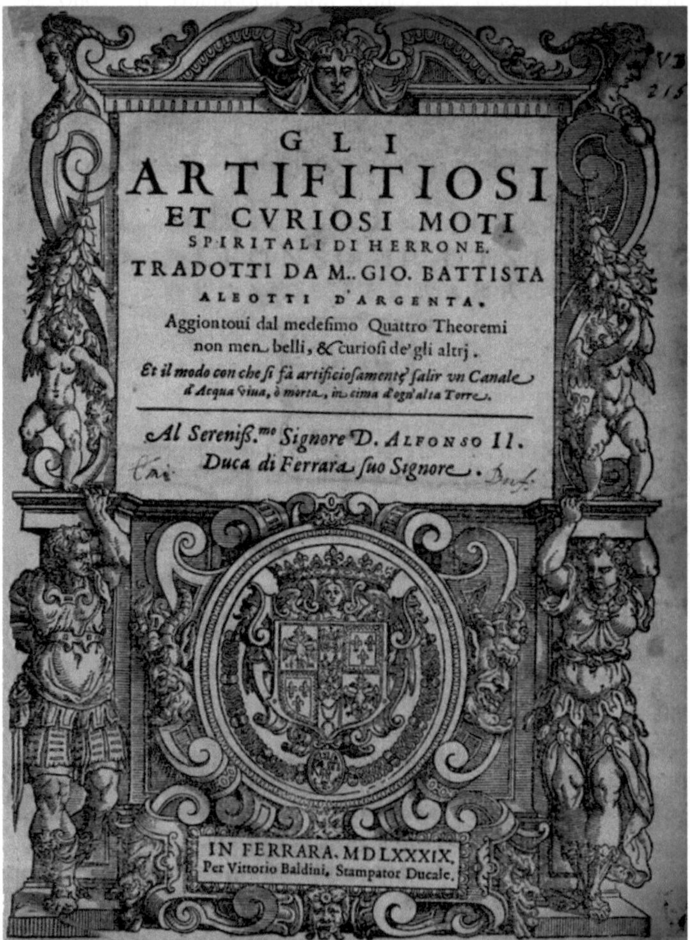

Fig. 2 Cover page of Heron’s “Gli Artifitiosi et curiosi moti…” translated by G.B. Aleotti in 1589

In 'Pneumatica' Heron describes the workings of several bird automatons. Fragments of Heron's writings were the first of the Greek works to be translated. These appeared for the first time in Latin in the work of Giorgio Valla (1501), followed by complete translations of his book into Latin by Federico Commandino, published posthumously in 1575. This work provoked a great interest among Renaissance scholars and was translated in 'Volgare' (italian vernacular) and published for the first time by Giovanni Battista Aleotti under the title 'Gli Artifitiosi et Curiosi Moti Spiritali di Herrone'. In its tranlsation Aleotti incorporated five ideas of his own. Other versions followed quickly, the best known being that by Alessandro Giorgi da Urbino of 1592 and 1595.

The 'Pneumatica' is a peculiar work, written in two volumes, which study, in 80 chapters, mechanical apparati, working with air, steam or hydraulic pressure. The first volume contains 43 chapters and the second 37 chapters. Heron begins with a theoretical consideration of pressure in fluids. Some of this theory is right but, of course, given the state of scientific knowledge at the time, some is quite wrong (Fig. 3).

There are, rather remarkably, descriptions of over 80 machines such as a fire engine, a wind organ, a coin-operated machine, and a steam-powered engine called

56 DELLI SPIRITALI

RISCALDATO UN VASO PIENO D'ACQUA far girare una Sfera vota ſu due Poli.
Theorema L.

IL riſcaldato Vaſo di acqua ripieno ſia A.B. la cui bocca ſia con diligēza turata con un coperto C. D. ſia dopoi con eſſo forato il piegato tubo E. F. G. del quale la eſtremità G. ſia con diligenza impoſta nella concaua Sfera H. K. & alla punta di queſto Diametro della Sfera ſia contrapoſto un polo L. M. piegato anco lui come il tubo E. F. G. conficcato nel coperto del vaſo C. D. & la Sfera habbia dui piegati tubi, l'uno al l'altro per Diametro oppoſti, & con eſſo forati, che con buchi ſi corriſpondino, & le loro piegature ſiano ad angoli retti, che auenirà, che riſcaldato il uaſo ſalirà il uapore nella Sfera per il tubo E. F. G. & caderà fuori per li piegati tubi, & aggireraſſi la Sfera con il modo, che alle uolte ſi uengono ragirare intorno artificioſi balli di Animali.

Fig. 3 Heron's Aeolipile described in "Gli artifitiosi et curiosi moti spiritali ..." published by Aleotti an 'aeolipile', which has much in common with a jet engine. It is described as a hollow sphere mounted so that it could turn on a pair of hollow tubes that provided steam to the sphere from a cauldron. The steam escaped from the sphere from one or more bent tubes projecting from its equator, causing the sphere to revolve. The 'aeolipile' can be considered the first known device to transform steam into rotary motion.

Gli Artifitiosi et Curiosi Moti Spiritali di Herrone

The book published by Aleotti is the first edition in vernacular of the famous work of Heron. Aleotti utilised the translation from Greek to Latin made by Federico Commandino, humanist and physician, who lived from 1509 to 1573. The translation of this work, due to its content and samples exposed, such as demonstrations that Heron gives of the use of water to play an organ or for the movement of stage machineries, will later provide fundamental basis for the design of the fantastic water games whereby Aleotti will grace the gardens which he designed, such as those designed for the Dukes of Ferrara in 1589, and to realize the scenografic machines that will make him respected and famous in his time as a scenographer.

Besides replacing the original title, Aleotti added a series of remarks, comments and additions that he titled "Four Theorems not less beautiful & curious than others. And the way to artificially raise a duct of living or dead water, at the top of every tall tower".

The first three 'Theorems' describe the running of many machines that, inspired by the theory shown and by the examples given by Heron, in an increasingly complex manner allow the coordinated motion of automatons which stage action with water jets and emission of sounds.

The first of these 'Theorems' (Fig. 4) is dedicated to a theatrical machine, or how to 'Make a dragon, who is the guard of the golden apples, fight Hercules with a sledgehammer, which he lifts as the Dragon hisses, and in the moment in which Hercules smites him in the head: to make it splash water in its face'. It is the description of a hydraulic machine employing siphons, funnels, tanks, watertight chambers, a ruler, a rocker switch and the combined use of the potential energy of the water, and water pressure, of the levers and the force of gravity allows, as the title indicates, to let Hercules and the dragon perform a series of predetermined movements, all accompanied by the emission of wheezing, sounds and splashes of water.

In the second theorem (Fig. 5) water games are even more complex. In this machine six automata (or possibly more) come into play that symbolize many rivers intent to pour out of their skins in a pool, around which they are arranged, and a triton that comes out from water emitting the sound of a trumpet (or shell). With loud and sound emergence the triton causes the interruption of the flow of water from the skins, and once served its purpose it can come back to dive. But with the demise of the triton skins return to dispense water and the game resumes.

Fig. 4 Aleotti's first theorem ("Gli artifitiosi et curiosi moti spiritali ...", 1589)

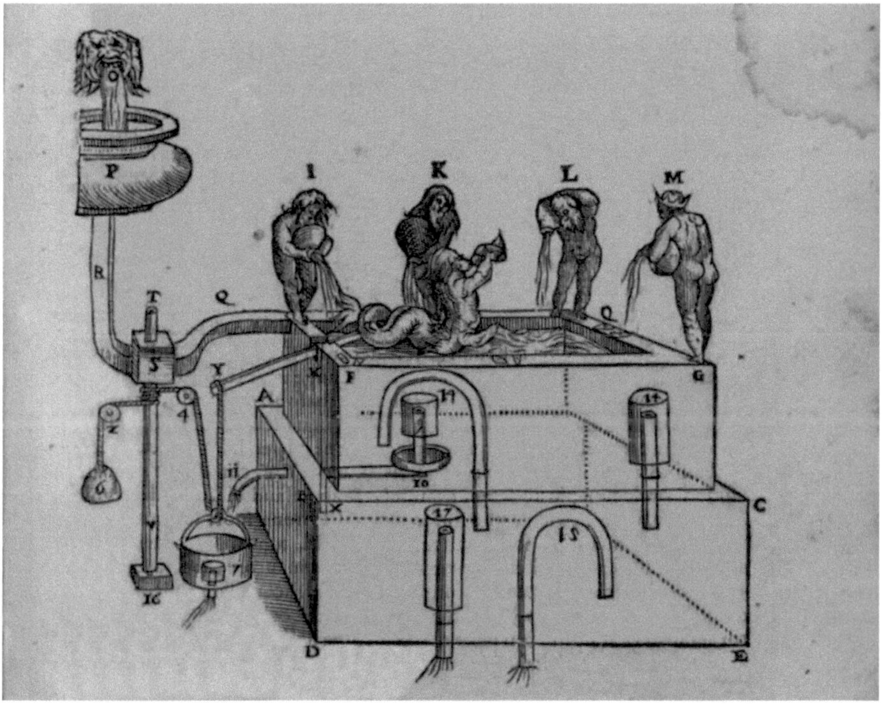

Fig. 5 Aleotti's second theorem ("Gli artifitiosi et curiosi moti spiritali ...", 1589)

In this machine the mechanisms are enriched in the hollow sphere described by Heron's Theorem IX and in the barrels used in the fire (Theorem XXV), in 'Spiritali' pipes (or diabetes) (Theorem II), in vessels that once filled and emptied of water can move ropes with counterweight through pulleys ('troclea') that regulate, in their turn, the movement of triton and cyclic opening of the valve that governs the entire scene.

In the case of the siphon, Heron employs diabetes (διαβήτης) to describe a particular device that consists of two tubes nested one inside the other, but whose walls are at a certain distance from each other. Moreover, the outer tube has its upper closure slightly raised with respect to the top air chamber, while the second, left open at both ends, allows only the flow of liquids.

The third theorem (Fig. 6) describes how to implement a single channel that simulates the glow of a forge and rules the movement of four craftsmen intent on beating the red-hot iron on an anvil to produce sparks. The mechanism is enhanced by an overshot water wheel that allows one to transmit, through some rows Π (Pi-greek) shaped fixed in its axle, the alternating movements that move the arm that delivers the iron smiths arranged around the anvil.

The fourth theorem (Fig. 7) addresses the issue of how to implement a system for cooling a room. Below the floor Aleotti indicates to realize two compartments, 5

Fig. 6 Aleotti's third theorem ("Gli artifitiosi et curiosi moti spiritali ...", 1589)

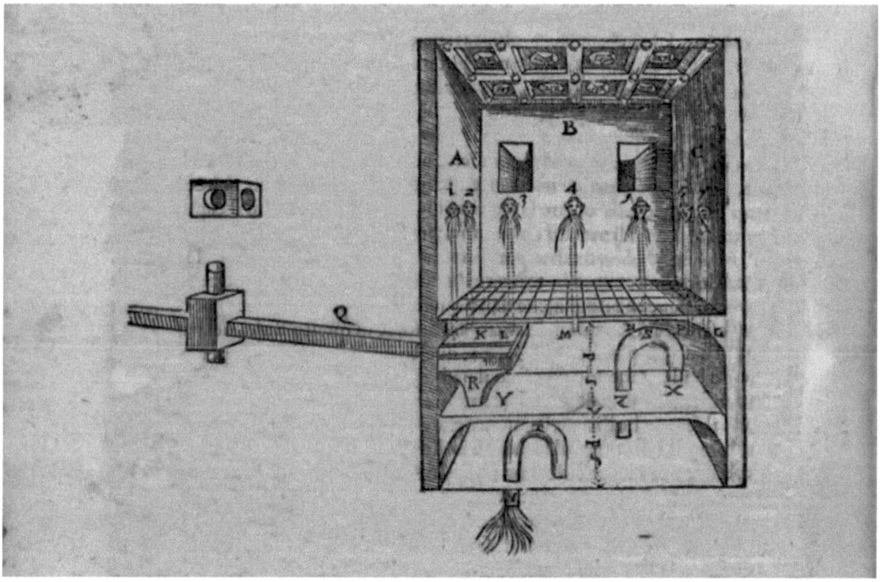

Fig. 7 Aleotti's fourth theorem ("Gli artifitiosi et curiosi moti spiritali ...", 1589)

feet tall each, and as wide as the room above, coated so as to be water and airtight. Two siphons arise: the first enables connecting the two rooms; the second acts as a drainage. Other tallest siphons connect the upper compartment with the room.

The system is powered by a pipeline through a tap and a gate valve fills, according to the needs, an "infundibulum" (tank funnel) located above the floor of the upper compartment. Its filling activates the water flow through the siphons and the simultaneous increase in air pressure that comes out from the compartment through the vents placed in the room to simulate the breath of the wind.

Aleotti wanted to add to the four previous theorems 'this way to raise a channel of living water to any great height', in order to allow the countries and all the cities in the plains to have the opportunity to be graced with "natural Fountains".

The first solution he proposed (Fig. 8), for this water-pump, uses the potential energy of water on an overshot wheel, with a diameter of five or six feet, one and half foot wide, and with fifteen semi-C-shaped buckets, made of oak ("wood that lasts a long time even if dipped in water").

Or he offered the alternative to realize an undershot wheel, very similar to that described by Vitruvius, immersed with its lower end in the watercourse so that its paddles could be driven by the velocity of the running water, with twelve long paddle ('scitale') capable, as Aleotti stated basing on experience given by the numerous mills along the Po river built in the same way, "if the 'scitale' will be very broad they will rotate the wheel with very little water".

The rotational motion of the water wheel is then converted in a reciprocating linear pistons compressor, by a crankshaft, resting on saddles ('Tapagni') made of

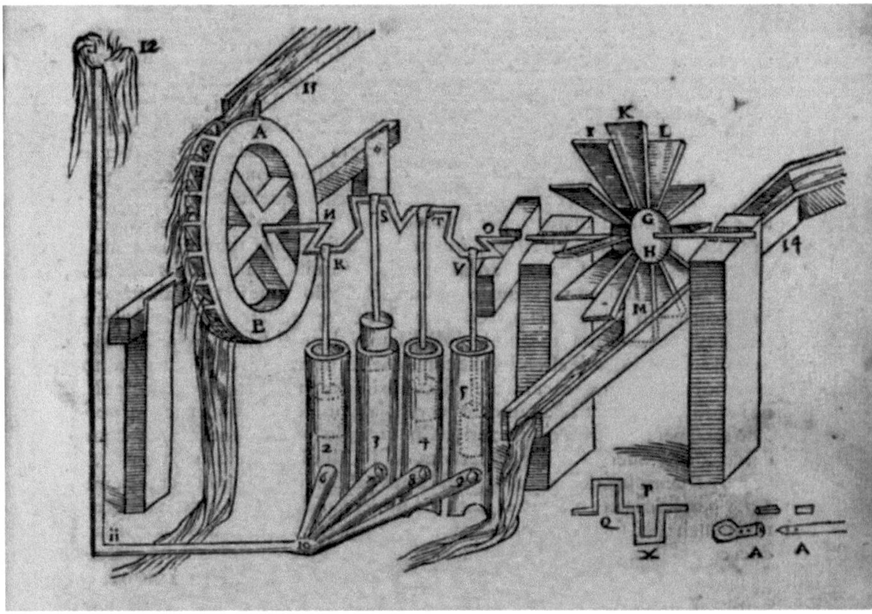

Fig. 8 Aleotti's "Way to raise a channel of living water to any height" ("Gli artifitiosi et curiosi moti spiritali ...", 1589)

bronze and with "crankpins". He reproduced the Ctesibian water pumping machine (Ctesibius 285–222 BC), quoted by Vitruvius (Marcus Vitruvius Pollio born c. 80–70 BC, died c. 15 BC), Vegetius (Publius Flavius Vegetius Renatus, a writer of the Later Roman Empire, late 4th—early 5th century) and Valturio (Roberto Valturio, Rimini, 1405–1475, Italian engineer and writer, author of the military treatise 'De Re militari'), made by "big ends", connecting bronze rods ('Anelli'), leather cylinders, pistons ('Modioli') and non-return valves ('Assarij' or 'Cartelle'), similar to those used to take out water from the ships in Roman age.

In the two alternative versions that Aleotti provides, the driving force could be given by one man (or actually a weak child) or by a horse taking advantage of the circular motions demonstrated by Aristotle (384–322 BC) in the 'Mechanica' (or 'Mechanical Problems'), able to transfer rotational motion, by a couple of toothed gears, to a reciprocating compressor.

In this book, as well as in his study and writings, Aleotti, demonstrates his deepening desire to increase and disseminate his knowledge and experiences. He promotes to the benefit of the world the knowledge of those sciences that had remained veiled under the Greek and Latin languages, at the same time enriching and updating them with the product of his theoretical speculations and practical applications. He does not conceal the hope and the ambition that every effort he expended in such work—as he clearly stated in the preface to Heron's book—will have the opportunity of worthy and systematic placement in "Horoscopij Acquatici" (Hydrologia). He claims to have already written the first four books, or even "I succeed with the satisfaction of the World, it gives me the courage to put the last hand to 'Theatro di tutte le Scienze, & arti' which I have no doubt that we ought to be more than roughly educated, those people who want to claim the name of the architect, or engineer, as today they are told".

The task of transmitting in the most clear and intelligible way his knowledge and his experience is given, as well as by the text in the vernacular that with a wealth of notes and comments describes the principles and operation of the machines, to detailed drawings in pseudo-assonometria or perspective showing the various parts which make them up, the movements they make and the forces that come into play.

As Rossi (Aleotti and Rossi 2000) highlights in the extensive introduction of the posthumous edition of 'Hydrologia', and especially as Aleotti in several passages of his extensive literature reiterates, the drawing is the primary tool for any architect. It is the technical language, the essential means of communication by which anyone wishing to profess to architecture is able to demonstrate his thoughts to clients and colleagues, but also to make himself understood by the workers who are subjected to him.

Della Scienza et dell'Arte del ben regolare le Acque

The 'Hydrologia', which was printed in 2000 thanks to the commendable work of Massimo Rossi with the title of "Della Scienza et dell'Arte del ben regolare le Acque di Gio. Battista Aleotti detto l'Argenta architetto del Papa, et del publico ne la città di Ferrara" had a long and troubled genesis, its imminent publication having been announced by the same Aleotti since 1600. But editorial and financial misfortunes made the materialisation of the print edition impossible during the life of the architect. Due to the lack, at that time, of any encoded treatise about hydraulics and hydraulic works, the 'Hydrologia' was conceived by its author as a work indispensable to provide a common basis of knowledge and understanding among the technicians who work in that sector. In it Aleotti deals with issues concerning the geography and the nature of rivers, construction of works for regiment or changing the riverbeds, reclamation of wetlands and flood repair, about the 'archimetro', the discussion of the operations of topographic surveying, measuring and leveling of land and water courses, without forgetting the law that relates to water courses and, in the sixth book, a lot of artificial and curious things in the field of water ('artificiose e curiose in materia d'acque') (Fig. 9).

In addition to the valuable and extensive treatment of hydraulics and hydraulic constructions, the treatise of Aleotti, whose disclosure and use through manuscript copies and transcripts will extend beyond the death of its author (Corradi 1717), is noted for the extensive description of the 'archimetro' and for a wide series of geometric measuring operations and survey ('geometriche operationi') that it is possible to achieve with that instrument, including a detailed instructions for its construction.

Fig. 9 'Measuring each distance of an inaccessible point' ("Della scienza et dell'arte del ben regolare le acque...", 2000)

The fourth book, that consists of 62 chapters and 19 'Propisitioni', in fact, is entirely devoted to the problem of survey and the construction of correct topographic and chorographic maps, mainly using the 'archimetro', the measuring instrument invented by Aleotti. It is a veritable handbook which aims to provide the proper instructions on how to work with that innovative technical instrument, capable, when used properly, to reduce the typical inaccuracies of instruments available at that age and to solve in an appropriate manner the usual problems for those people who must "show with art and intelligence" the territories on which interventions, projects and works should be developed.

In the preparation of his treatise, both as regards hydraulics, and as regards the survey, Aleotti refers to the teachings of Archimedes, of Heron, as well as the work of Aristotle, to whom he refers, with the judgment about the perfection of the number three (De Caelo et Mundo, First Book, 350 BC), to enunciate the perfection that is in triangular measurements obtained by the intersection of the lines, or even to Euclid's Elements (Chapter IX, 'through the doctrine of the nineteenth proposition of VII of 'Elementi'), as well as to the teachings of Pythagoras (570–495 BC), Archimedes and Apollonius of Perga (262–190 BC).

In the preface Aleotti, in fact, emphasizes the importance and the benefit that geometry may dispense to professors of the 'science of the smallest description of the countries' (chorography) through the use of the square, ruler and compass, further stating that geometry is, nevertheless, one of the mathematical disciplines in which we consider the continuous quantities, measuring surfaces and all kinds of solids, all lengths, widths, heights and depths. In this field, his reference is Vitruvius —"the most learned and erudite of geometry"—whose work is a witness, as he says, in particular the sixth and eighth books. Starting with a discussion of the measurement operations using the intersection ahead, Aleotti addresses a large series of important surveying and representation of a plan of a province, of how to draw countries and to find the distances between places, to measure and to draw the surfaces, how to find the side 'tetragonico' (the square root) of a number, to zoom in or out a drawing, to measure distances inaccessible in various operating conditions, thanks to the different methods allowed by the 'archimetro', as well as to measure the heights of any body or the depth of a water hole, how to draw a straight line parallel to a given line or perform leveling operations, providing a broad overview of the case study and examples about it, but even to give most detailed and thorough guidance on how to measure 'by the degrees of the sphera circles, which is the distance between any two cities, or places on the universal globe of Earth', i.e. the distance between two points on a spherical surface (Fig. 10).

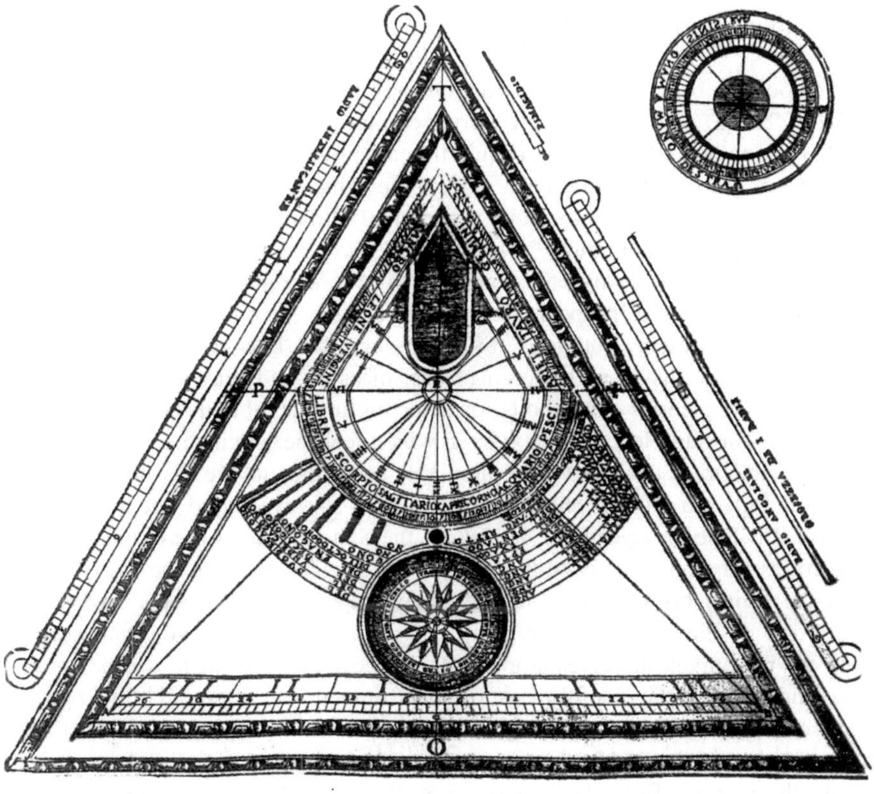

Fig. 10 The 'Archimetro' ("Della scienza et dell'arte del ben regolare le acque...", 2000)

Most of the steps are possible thanks to the measuring instrument that Aleotti has designed and developed, being able to build up, by his own speculations and thanks to his own experiences in surveying, starting from the existing knowledge on measuring instruments, perfecting and improving the accuracy in the operational phases. As Aleotti explains, "it is a geometric instrument invented by us many years ago, because only with this you can have all operations you need and to use them is necessary; and we have called it 'archimetro', because among all the other geometric instruments we have found out that this name is not unbecoming". The name 'archimetro' was adopted almost simultaneously by Ostilio Ricci (1540–1603) to define a new topographic compass of his own invention (Archimetro Florentine) and the shape of the instrument could probably derive from the similar 'trine balistario' of Giovanni Fontana (1395–1454). The 'archimetro' was composed of a triangular tablet with various measurement rules, mounted on a stand and provided with a diopter for sighting the places in the distance, and a magnetic compass. It consists of the lines equinoctial sundial, the circle and the horizontal circle, divided according to eight ordinary winds, besides two 'radij' (rules) and 'pinnaccidij' (pivots) to which the Aleotti devotes a specific chapter of the book for describing how it works and how to build it. In addition to building the 'archimetro', in the fourth book, Aleotti provides instructions for building a quadrangular compass useful to 'bring up on the paper the surveyed plan of the country that you'll like to draw', or even he explains how to equip the 'archimetro' of a horizontal clock, how to build a night watch on the back side of the same instrument, which would not otherwise be used, or how to make a perpetual almanac, on any of its sides that would otherwise remain empty. (Fig. 11)

To illustrate how to use the 'archimetro' for a topographic survey, Aleotti explains, by means the 'proposizioni' how to use the 'archimetro' as level, how to survey and represent countries by applying the intersection of the lines, calculate/measure distances between places, measure and draw the surface of the territories, a long series of methods for detecting inaccessible distances, measuring heights, such as land leveling, or even astronomical measurements needed to build gnomons and to know what time at night, according to the invention of Peter Appiano, and in a chapter shows, the data of a survey carried out for the Duke Alfonso II d'Este on December 4, 1596 in the San Alberto at the mouth of the Po Primaro, as training to be used to trascript the data for drawing the map.

In the fourth book Aleotti, besides providing a comprehensive and detailed description of the many measurement and leveling tasks that you can carry out using the instrument, he gives ample proof of how the instrument can be enriched and completed by other useful measuring instruments, but above all how to put into practice the knowledge of geometry which, as he says in the preface, the exercise of the profession of the architect, who must also take care to drain, reclaim, irrigate, defend the country from the waters, as well as monitor the progress and divert rivers, you can not do without being able to detect a plant, measure distances or inaccessible to know the trends and slopes of the rivers. And to that end it pays to use 'archimetro', an excellent tool for surveying and leveling of the water (Fig. 12).

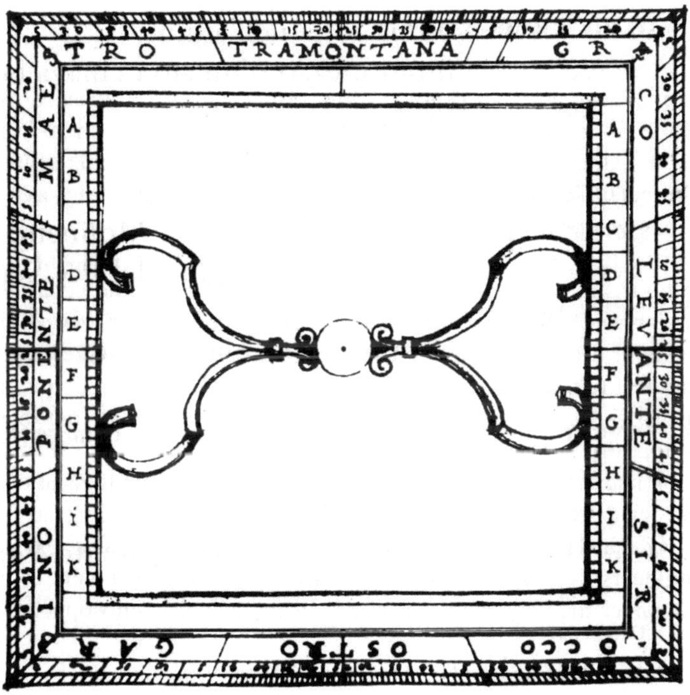

Fig. 11 The quadrangular compass ("Della scienza et dell'arte del ben regolare le acque...", 2000)

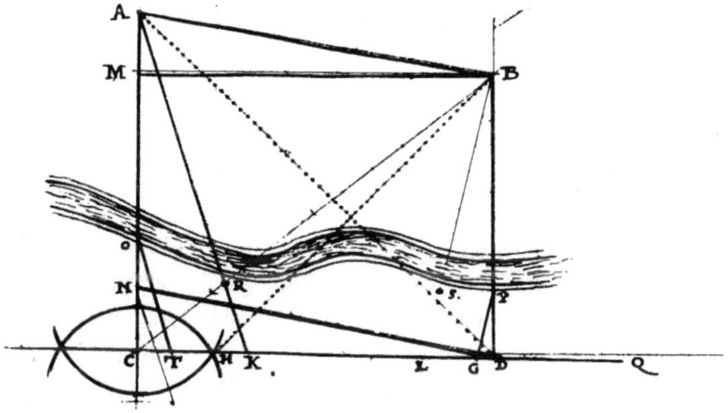

Fig. 12 'To draw a parallel straight line' ("Della scienza et dell'arte del ben regolare le acque...", 2000)

On the Circulation of Works

The work of 'Pneumatica' of Heron, as it has been said, was translated by Federico Commandino from greek to latin and published, posthumously in 1575 in Urbino. From here an extraordinary fortune of the text arises during the sixteenth and early seventeenth centuries, which intersects the emergence of mechanics as a discipline claiming its own special status.

As reported in Aleotti's supplication to Alfonso II d'Este, Duke of Ferrara, that opens the book on "Gli Artifitiosi et Curiosi Moti Spiritali", he began to work on the translation and editing of the book in 1586 when, during the period of convalescence after a serious illness that struck him, he was not able to practice but devoted himself to the theory. He published the book in 1598 in Ferrara at Vittorio Baldini. In 1615 Aleotti completed the "Additional Notes" to the four theorems that already at that time he had established and put in an appendix to Heron's text.

The book had later editions (i.e. Bologna, 1647) and some translations. In 1680 it was translated into Latin and published in Amsterdam; in German it was published in 1688 in Frankfurt, from Johann Wilhelm Ammon. The text was spread among engineers as well as among humanists for the establishment of a technical know-how under the banner of the 'return of the ancient philosophers' (Trabucco 2010).

As we said, in the intentions of Aleotti the additions he made to the text of Heron were going to be the sixth and final book of his 'Hydrologia'. This is, maybe, his most important work, whose history is marked by hardships and vicissitudes that do not allow Aleotti to publish it during his lifetime. The manuscript of the first volume appeared for the first time in 1598, and it had other issues, in 1604, in 1622, in 1630 and finally in 1632. In 1600 appeared the manuscript for the second book of 'Hydrologia', followed by subsequent editions in 1602, in 1605, in 1616, in 1626 and again in 1627.

Modern Interpretation of Main Contributions

The second book of 'Hydrologia', as well as the other five, unfortunately remain unpublished, after they had come to miss funding firstly promised for publication.

The complete work was published in 2000 by Massimo Rossi, such as transcription based on the manuscripts preserved in the city library Ariostea of Ferrara and the Biblioteca Estense in Modena (Aleotti and Rossi 2000).

Complex and articulated, as well as not easily extricable, is the story that accompanies the manuscript aleottiano and the three transcripts that reached us. Over the centuries, several authors have reported, described and commented on the 'Hydologia' (Corradi 1717; Cittadella 1847; Strocchi 1926; Mainprice 1967) or some of its parts, such as the reconstruction of the practice of a cartographic survey made by Cazzola (1988).

But more than his written work, the most important heritage left and tansmitted by Aleotti is that of his works. If on one hand, we do not know if and what has been the contribution of 'Hydrologia' in particular or some of the topics treated in it on his contemporaries and on the generations immediately following, Aleotti's cartographic works have witnessed for many years the value of the detector expert and skilled cartographer. The mathematician Domenico Corradi, in his publication on "Harmful effects that will be produced if we put Reno into Lombardy's River Po" (1717), in the reconstruction of the profile of the Adriatic coast in different historical eras, relies on the discussion made byAleotti in 'La Scienza'. But the Aleotti's cartographic drawings are real role models for engineers who will work on the territory of Ferrara in the following centuries, as well as one of its most important cartographic works the 'Corografia of the State of Ferrara' that was chosen by Abraham Ortelius (1601) to be hosted in his work the 'Theatrum Orbis Terrarum' as a worthy representation of the Duchy of Ferrara, in the first modern atlas, that remains one of the most wonderful and popular works of the geography of the seventeenth century.

In recent years important seminars and studies were promoted by various institutions and university departments which have contributed, especially from a historical point of view, to re-profile and deepen not only the articulated, complex and rich figure of the architect from Argenta, but also the historical, social, technical and cultural environment in which he was formed and operated. In 1996, a conference dedicated to Giambattista Aleotti and the Renaissance engineers have highlighted reports that intervened in those years between science and technological innovations, that took shape and to which Aleotti gave an important and significant contribution.

Bibliography

AA VV (1994) Atti relativi al Seminario di studi su G.B. Aleotti (1546–1636)/a cura del Dipartimento di Matematica dell'Università di Ferrara, 4 Febbraio 1994

AA VV (1994) Atti relativi al Seminario di studi su G.B. Aleotti (1546–1636)/a cura del Dipartimento di Discipline Storiche dell'Università di Bologna, 25 Novembre 1994

AA VV (1995) Atti relativi al Seminario di studi su G.B. Aleotti (1546–1636)/a cura dell'Istituto di Studi Rinascimentali di Ferrara, 26 Maggio 1995

AA VV (1996) Atti relativi al Seminario di studi su G.B. Aleotti (1546–1636) e gli ingegneri del Rinascimento/a cura del Dipartimento di Matematica dell'Università di Ferrara, 6 e 7 Dicembre 1996

Adorni B (1974) L'architettura farnesiana a Parma (1545–1630), Battei, Parma

Aleotti GB, Rossi M (a cura di) (2000) Della scienza et dell'arte del ben regolare le acque di Gio. Battista Aleotti detto l'Argenta architetto del papa, et del publico ne la città di Ferrara, Panini, Modena

Argoud G, Guillaumin J-Y (1997) Les pneumatiques d'Héron d'Alexandrie, Université de Saint-. Etienne, Saint-Etienne

Berselli A (1977) Contributi sugli aspetti idraulici, architettonici e teatrali dell'Aleotti, in Storia dell'Emilia-Romagna, University Press, Imola

Boas M (1949) Hero's Pneumatica. A Study of its Transmission and Infiuence, Isis, vol XI

Cazzola F (1988) Problemi d'acque e tecnica cartografica: l'"Hidrologia" di G.B. Aleotti, in Schifanoia, n. 6, Panini, Modena

Ceccarelli M, Cigola M (2001) Trends in the drawing of mechanisms since the early Middle Ages. J Mech Eng Sci 215, pp 269–289

Cigola M, Ceccarelli M (1995) On the evolution of Mechanisms drawing. In: Proceedings of IXth IFToMM World Congress Politecnico di Milano, vol 4, 1995, pp 3191–3195

Cigola M (2012) In praise of parallel theories: Descriptive Geometry and Applied Mechanics. In: Carlevaris L, Filippa M (eds) In praise of theory. The fundamentals of the disciplines of representation and survey, Gangemi, Roma, 2012, pp 39–46

Cittadella LN (a cura di) (1847) Dall'interrimento del Po di Ferrara e divergenza delle sue acque nel ramo di Ficarolo. Discorso inedito di Giambattista Aleotti argentano, premesse le memorie per servire alla biografia dell'autore, Domenico Taddei, Ferrara

Corradi D (1717) Effetti dannosi che produrrà il Reno se sia messo in Po di Lombardia [...], Soliani, Modena

Fiocca A (a cura di) (1998) Giambattista Aleotti (1546–1636) e gli ingegneri del Rinascimento, Olschki, Firenze

Giardina GR (2003) Erone di Alessandria. Le radici filosofico-matematiche. Della tecnologia applicata. Definitiones. Testo, traduzione e commento, CUECM, Catania

Keller AG (1967) Pneumatics, Automate and the Vacuum in the Work of Giambattista Aleotti, in The British Journal of the History of Science, III, 12, London

Mainprice JL (1967) Aleotti's Idrologia, in The British Museum Quarterly, vol XXXII, nn. 1–2

Moretti W (1996) Aspetti letterari dell'Idrologia di GB. Aleotti, in Atti del Seminario su GB. Aleotti (1546–1636), a cura del Dipartimento di Matematica dell'Università di Ferrara, Ferrara

Ortelius A (1601) Theatrum Orbis Terrarum

Raimondi S (2004) Giovan Battista Aleotti. Un grande personaggio del Cinque-Seicento ferraree tra arte, scienza e letteratura, Ariostea, Sabbioncello San Pietro

Strocchi G (1926) Cenni sul codice autografo cinquecentesco inedito di GB. Aleotti (detto l'Argenta) e sulle opere di questi, Tipografia F. Lega, Faenza

Torlontano R (1993) Le "legi all'architettore necessarie secondo la mente di Vitruvio" di Giovan Battista Aleotti, in OPUS Quaderno di storia, architettura, restauro—n. 3/1993

Trabucco O (2010) L'opere stupende dell'arti più ingegnose: la recezione degli Pneumatika di Erone Alessandrino nella cultura italiana del Cinquecento, Olschki, Firenze

Giovanni Pomodoro (XVI Century)

Stefano Brusaporci

Abstract Giovanni Pomodoro was a mathematician, astronomer and land-surveyor, and an experienced instruments builder, who lived in the XVI century. His figure and his name cannot be split off from his only known work: the treatise "Geometria Prattica" [Practical Geometry], published posthumously in 1599. The treatise has a constant practical aim and in all probability in this aspect lies the fame and fortune of the work through the centuries: It is "modern", shaped as a "handbook", more than as an exhaustive "treatise", and it favoured the diffusion of scientific notions and solutions between clients and common people, according to the process of knowledge diffusion that would be implemented between the XVI and the XVII centuries. In particular it is thanks to the numerous tavolas dedicated to the "surveyor's cross" that the work has always been a reference in the study of historical surveying methods and tools.

Introduction

The "Geometria Prattica" is the main source of information on Pomodoro, in particular contained in the Foreword and in the Introduction, respectively written by Pomodoro's brother Pietro and by Giovanni Scala. The scarcity of biographical information on Pomodoro is a counterpoint to the fame of his work.

The work had six editions, all published in Rome: 1599, 1603 (reprint of the previous one with a new frontispiece); 1624, 1667, 1691 (reprint of 1667 edition), 1772 (Fig. 1).

S. Brusaporci (✉)
Department of Civil, Construction-Architectural and Environmental Engineering, University of L'Aquila, via G. Gronchi 18, 67100 L'Aquila, Italy
e-mail: stefano.brusaporci@univaq.it

© Springer International Publishing Switzerland 2016

M. Cigola (ed.), *Distinguished Figures in Descriptive Geometry and Its Applications for Mechanism Science*, History of Mechanism and Machine Science 30, DOI 10.1007/978-3-319-20197-9_9

Fig. 1 Frontispiece of the 1667 edition of "Geometria Prattica"

It is configured as a four-handed work because, due to Pomodoro's premature death, Giovanni Scala completed the work with the description of the drawings and with explanations.

The analysis of the treatise points out the competence of Pomodoro in geometry and, mainly, his abilities in surveying and mechanics: the numerous and detailed drawings of mechanisms demonstrate his skills as a designer, manufacturer and user of measuring instruments.

Giovanni Scala was a mathematician, military engineer and architect from Friuli, operating between 1547 and 1599, and author of the treatise "Delle fortificazioni" [On Fortifications], published in 1596. However this work does not exceed the fame of Pomodoro's and Scala is remembered primarily as the commentator of "Geometria Prattica"'s Tavolas.

Biographical Notes

Giovanni Pomodoro was born in Venice, in the XVI century, and died before 1599a, b, c, previous to the publication of his treatise. Information can be gleaned from the Foreword and the Introduction of "Geometria Prattica": the first written by his brother Pietro Pomodoro, and the second by Giovanni Scala.

According to the text, Scala may have known Pomodoro, at least for fame, and describes him as «[…] excellent in Arithmetic and Geometry, and famous in his age […]». And later in the treatise, commenting on the Tavola XXXVIII, he writes: «The author is bright in figures on practical surveying indications, because he led for many years the art of measuring, and being one of the most surveying experts at his age, it's clear that composing this work he has not forgotten anything, that was necessary to the art of measure […]». From this information we can understand that Pomodoro was a practitioner of geometry and of architectural and land surveying. The notation "famous in his age" suggests the portrait of a person active in his cultural context, and in particular in the Venetian one, at that time the most scientifically and culturally alive. This is supported by a notation that can be derived from analysis of the Tavola XXVIII: the caption of the drawing—where the calculation of the surface of an oval is solved—reads: «The oval figure of Alberto Durero [Albrecht Dürer], practically measured, measures 680 5/42 sqr.» (Fig. 3). Namely it gives the solution of a precise measuring problem and cites explicitly Albrecht Dürer (1471–1528), confirming that Pomodoro was a cultured scholar and a professional involved in his historical context.

If it is certain that Pomodoro died before 1599, i.e. before the first edition of his treatise, in general it is very difficult to define univocally the timeframe in which he operated. Traditionally, his activity is ascribed to the second half of the XVI century, but Scala writes in 1599 that «[…] the work remained, as I said, almost useless and abandoned […]» and further: «therefore for many years it has been buried, although M. Pietro Pomodoro, brother of M. Giovanni, tried […] to find a virtuous person, who wanted to work on it and fulfil what was missing, and explain what has been realised, and yet he (according to what he told me) didn't find anyone who wanted to take charge of it […]». Since the difference of age between the two brothers is not known, nor how old Piero was at the time of publication of the treatise, the worlds of Scala could bring forward for a few years the timeframe of Pomodoro's activities. However, it is unlikely, for chronological reasons, that Pomodoro may have personally known the Dürer, in the occasion of one of his Venetian stays (1494–1495 and 1505–1506). The fact is that Pomodoro lived in the Veneto region of the mid-sixteenth century, place of the highest scientific and intellectual vitality, contemporary of personalities such as Palladio (1508–1580), Daniele Barbaro (1514–1570) and Galileo Galilei (1564–1642), who played an important role in the culture of the time.

The "Geometria Prattica"

The treatise "Geometria Prattica", in its overall configuration, is the synthesis of the work of two authors: Giovanni Pomodoro drew the tavolas—engraved on copper plates—and Giovanni Scala wrote the text, to explain and comment on the drawings. Pomodoro left some drawings quite unfinished. Scala also added some tavolas with the intent to give completeness to the treatise. The whole work consists of 51 tavolas, of which 44 were made by Pomodoro, and 7 by Scala. According to the editions, the tavolas are preceded by the written body of Scala or are alternated with the text. The textual pages have a layout with two columns, sided by Arabic numerals referring to the drawings in the tavolas. Drawings present a particular graphical quality, allowed by the copper engraving technique.

The work's structure is didactic. Many times the examples are repetitive, sort of tutorials: in fact very often drawings have measures, in particular the surveying ones, and the problems are developed until the solution. Moreover tavolas present useful captions.

For example in the Tavola VI Pomodoro rests on examples based on the distributive propriety of multiplication over addition; again Scala in the Tavola IX inserts a paragraph where he explains the meaning of "integer measure" (integer number) and "broken measure" (rational number). These notations suggest that the "Geometia Prattica" was addressed to an audience not accustomed to mathematical terms.

The treatise opens with the description of the principal instruments for drawing, surveying and military planning. Scala recommends that: «they have to be realised of appropriate size, so they can be put all in a casing, in a sheath, or in a case [...]» (Fig. 2).

Pomodoro presents a repertoire of operational tools for field operations: drawing and surveying instruments: compass, ruler, square ruler, penknife, pencil lead, stiletto (to draw white lines, i.e. without ink), drawing pens, file for pinning pen; plumb-bob level; gun compass (to measure the diameter of cannon's mouth and of cannonballs); other different kinds of compasses; hinged rules with goniometer and magnetic compass (to be adapted to both exterior and interior edges of the walls, to measure the angles or to "make an impression" and translate them on the sheet to draw plants); the surveyor's cross; the geometric square with quadrant (called "geometric quadrant" by Pomodoro, to measure distances, depths and lengths, with graduated alidade with viewfinder and a quarter of a circle inside the square).

There follows a series of tavolas (Tavolas II–XXX) on geometrical definitions and operations, useful to solve surveying problems: here they are formulated in an abstract way, and then they will be developed and explained with applicative examples (Fig. 3). The analytical solution of indirect measurements is substantially based on the use of the Pythagorean Theorem and on the theory of similar triangles. This is the one presented in the VI Book of the "Elements" of Euclid (IV–III century).

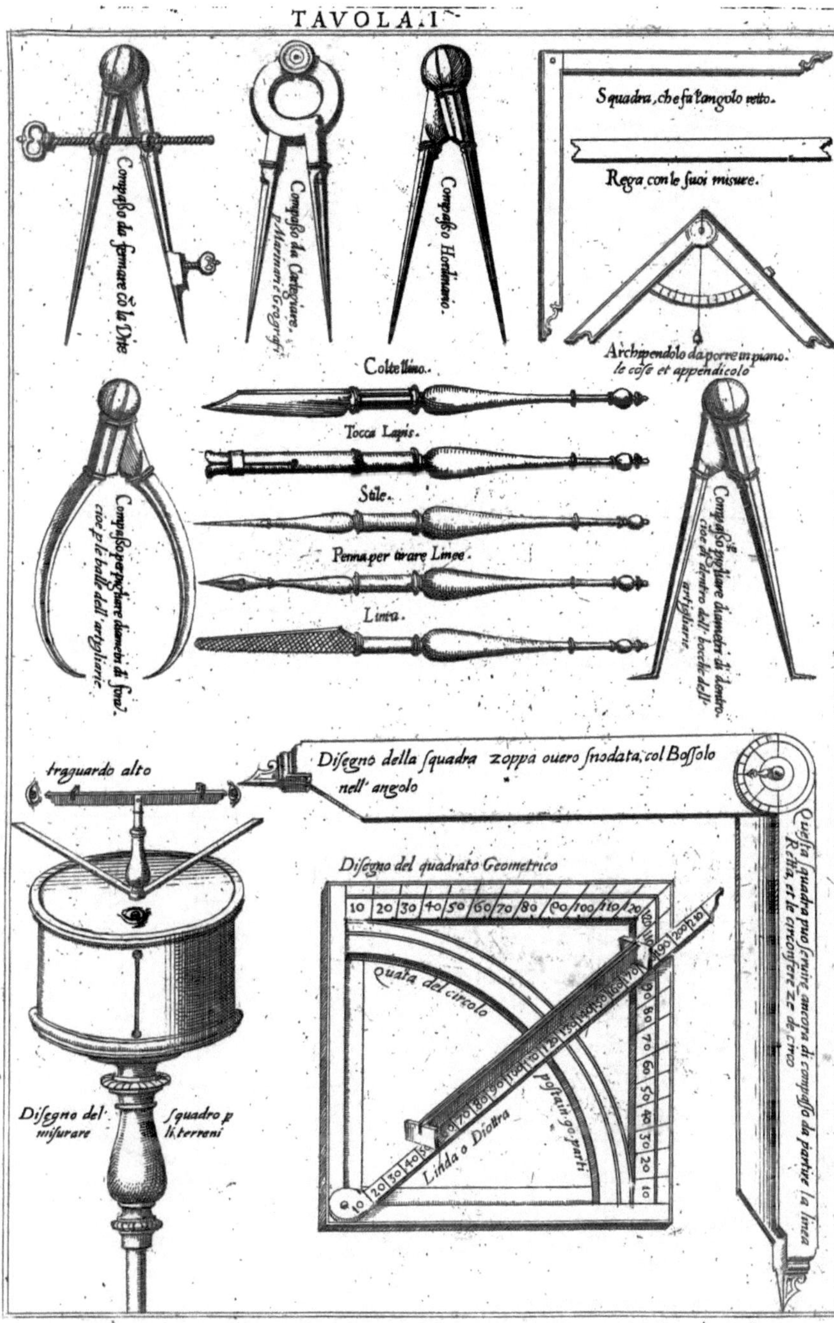

Fig. 2 Tavola I: drawing and surveying instruments

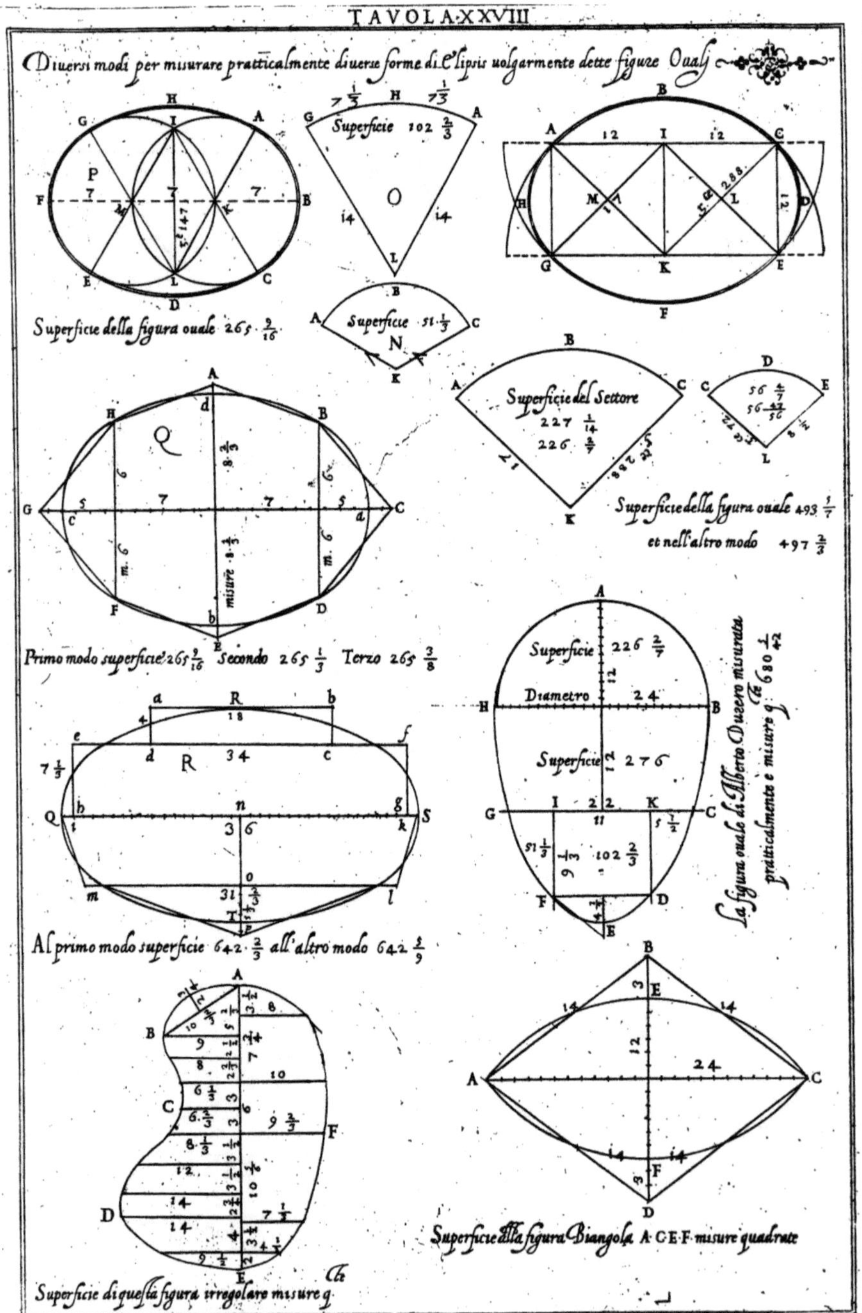

Fig. 3 Tavola XXVIII: area calculation of elliptical, oval and irregular surfaces

In the following tavolas we enter fully into the applicative part of Pomodoro's treatise. The Tavolas XXXI–XXXIX are focused on the use of the surveyor's cross (Figs. 4, and 5); the Tavola XL describes a square with mobile angle for planimetric survey (Fig. 10); the Tavolas XLI–XLIII deal with problems of distance measurements solved by using proportions between similar triangles (Figs. 6, and 7); the same at the Tavola XLIV, to calculate the heights (Fig. 8). The fundamental approach to the problems of measurement are the following ones: (1) definition of distances either directly or through mathematical calculations, based on the similarity between triangles; (2) calculation of areas by dividing or inscribing them with simple geometric shapes, such as triangles, right triangles, squares and rectangles. The reading of Scala's text about drawings n. 4, 5, 6 of the Tavola XXLIII suggests that he doesn't seem to have understood the full meaning of the drawings, which are to be read in succession: it's the measurement of a distance with a not practicable interposed land; the way presented by Pomodoro is the same as that first described by the Arab astronomer Al-Biruni in 1029 for the measurement of a river width (Fig. 7). About the drawings in the Tavola XLIV, it is interesting that Pomodoro uses hinged staffs for operations usually realised with the geometric square in other treatises (Fig. 8); about sun shadows and mirrors, their use can be referred to the "Optics" by Euclid (IV–III century). Similar operations with mirrors, squares, staffs, wooden triangles, and geometric squares are described in the "Geometria incerti auctoris" (972–1003) ascribed to Gerbert d'Aurillac—future pope Silvestro II (ca 950–1003).

The Tavolas of Pomodoro are followed by seven tavolas by Giovanni Scala, added with the intent to integrate the contents of the "Geometria Prattica", in particular with the explanation of the calculation of volumes of solids, constructive elements and buildings. Scala's construction expertise is underlined by the recommendations concerning construction yard aspects, or by the comments on building characteristics of foundations and retaining walls.

Scala's astronomical interests are evident in many examples, in particular the ones used to exemplify kinds of curved lines and their intersections: he cites the sun path between the tropics, the sundials, the relative position between the sun and the moon and the observer on the earth, the intersection of the «Meridian with the Horizon, with the Equinox, with the Tropics, and with the Arctic and Antarctic circles», the "twisting" form of comets' tail.

A philological notation: Scala points out expressly the authorship of his additions. And Losi, editor of the publication in 1772, writes in his Preface: «I do not want to leave more to warn the public, that this new reissue is totally similar to the first. The usual and right complaint of literates, that in the reprinting the works of ancient authors—with the right aim of improving them—often they are damaged, was the cause for which I did not want any mutation».

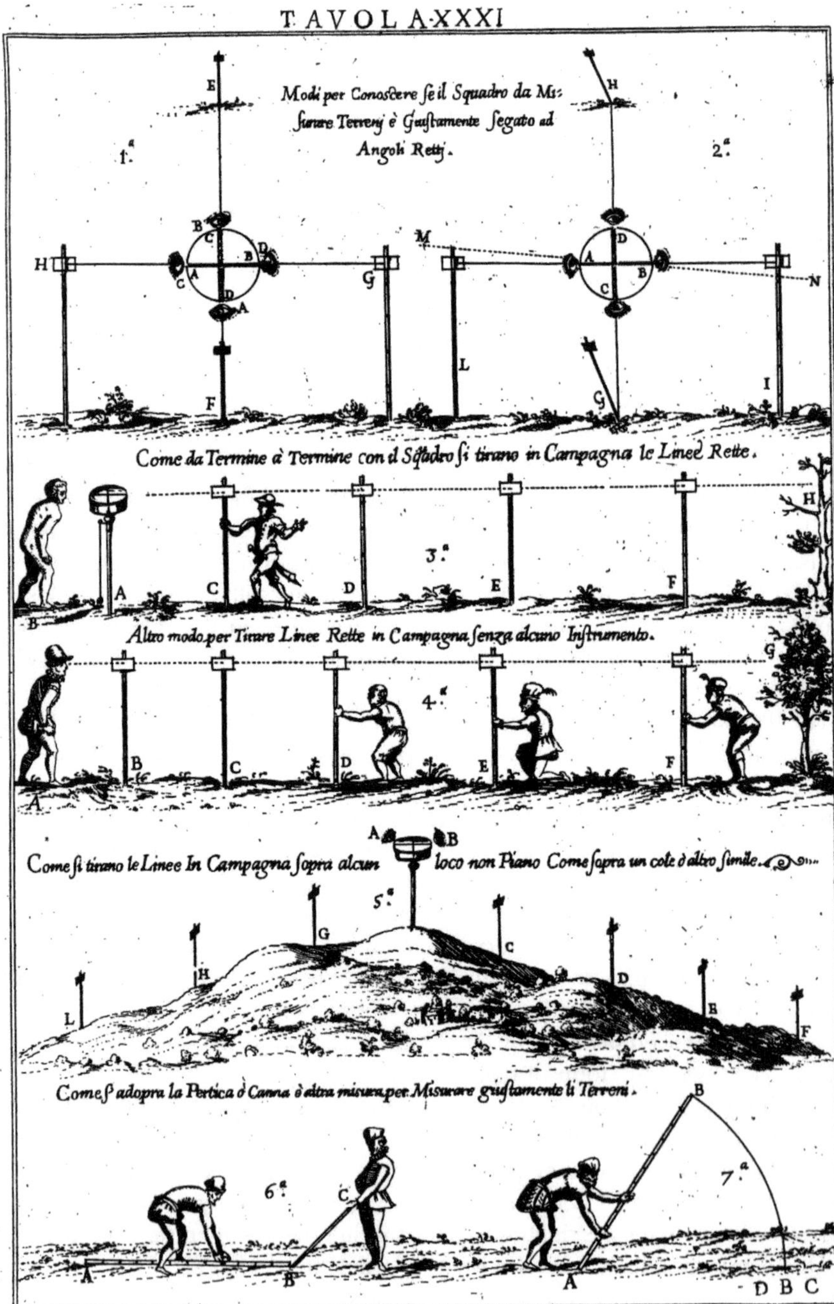

Fig. 4 Tavola XXXI: the use of the surveyor's cross

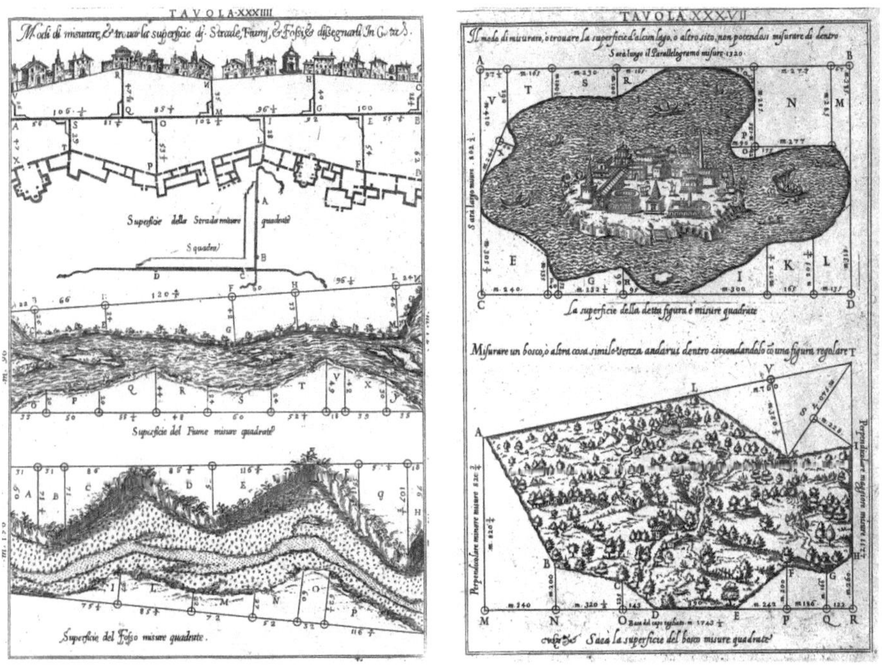

Fig. 5 Tavolas XXXIV and XXXVII: drawings explaining the geometrical methodology for land surveying using the surveyor's cross

On the Circulation of the Work

The work of Pomodoro is configured as a treatise of geometry applied to surveying, where the theoretical dissertation is related to numerous practical applications.

According to the wide space reserved for the description and use of the surveyor's cross, the "Geometria Prattica" is usually mentioned in writings that deal with this instrument. Such as: Niccolò Carletti, "Istituzioni di architettura civile" [Institutions of civil architecture] (1772), that cites the XXXII, XXXVII e XXXVIII Tavolas of Pomodoro. Lodovico Perini, "Trattato della pratica di geometria" [Treatise on the practice of geometry] (1751), where some illustrations are similar to Pomodoro's ones; Angelo Comoli, "Bibliografia storico-critica dell'architettura civile ed arti subalterne" [Historical-critical bibliography of civil architecture and subordinate arts] (1791), where he says that the treatise of Pomodoro is a «work useful to the architects, and indeed to all the artists» (Comoli 1791, 79). And again the "Geometria Prattica" is cited by: Luigi Marini in his comment to "Architettura militare di Francesco de' Marchi" [Military Architecture of Francesco de Marchi] (1811); Mariano d'Ayala, "La bibliografia militare d'Italia" [The Italian military bibliography] (1854); Stefano Grillo, "Prolusione al corso di Geodesia" [Prolusion for the course of Geodesy] (1861); "School Science and Mathematics", vol. 10,

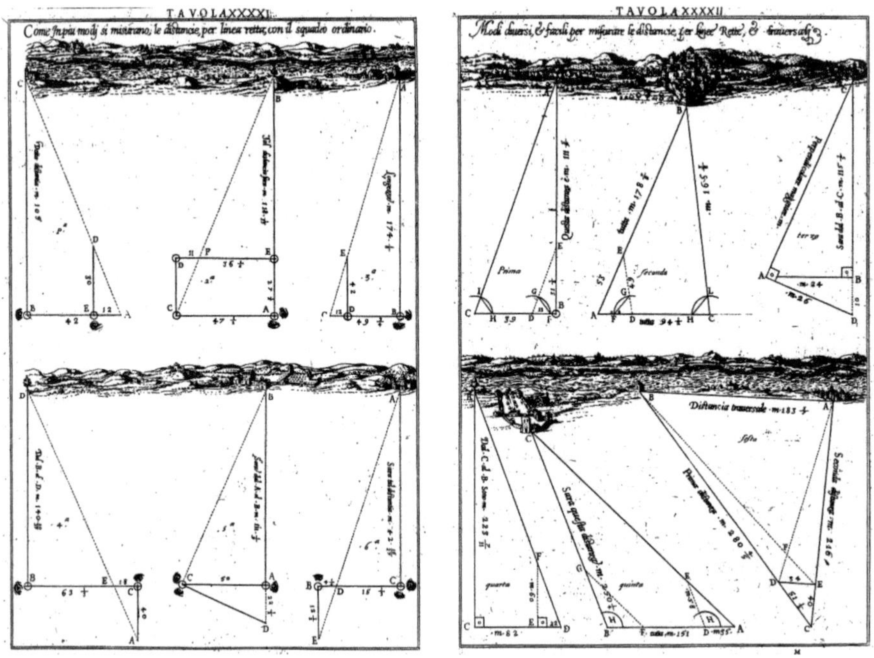

Fig. 6 Tavolas XLI and XLII: distances measurement using right-angled triangles and similar triangles

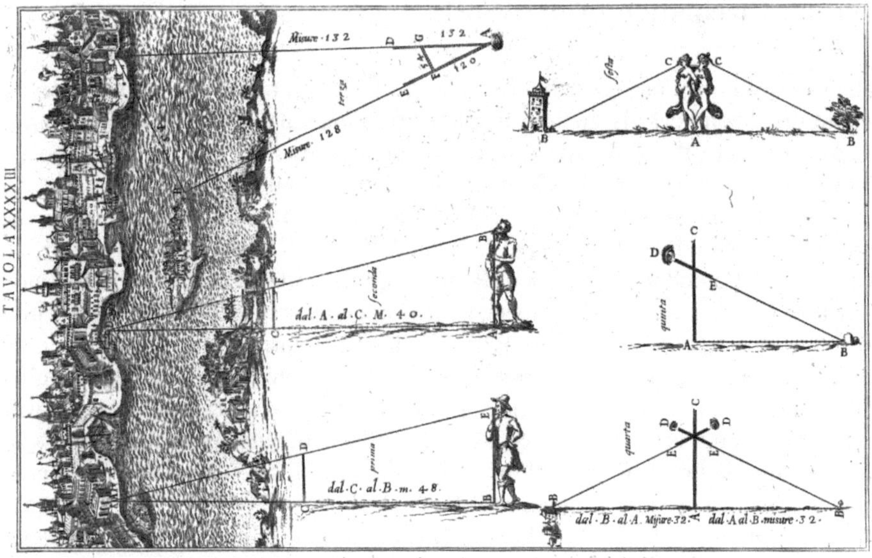

Fig. 7 Tavola XLIII: measurements through staffs

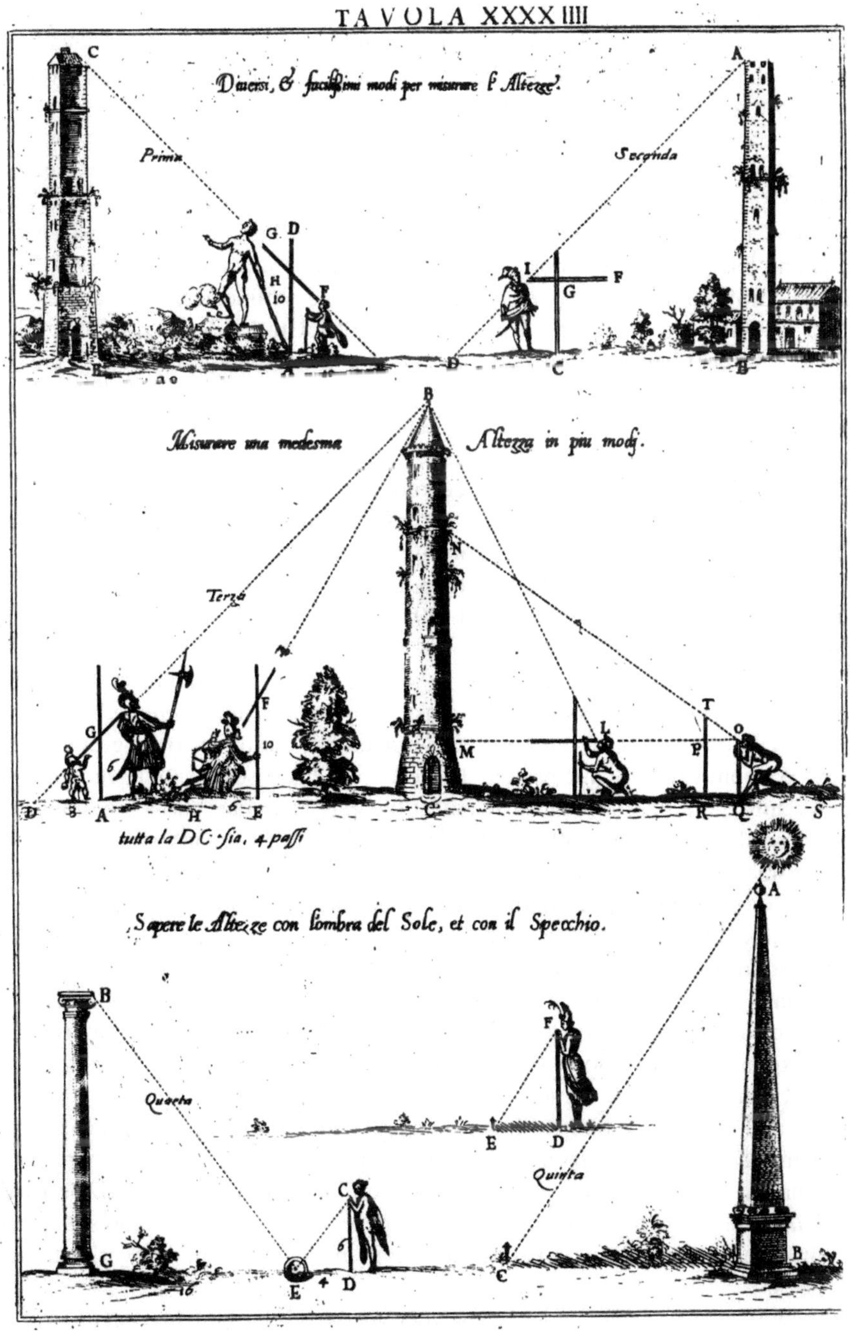

Fig. 8 Tavola XLIV: measurements using staffs, shadows and mirrors

Texas Tech University (1910); Giuseppe Boffito in his treatise on historical scientific instruments (1929); Federigo Enriquez and Giorgio Diaz De Santillana in the "Storia del pensiero scientifico" [History of scientific thought] (1932); Edmond Kiely, "Surveying instruments: their history and classroom use" (1947).

More recently the "Geometria Prattica" is mentioned almost systematically, whenever it is debated on history of surveying. For example, we remember: "Cultura, scienze e tecniche nella Venezia del Cinquecento" (1987); Enrico Gamba and Vico Montebelli, "Le scienze a Urbino nel tardo Rinascimento" (1988); Mario Docci and Diego Maestri, "Storia del rilevamento architettonico e urbano" (1993); Daniela Stroffolino, "La città misurata. Tecniche e strumenti di rilevamento nei trattati a stampa del Cinquecento" (1999). In 2007 Uta Lindgren cites Pomodoro among the renaissance authors employing the triangulation method (Lindgren 2007).

The surveyor's cross described by Pomodoro in the Tavola I has a "superior viewfinder" mounted on the cylinder (Fig. 2). Perhaps it would make it possible to trace lines at 45° or with different directions, for example parallel lines to use the principles of similar triangles. Instead, the instrument represented in Tavola XXXI, consists only of a cylinder with four intaglios (Fig. 4).

The surveyor's cross—first cited by Lazesio (1470?–1542) in 1527—derives from the roman "groma" (Rossi et al. 2011), and Tartaglia (1499?–1557) in his treatise "Quesiti et inuentioni diuerse de Nicolo Tartaglia" [Issues and different inventions of Niccolo Tartaglia] (1554) represents a cross similar to the one of classical age. Rather the Pomodoro's cross is cylindrical but, differently from the ones of Ryff or Oddi, it has an upper viewfinder, not dissimilar from groma's arm. However, compared to the groma, Pomodoro's cross seems to have only one plumb, unlike the traditional instrument of the classical era with four plumbs. The cross described by Ryff (1500 ca–1548) in 1547 has six intaglios, to draw lines even at 45°; similarly the cross by Muzio Oddi in his treatise of 1625. Again in 1840 the cross by Giuseppe Antonio Alberti has four intaglios, but with a magnetic compass fixed on the cylinder's top (Fig. 9).

Pomodoro's insistence on using the cross derives from his idea that this instrument—manoeuvrable and easy to use—can favour the simplification of measurement's problems, by the calculation of areas of rectangles or right triangles. Scala explains: «Up to now we have been dedicated to the study of regular and irregular figures, and we demonstrated in how many ways they can be described, divided and elaborated, both with the compass and with the numbers, but now for going farther, it's necessary to deal with the practice of the measurement in the field, that is the aim of all our efforts. But there is an instrument for land surveying, called a cross, useful to reduce places and maps to small and smart figures, therefore first it's necessary that I illustrate how it's made and how it works to take measures»

Likewise Oddi writes a few years after in the Preface of his treatise on the surveyor's cross: «[…] it's most appropriate and suitable among other instruments for the simplicity in building it and for the facility in using it […] And sometimes it requires little struggle, but the advantage is that you use only one instrument—

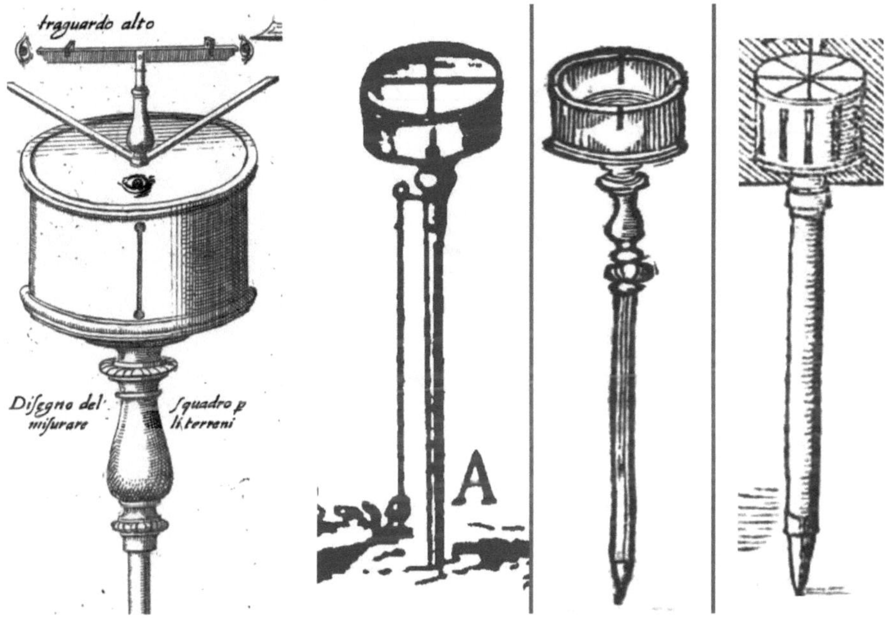

Fig. 9 Surveyor's cross by Pomodoro (1599), Ryff (1547) and Perini (1751). In particular Pomodoro presents two kind of instruments, the first one with an upper viewfinder, perhaps to trace lines at 45° or with variable directions

simple and rough—to realise quite all operations that are usually done with many tools, made with finesses and ability» . The secret of the surveyor's cross lies in its ease of use and of construction, and these are the reasons of its fortune in the treatises.

With regard to the methods of measuring adopted by Pomodoro, we observe that he doesn't propose measurements "on forward intersection", i.e. according to the graphic method based on determination of the vertex of a triangle, knowing the base and corners on it—a method already described and used by Alberti (1404–1472) in his "Descriptio Urbis Romae". The reason is that the surveyor's cross does not allow one to measure angles—except the future version of the nineteenth century with magnetic compass. Therefore also in the examples in Tavolas XLI and XLII, resorting to the use of parallel or perpendicular lines, Pomodoro solves the measurement problems using geometric—mathematical methods, i.e. establishing proportions between similar triangles.

However the "Geometria Prattica" also deals with other tools.

In particular in the Tavola XL the "lame square" is presented, that is made by hinged rules with compass, to survey interior and exterior angles of buildings (Fig. 10). The drawings are accompanied by a surveying notebook example. This is of particular interest because it indicates that it was usual to measure angles, and it's also written in the comments of Scala. This procedure may be the same as that

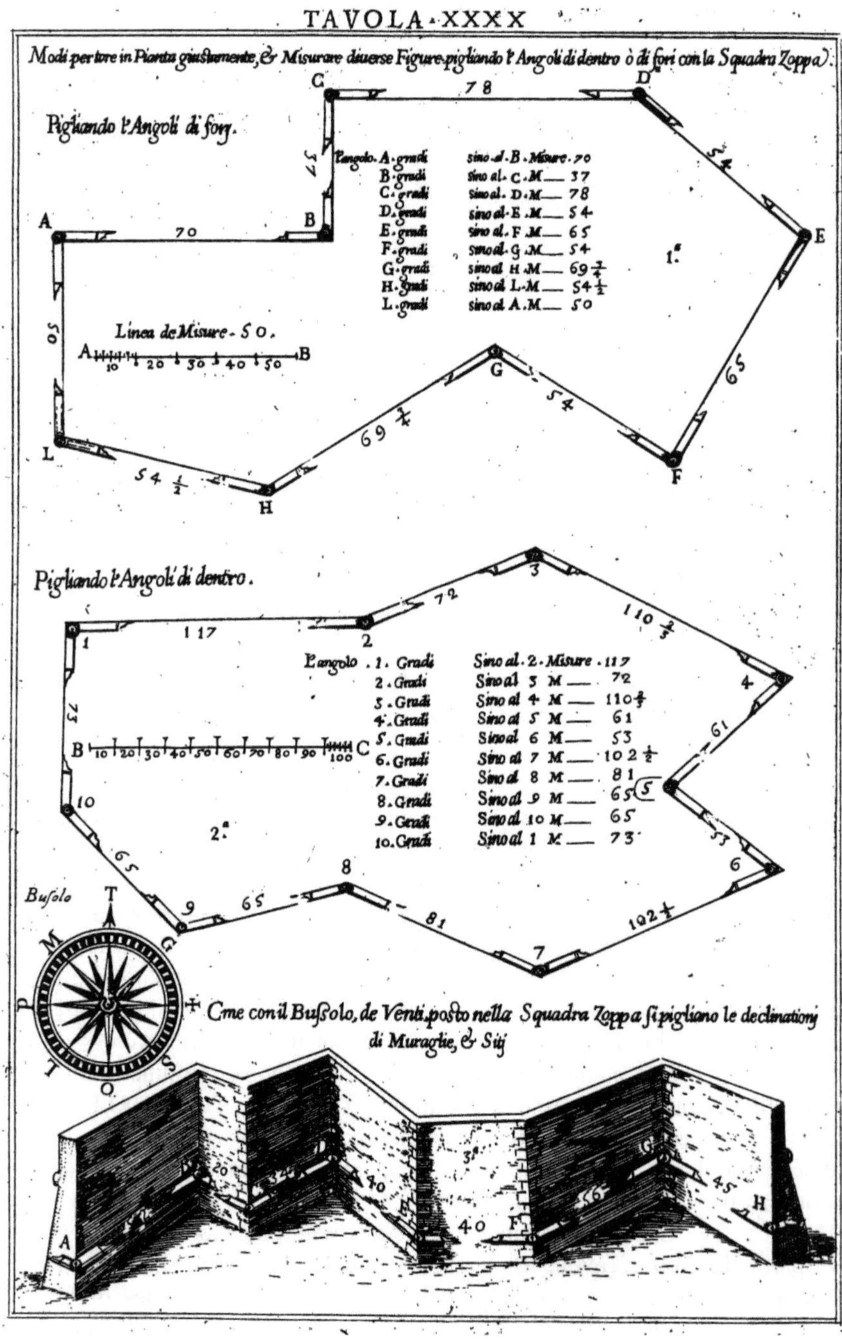

Fig. 10 Tavola XL on the use of the "lame square", i.e. hinged rules with compass to survey interior and exterior angles of buildings. The drawings are accompanied by a surveying notebook example, where the length of traverse's sides and the angles are recorded

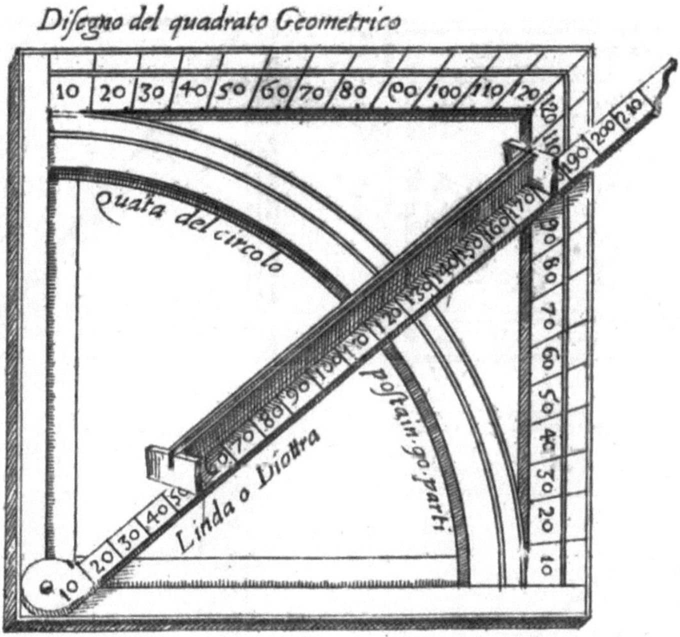

Fig. 11 The geometric square represented in the Tavola I

suggested by the land-surveying notes for Imola's map—often attributed to Leonardo da Vinci: the survey is realised with closed concatenated traverses, where the lengths of blocks' sides are measured but not the angles. This would suggest that the surveyor did not record the measures of angles but directions in relation to the cardinal points, using an instrument with magnetic compass, and re-drawn them directly on the paper (Cozzi 1987; Docci 1987). This procedure, based on the direction's recording—i.e. assessed by referencing them to the direction of the winds, is an application of the land surveying of a technique that has long been present in the navigation and it is explained in "Ludi rerum mathematicarum" by Alberti (1450–1452) and in the treatises by Gemma Frisius (1533), Ryff (1547), Bartoli (1564).

Finally, a consideration on the drawing of the geometric square represented in the Tavola I (Fig. 11): the drawing is of large dimensions compared to the overall layout of the tavola, and it presents an accurate description. This one by Pomodoro is a refined instrument, that can be compared—and perhaps more elaborated—to the ones presented by Ryff (1547), Peverone (1558) or Bartoli (1564) (Fig. 12). Unique notation: it has no plumb, so that the quarter of the graduated circle, inside it, does not seem to be usable, similarly to the one of the "quadrant of the circle" (Centofanti and Brusaporci 2013). Bearing in mind the importance that Pomodoro attributes to this drawing, and the fact that the geometric square allows calculation of the distances using proportions between similar triangles—i.e. according to the

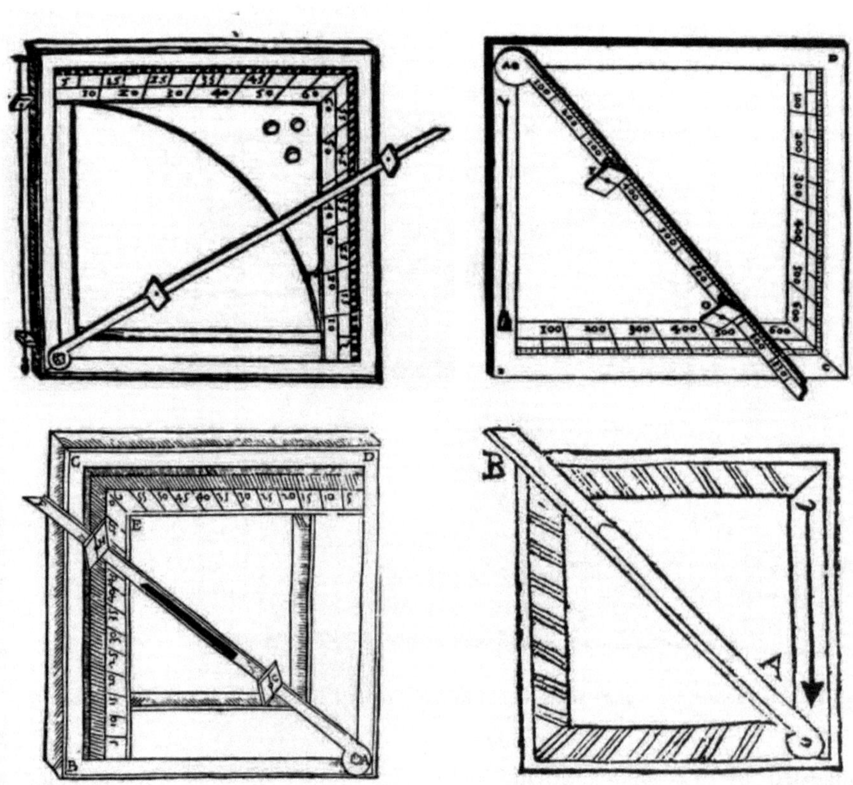

Fig. 12 The geometric square by Ryff (1547), Peverone (1558), Bartoli (1564), Fonticulano (1597)

method used and explained systematically by Pomodoro, also in T Tavola XLII with applicative examples—the fact that often the "Geometria Prattica" is remembered only for the use of the surveyor's cross is considered improper and reductive. Considering the nature of "unfinished" work of the "Geometria Prattica"—evident from various partially incomplete drawings—and for Pomodoro's graphical attention in illustrations, it could be argued—but it is purely a hypothesis—that the work should be provided for other tavolas, explanatory of the geometric square.

The fame of the Treaty—and probably also the one of Pomodoro—is demonstrated by the fact that in 1600—that is only one year after publication of the "Geometria Prattica"—it is widely cited by Giovanni Vasari the Younger (1562–1625) in his manuscript "Raccolto […] di varii instrumenti per misurare con la vista" [Collection of various tools to measure with the view]: of the whole 76 folios—each one dedicated to one instrument and procedure—the folios 8, 9, 10, 11, 12, 13 are referred to the "Geometria Prattica". In particular Vasari re-draws (mirrored, because he re-traced them) Pomodoro's illustration of the surveyor's cross in

Tavola I, sided by the first and second T Tavola XLI (examples of measurements using the cross); the fifth and sixth drawings in Tavola XLII on measurements with similar triangles; the second and third illustrations in Tavola XLIII concerning graduated staffs to measure distances; the drawings of Tavola XLIV on staff combination and the use of mirrors and shadows to measure heights (Camerota 1996).

Modern Interpretations

The second half of the XVI century was a period of cultural development and codification of knowledge, as well as of transformation of communication methods, in particular thanks to the proliferation of large printing, especially in Venice, Bologna and Rome.

With the printing in 1545 of the Italian version of "Euclid's Elements", published in Venice and edited by Niccolò Tartaglia (1499?–1557), geometry becomes a scientific fundamental reference for knowledge and for the study of reality, and therefore the basis for any action aimed at transforming Nature. Therefore, the study of geometry is seen not only as an independent speculative activity, but also as a tool for solving everyday problems, a means of urban and land knowledge, the basis for architectural design and fortress building.

In particular, the concept of "measure" plays a central role, as a tool for certain knowledge, domain, management and transformation of the world. Consequently the measuring instruments become very important, essential equipment for a scientific "quantization", but only if they are properly built and used according to the principles of Euclidean geometry. Along this line, land and architectural surveying, until now ascribed to individual initiatives—and often directed to the knowledge of the remains of the classical era—expand their horizons and become a professional practice.

Consequently to this "Copernican revolution", the treatises—following the new Renaissance science—are no longer based on a contemplative knowledge, but on an operational and functional one, and they are tested on empirical observation. Therefore, numerous publications assume the character of general theories, useful for application, often based on personal experiences. These works can interest and be addressed to different kinds of people—mathematicians, master builders, engineers, military, etc.—in particular by providing a method: that is giving a mathematical basis to operate with awareness. In this sense, usually aim of the treatises is not be operational guidelines or a practical compendium, because geometry is intended as a means to understand and solve problems, with the highest degree of accuracy.

The treatise writers—being scientists—combined theoretical, technical and practical interests. The ones dealing with surveying propose a set of important scientific issues, but usually with an "aristocratic" attitude (Maestri 2001, 47): it is clearly declared by Ieronimo Pico Fonticulano (1541–1596) in his treatise on geometry published in 1597: «If I describe the practice of fools, simply so that every mediocre talent can be used, I would do wrong to professors because I would facilitate the ignorant [...] who have only a naked practice that they do not know is right or not, but want to show that they do know, and presumptuously become professors» . Focusing on the background, it is important to understand the meaning of the "Geometria Prattica" in relation to its historical cultural context, because Pomodoro conceives a work fundamentally different from the treaties of the period and in accordance with the most innovative experiences.

The "Geometria Prattica" has a general structure that reflects one of the most widespread treatises of the period: a first theoretical part, focused on Euclidean geometry, and a second applicative one. However, unlike scientists who write treatises where problems are seen as consequent practical applications of geometry, Giovanni Pomodoro changes the point of view and creates a "handbook", aimed at concrete and rigorous solving of problems. Thus Pomodoro's work, as highlighted by the title "Geometria Prattica", is configured as a guide, offering methods and techniques to be applied in daily operations. And perhaps it is in this "slant"—which could be called "vulgar" and which would not escape his contemporaries—that there is a real reason why—as stated by Scala—there was nobody willing to comment on Pomodoro's tavolas- if not Scala himself but other mathematicians and fortress designers interested in professional practice.

And Pomodoro also deviates in operating approach from the tradition of the treatises, with the stated choice of the use of simple tools and methods. In other treatises the "refinement" and the "complexity" of instruments—that require the full understanding of geometric principles—seems to be based on complacency: you remember the "Trattato del radio latino" of 1586 by Latino Orsini (XVI century), or many pages dedicated to the geometrical square, to the quadrant of the circle or to the asymmetrical square in the various treatises.

More generally, the good fortune of Pomodoro's treatise is in offering to his contemporaries a work suited to the needs of clients—increasingly numerous and demanding, useful for practical administrative needs, simple in format and smart in contents, because the management of the problems could not be ascribed only to the "intelligence" of a small number of academics.

The fortune of "Geometria Prattica" also lies in another aspect: Pomodoro's drawings have a peculiar graphical quality, being a smart visual guide for professionals. Pomodoro has a pragmatic approach, not from ignorance—and in fact he proves to have knowledge of geometry and analytical calculation, solving problems even mathematically complex, but he presents a different approach, more "modern" than the one diffused in the sixteenth century—you consider that the XVI century is finishing up. And the precision of Pomodoro's work shows his ambition to explain the surveying discipline to all people.

The "Geometria Prattica" is different from many other coeval treatises also for another reason: it does not follow a dialogic structure—also present in Galileo's "Dialogo sopra i massimi sistemi" [Dialogue Concerning the Two Chief World Systems] (1632)—and sometimes even declined in the treatises of buildings, often organized into questions and solutions—such as in Tartaglia (1499?–1557).

Pomodoro's work is colloquial, of course for reasons dictated by the "Geometria Prattica" way of composition, with Scala's explanations to Pomodoro's tavolas. And this didactic model seems to be similar to the one already used by Serlio (1475–1554?) and by Palladio (1508–1580).

The culture of the Renaissance expands outside of the universities, precisely thanks to the treatises. Academies prosper, favouring the diffusion of knowledge, with their societies not restricted only to scientists. The Academies, more than the Universities, become the core of intellectual life, a place of expression and development of the "dialogue" between minds. There is a process of erudition's diffusion—and even of "dilution", and in this context the "Geometria Prattica" operates in consonance. In fact, it is the first handbook in the history of surveying: according to an "aristocratic" point of view, it produces a "vulgarization" of the notions, but for example fully in trend with a line that has founded fortune —"mutatis mutandis"—in the series of small size volumes of Greek, Latin and Italian classics, disseminated in the early sixteenth century. This is in accordance with the new climate of the Post-Reformation, to which the policy of the Republic of Venice had consciously adhered to stabilize its role as a continental power: Post-Reformation «able to weld the culture of the dominant classes to the one of the lower ones, to realize the most complete process of ideological homogenization that the Church had ever made» (Cozzi 1987, 25).

Bibliography

Pomodoro's Work Publications

Pomodoro G (1599) "Geometria prattica tratta dagl'Elementi d'Euclide et altri auttori da Giouanni Pomodoro venetiano mathematico eccellentissimo descritta et dichiarata da Giouanni Scala mathematico. Nella quale si uede in 50 tauole di rame scolpito tutto quello che ad un buon giometra s'appartiene di sapere et porre in uso. Opera non meno uttile che necessaria, a' misuratori di terreni, di fabriche, et altri simmili, ma' in'oltre ancora a, geografi, cosmografi, architetti ciuili, et milittari, a' bombardieri, soldati privati, a' capitani, mastri di campo, et a' qual si voglia altra persona virtuosa", Stefano Paolini, Roma

Pomodoro G (1599) "Geometria prattica dichiarata da Giouanni Scala sopra le tauole dell'ecc.te matthematico Giouanni Pomodoro tratte d'Euclide et altri authori. Opera per Generali da guerra Capitani Architetti Bombardieri e Ingegneri Cosmografi nonché per odinari professori di misure", Stefano Paolini, Roma. [reprint 1603 with new frontispiece]

Pomodoro G (1624) "La geometria prattica di Gio. Pomodoro venetiano cauata da gl'Elementi d'Euclide, e d'altri famosi autori, con l'espositione di Gio. Scala matematico. in cinquanta tauole, scolpite in rame, dalle quali con facilità possono apprendere tutte le cose, che al buon geometra appartengono. Opera necessaria à Misuratori, ad Architetti, à Geografi, à Cosmografi, à Bombardieri, à Ingegneri, à Soldati, & à Capitani d'Eserciti", Giovanni Angelo Ruffinelli, Roma, [printing office "Andrea Fei stampator ducale", Bracciano 1623]

Pomodoro G (1667) "La geometria prattica di Gio. Pomodoro venetiano cauata da gl'elementi d'Euclide, e d'altri famosi autori, con l'espositione di Gio. Scala matematico. Ridotta in cinquanta tauole, scolpite in rame, dalle quali con facilita si possono apprendere tutte le cose, che al buon geometra appartengono. Opera necessaria à Misuratori, ad Architetti, à Geografi, à

Cosmografi, à Bombardieri, à Ingegneri, à Soldati, & à Capitani d'Eserciti", Gio. Batta de Rossi milanese in piazza Nauona, Roma [in printing office of "Moneta", Roma 1667]
Pomodoro G (1667) "La geometria prattica di Gio. Pomodoro venetiano cauata da gl'elementi d'Euclide, e d'altri famori autori, con l'espositione di Gio. Scala matematico. Ridotta in cinquanta tauole, scolpite in rame dalle quali con facilità si possono apprendere tutte le cose che al buon geometra appartengono. Opera necessaria à Misuratori, ad Architetti, à Geografi, à Cosmografi, à Bombardieri, à Ingegneri, à Soldati, & à Capitani d'Eserciti", Matteo Gregorio Rossi romano in piazza Nauona, Roma [reprint 1691 in printing office of " Moneta"]
Pomodoro G (1772) "Geometria pratica di Giovanni Pomodoro veneziano ridotta in tavole cinquantuno con le spiegazioni di Giovanni Scala matematico", Carlo Losi, Roma 1772 [in printing office of "Generoso Salomoni"]

Other References

Alberti GA (1840) "Istruzioni pratiche per l'ingegnere civile", Borroni e Scotti, Milano
Alberti LB (2005) "Descriptio Urbis Romae", 1443–1448. In Boriaud JY, Furlan F (eds) "Leonis Baptistae Alberti. Descriptio Vrbis Romae. È dition critique", Leo S. Olschki Editore, Firenze
Alberti LB (1960–1973) "Ludi rerum mathematicarum", 1450–1453, in Grayson C (ed.), "L.B. Alberti. Opere volgari", Laterza, Bari, 1960–1973, III, pp 133–173
Al-Biruni (1934) Book of instructions in the elements of the art of astrology, 1029. In: Ramsay Wright R. (ed) Luzac, London
Bartoli C (1564) "Del modo di misurare le distantie", Sebastiano Combi, Venezia
Boffito G (1929) "Gli strumenti della scienza e la scienza degli strumenti", Seeber, Firenze
Camerota F (1996) "Introduzione", in Vasari il Giovane G., "Raccolto fatto dal Cav:re Giorgio Vasari: di varii instrumenti per misurare con la vista", Giunti, Firenze
Carletti N (1772) "Istituzioni d'architettura civile di Niccolò Carletti Filosofo, Professore di Architettura, accademico di merito di S. Luca", stamperia Raimondiniana, Napoli
Ceccarelli M, Cigola M (2001) Trends in the drawing of mechanisms since the early Middle Ages. J Mech Eng Sci 215:269–289
Centofanti M (2001) "Agrimensura". In: Maestri D (ed) "Essendo la geometria origine e luce di molte scienze et arte", Fondazione Cassa di Risparmio della Provincia dell'Aquila, L'Aquila, pp 155–164
Centofanti M (2001) "Strumenti e metodi per il rilevamento". In: Maestri D (ed) "Essendo la geometria origine e luce di molte scienze et arte", Fondazione Cassa di Risparmio della Provincia dell'Aquila, L'Aquila, pp 123–154
Centofanti M, Brusaporci S (2013) "Surveying methods and instruments in the sixth book of Ieronimo Pico Fonticulano's Treatise on Geometry (1597)". In: Pisano R, Capecchi D, Lukešová A (eds) "Physics, astronomy and engineering. critical problems in the history of science and society", Scientia Socialis Press, Šiauliai, pp 177–184
Cigola M (2012) In praise of parallel theories: Descriptive geometry and applied mechanics. In: Carlevaris L, Filippa M (eds) In praise of theory. The fundamentals of the disciplines of representation and survey, Gangemi, Roma, pp 39–46
Cigola M, Ceccarelli M (1995) On the evolution of Mechanisms drawing. In: Proceedings of IXth IFToMM world congress, Politecnico di Milano, vol 4, pp 3191–3195
Comoli A (1791) "Bibliografia storico – critica dell'architettura civile ed arti subalterne", vol III, Il Salvioni, Roma
Cozzi G (1987) "La politica culturle della Repubblica di Venezia nell'età di Giovan Battista Benedetti". In: "Cultura, scienze e tecniche nella Venezia del Cinquecento", Istituto Veneto di Scienze Lettere ed Arti, Venezia, pp 9–28
Cultura, scienze e tecniche nella Venezia del Cinquecento (1987) Istituto veneto di scienze lettere ed arti, Venezia
D'Ayala M (1854) "La bibliografia militare d'Italia", Stamperia reale, Torino

Docci M (1987) “I rilievi di Leonardo da Vinci per la redazione della pianta di Imola”. In: Benedetti S, Miarelli Mariani G (eds) “Saggi in onore di Guglielmo De Angelis d’Ossat”, Multigrafica, Roma, pp 29–31

Docci M, Maestri D (1993) “Storia del rilevamento architettonico e urbano”, Laterza, Roma-Bari

Enriguez F, Diaz De Santillana G (1932) “Storia del pensiero scientifico”, vol I. Treves Treccani Tumminelli, Milano – Roma

Fonticulano IP (1597) “Geometria”, Giovanni Francesco Delfini, L’Aquila

Gamba E, Montebelli V (1988) “Le scienze a Urbino nel tardo Rinascimento”, Quattro Venti, Urbino

Gemma Frisius R (1533) “Libellus de locorum describendorum ratione”, Antwerp

Gerbert d’Aurillac (1963) “Geometria incerti auctoris”. In: Bubnov N (ed) “Gerberti postea Silvestri II papae Opera Mathematica (972–1003)”, Georg Olms Verlagsbuchhandlung, Hildesheim

Grillo S (1861) “Prolusione al corso di Geodesia”. In: “Giornale dell’ingegnere – architetto ed agronomo”, IX, Stabilimento Saldini, Milano

Kiely ER (1947) Surveying instruments: their history and classroom use. The Columbia University Press, New York

Lazesio (de) Feliciano F (1527) “Libro di arithmetica & geometria speculatiua & praticale”, Francesco di Alessandro Bindoni, & Maphoo Pasini, Venezia

Lindgren U (2007) Land Surveys, Instruments, and Practitioners in the Remaissance. In: Woodward D (ed) The History of Cartograpy—volume 3. Cartography in the European Reinasseance”, University of Chicago Press, Chicago 2007, pp 477–508

Lyons HG (1927) Ancient surveying instruments. Geograph J LXIX:132–143

Maestri D (ed) (2001) “Essendo la geometria origine e luce di molte scienze et arte”, Fondazione Cassa di Risparmio della Provincia dell’Aquila, L’Aquila

Marini L (1811) “Architettura militare di Francesco De’ Marchi illustrata da Luigi Marini”, Roma

Nappo T (ed) (2007) “Indice biografico italiano”. In: Saur KG, vol 10. Munchen

Oddi M (1625) “Dello squadro”, Bartolomeo Fobella, Milano

Orsini L (1583) “Trattato del radio latino”, Vincentio Accolti, Roma

Palladio A (1570) “I quattro libri dell’architettura”, Dominico de’ Franceschi, Venezia

Perini L (1751) “Trattato della pratica di geometria…”, Giuseppe Berno, Verona

Peverone GF (1558) “Due breui e facili trattati, il primo d’arithmetica: l’altro di geometria ne i quali si contengono alcune cose nuoue piaceuoli e utili, si à gentilhuomini come artegiani. Del sig. Gio. Francesco Peuerone di Cuneo”, Gio. di Tournes, Lione

Pomodoro G (1599) “Geometria prattica tratta dagl’Elementi d’Euclide et altri auttori da Giouanni Pomodoro venetiano mathematico eccellentissimo descritta et dichiarata da Giouanni Scala mathematico”. Stefano de Paulini, Roma

Promis C (1874) “Biografie di ingegneri militari italiani dal sec. XIV alla seconda metà del XVIII”, Fratelli Bocca, Torino

Rossi C, Ceccarelli M, Cigola M (2011) “The groma, the surveyor’s cross and the chorobates. In-deph notes on the design of old instruments and their use”. In: “Disegnare Idee Immagini” 42:22–33

Ryff WH (1547) “Der furnembsten notwendigsten der gantzen Architectur angehörigen Mathematischen und Mechanischen Künst eygentlicher Bericht und verstendliche unterrichtung”, Petreius, Nurnberg

Scala G (1596) “Delle fortificazioni”, Roma

School Science and Mathematics (1910) Vol. 10, Texas Tech University, Lubbock (Texas)

Serlio S (1584) “I sette libri dell'architettura”, Venezia

Stroffolino D (1999) “La città misurata. Tecniche e strumenti di rilevamento nei trattati a stampa del Cinquecento”, Salerno editrice, Roma

Tafuri M (1987) “Daniele Barbaro e l acultura scientifica veneziana del ‘500”. In “Cultura, scienze e tecniche nella Venezia del Cinquecento”, Istituto Veneto di Scienze Lettere ed Arti, Venezia, pp 55–84

Tarozzi G (1985) "Gli strumenti nella storia e nella filosofia della scienza", Istituto per i beni artistici, culturali, naturali della Regione Emilia-Romagna, Bologna

Tartaglia N (1550) "La Noua scientia de Nicolo Tartaglia con una gionta al terzo libro", Nicolo de Bascarini, Venezia

Tartaglia N (1554) "Quesiti et inuentioni diuerse de Nicolo Tartaglia", Nicolo de Bascarini, Venezia

Vagnetti L (1970) "Cosimo Bartoli e la teoria mensoria nel secolo XVI. Appunti per la storia del rilevamento". In: "Quaderno dell'Istituto di Elementi di Architettura e Rilievo dei Monumenti di Genova" 4:111–164

Vasari il Giovane G (1600) "Raccolto fatto dal Cav:re Giorgio Vasari: di varii instrumenti per misurare con la vista"

Jacques Ozanam (1640–1718)

Cristina Càndito

Abstract The life of Jacques Ozanam has been conditioned by the fact that he was the second-born in a well-off land-owning family, under a regime where only the first-born could inherit the family wealth. His father for this reason pushed him into clerical studies, thereby prohibiting him from carrying out scientific studies during his training. His joint interests in teaching and research lead him to not just explain with clarity and simplicity the scientific discipline that he delved into in his texts, but to also arrive at original results which demonstrate, for example the use of a method of measurement applied to perspectives and the illustration of a clever new machine using human propulsion.

Introduction

The period of Ozanam's scientific production, between the XVII and the XVIII century, was characterized by a persistent unity of science, theory, and practice in a jointly-held mathematical matrix. Each one's applicative branch often assumes only sporadically an autonomous dimension and, at times, this is characterized by a less systematic treatment, conceived to meet practical necessities.

Some scientific sectors, as is known, have undergone conspicuous methodological revolutions. In the field of mechanics, for example, the premise is provided by the contributions made by Guidobaldo del Monte (1577) and Galileo Galilei (1600; see Ceccarelli 1998). In the sector of perspectives an evolution had already been conceived, introducing the elements to infinity, and allowing for the consideration of the unity of projections thanks to the contribution of Girard Desargues in 1639 and 1640.

Work by Ozanam must be inserted into this scientific context, of which he was aware, while making the effort to apply them in accordance with the objectives of

C. Càndito (✉)
Department of Sciences for Architecture, University of Genoa,
Stradone S. Agostino 37, 16123 Genoa, Italy
e-mail: candito@arch.unige.it

© Springer International Publishing Switzerland 2016

M. Cigola (ed.), *Distinguished Figures in Descriptive Geometry and Its Applications for Mechanism Science*, History of Mechanism and Machine Science 30, DOI 10.1007/978-3-319-20197-9_10

his writings, which are very diverse: sometimes general and explanatory treatment, but in other cases focused on providing practical instruction in a specific sector.

In the field of mechanics and descriptive geometry Ozanam concentrates on the systemization of knowledge in diverse sectors and on the attempt to find simplifying methods, even if they are rigorous in their scientific foundation. This is demonstrated by the conspicuous presence of experimental explanations and applicative methods, characterizing the production of a scholar whom, already at a precocious age, was involved in the investigation of scientific knowledge with notable continuity and rigor.

Biographical Notes

Jacques Ozanam was born on June 16, 1640 in Bouligneux (the Princedom of Dombes, Ain, France). His family had made a name for themselves in the local administration, and Jacques was remembered as the "honneur de sa Dombes" by the sovereign princess of the princedom (Monfalcon 1862).

A recent family history (Ozanam 2011) permits the identification of his parents as Jacques Ozanam (1600–1660) and Euphémie Prost (who died in 1674). It has been said that the family, of Jewish origin, converted to Catholicism in 607, due to the influence of St. Desiderius of Vienna.

As a second son, Jacques was pointed towards a career in the clergy in spite of his precocious interest in mathematics and chemistry. His first biographer (Fontenelle 1719) in fact tells of Jacques' studies of the movements of the stars in the night sky at age 10 or 12, seen from the garden of his father's home. Another testimonial to his precocity is seen in his compilation, at age 15, of a mathematical study, which is still a manuscript.

His father tried to give Jacques the right education, and in fact his training was carried out at the Jesuit College in Reims. His theological studies however, were abandoned upon the death of his father, after only 4 years.

Ozanam, it seems, took private lessons in mathematics in Dijon from the Jesuit mathematician, Jacques de Billy (1602–1679), who later (1665–1668) would take over the position as mathematics professor at the Jesuit College of Dijon. According to several sources, however, Ozanam's true master may have been Claude François Millet Dechales (1621–1678) who taught mathematics and its applications in navigation and architecture at the Jesuit Colleges of Chambery and of Lyon. According to this hypothesis, the reference to de Billy is only due to the fact that Ozanam was inspired by his study methods (Niceron 1739).

In Lyon, Ozanam gave free lessons in mathematics until his financial conditions obliged him to ask for fees—his entire inheritance went, in fact, to his older brother upon the death of his father. In 1670 he was in charge of the publication of Adriaan Vlacq's logarithms and trigonometry tables which were said to be very accurate.

An incident has been documented in Lyon where Ozanam loaned money to two foreign friends of Henry D'Aguesseau (1636–1716), the father of the Chancellor of

France, Henry François D'Aguesseau (1668–1751). Following upon this generous offer, sometime around 1670, Ozanam was invited to the Chancellery in Paris, where he was able to give private lessons.

At first, when he arrived in Paris—perhaps related to his displeasure with his family—Ozanam started gambling, and had friends of bad reputation. But in 1673 he published his first treatise on sundials, and then in 1674 married, returning to a normal life.

With his high-born wife, Françoise Prevost (who died in 1701) Ozanam rented a house on Rue des Colombiers near Saint Sulpice. They had 12 children, most of whom died as infants. Without financial worries—thanks to his teaching earnings, in addition to his wife's rich dowry—Ozanam was able to dedicate himself to the studies which are documented by his publication in these years of numerous texts.

In 1701 he became a widower and—upon the War of the Spanish Succession which kept many students away from Paris—Jacques once again began to have economic difficulties. He moved into a more modest home on Rue de Seine. Then in 1708, after his sons, Jacques (1676–?) and Ignace (1695–?), left home, he set up house with a faithful servant in a two-bedroom apartment by the Abbey of Saint-Germain-des Prés.

He was admitted as "Élève géomètre" to the Royal Academy of Sciences in 1707 (Institut de France 1939), and on February 18, 1711 became an "Associé mécanicien".

By now we are certain that Jaques Ozanam died in Paris on April 3, 1718 (Giraud 2010) and not in 1717, as was erroneously stated in the printing of the 1719 eulogy presented to the Academy of the Sciences by Bernard le Bovier Fontenelle (1657–1757).

It is also necessary to point out that there is an almost total lack of iconographic documentation on Jacques Ozanam, with the exception of a presumed portrait (Fig. 1) drawn in red chalk on the cover sheet of the first of the two volumes of the second edition of "Récréations mathématiques et physiques" (Paris, Jean Jombert, 1696). The text, owned by Joseph Ehrenfried Hofmann (1900–1973), then passed to the Institute for the History of Natural Sciences (Ludwig Maximilians University, Munich) and now, after 2012, can be found in the rare book collection of the University Library of Munich (W 8 Vetus 1771(1). The presumed portrait was never published, with the exception of its appearance in an article on Ozanam's algebraic studies (Hofmann 1969).

Review of Main Works

Ozanam's writings are dedicated to mathematics, with emphasis placed on subjects related to arithmetic and to applications such as practical geometry, gnomonics, cartography, descriptive geometry, mechanics, and military engineering.

Among the studies undertaken by Ozanam there is a 380 page unpublished manuscript on the arithmetic of Diophantus of Alexandria (circa 200–circa 284).

Fig. 1 Anonymous, Ozanam portrait, "Récréations mathématiques et physiques" (Paris, Jean Jombert, 1696, leaf sheet). Volume conserved at the University Library of Munich, Ludwig-Maximilians-Universitaet; shelf mark W 8 Vetus 1771(1)

The original Alexandria text contains 130 problems which provide the solutions to determinate and indeterminate equations, but only six of the 13 original books have come down to us through Arab translations.

The first text that Ozanam published in 1673 is concerned with gnomonics. Its amplified second edition, dating to 1685, comes with schematic tables and is introduced by a general theorem and by 13 lemmas. The treatment is developed across four chapters. The first describes the hour lines and astronomical annotations for sundials built on a horizontal plane. The second deals with sundials on a vertical plane, while the third discusses sundials on inclined planes. The fourth and last chapter—on odd sundials—has eight problems describing a portable sundial, differing types of universal clocks (elliptical, 90°, hyperbolic, and parabolic), an astrolabe, and sundials that use refraction or reflection.

A study which bears testimony to Ozanam's interest in applications is "Géométrie Pratique" (1684). It begins with an exposition of the principals and the definitions of geometry, backed by 23 elementary problems for application. The first part is dedicated to plane trigonometry, the second to problems of length, the third to problems related to surfaces, and the fourth (and last) of those related to solids. It is structured as a collection of reoccurring themes in the author's studies.

Fig. 2 Adriaan Vlacq, Jacques Ozanam, "Tabulae sinuum, tangentium et secantium et logarithmorum", 1670: title page

Another element which is frequently found in Ozanam's works is constituted by trigonometry and logarithm tables. This demonstrates his intention to provide instruments that are precise, but easy to use for calculations. They can also be used for teaching. His 1685 "Tables des Sinus, Tangentes et Sécantes" are considered to be very accurate when compared to the better known works of the time. In particular, they remind one of the studies undertaken by Henry Briggs (1561–1631), published in "Arithmetica Logaritmica" (London 1624) which Adriaan Vlacq (1600–1667) then integrated (Gouda 1628). Vlacq's tables were published in a posthumous version by Ozanam himself in 1670 (Fig. 2). The subject then found its own systemic exposition in "Nouvelle Trigonométrie" which Ozanam published in 1697.

The subject of conic sections, which is of great theoretical and practical interest, was developed by Ozanam in his brief "Traité des lignes du premier genre" in 1687. For further in-depth studies he refers to the work by Philippe de la Hire (1640-1718) from 1673 (Fig. 3). Ozanam discusses the parabola, the ellipse, and the hyperbola, with the intention of supplying the foundations for solutions of equations of a greater degree than the second by the use of conic sections (Fig. 4). He proposes

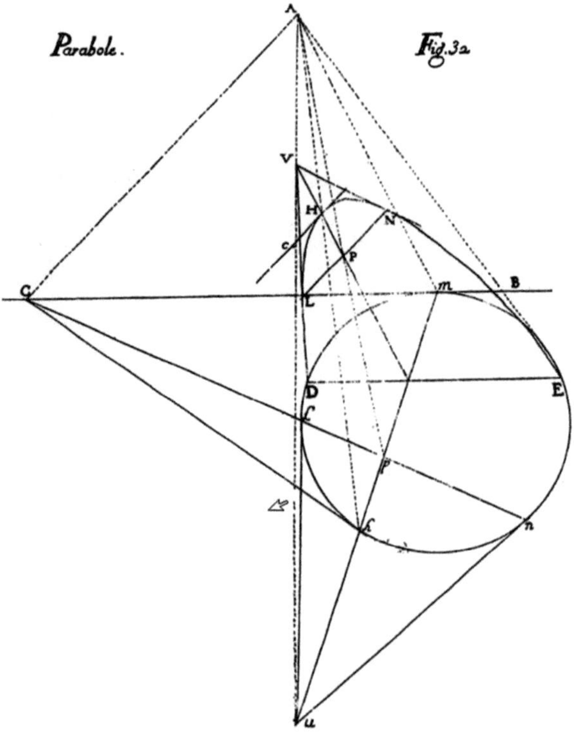

Fig. 3 The parabola of Philippe de la Hire. "Nouvelle Méthode en Géométrie pour les sections des superficies coniques", 1673, Fig. 32

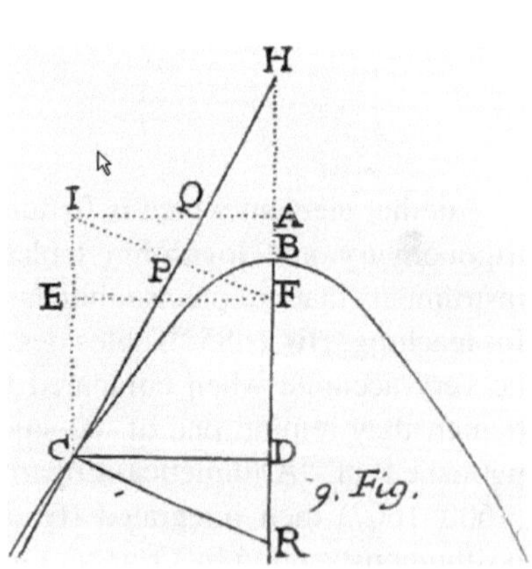

Fig. 4 Jacques Ozanam, "Traité des lignes du premier genre", 1687, Fig. 9

differing problems related to the relationship of certain geometrical entities with cones, as for example the discovery of a straight line which forms a right angle with a parabola ("Traité des lignes du premier genre", 1687, Fig. 9).

Ozanam suggests a general method for constructing the conics by the use of ruler and compass. In the introduction of the treatise, which anticipates the publication at the same time of another manuscript, the "Traité des lieux géométriques expliqués par une méthode courte et facile" integrates the previous one by supplying notions of elementary geometry which include conic sections. During the same year, he also published his "Traité de la construction des équations, pour la solution des problèmes indéterminés", which completed the whole picture of a subject which involved diverse branches of mathematics.

Ozanam also dealt with the "Traité de la sphère et son usage" (Paris, 1606) by Jean Boulenger (?–1636) proposing a corrected and amplified edition in 1688, as the realization of his desire to accommodate the knowledge provided by famous preceding works. Once again for Boulenger, Ozanam takes care of the publication of "Géométrie pratique" in 1691.

Among the publications of preceding texts, one must also remember that on Euclidian geometry by Jesuit Claude François Millet Dechales ("Les Éléments d'Euclide", Lausanne, 1678), which could be identified with his teacher of mathematics. This text, published in 1735, came out in several editions, among them being a version by Jacques Audierne in 1753 and an Italian translation in 1785.

In 1688 Ozanam provided an ample explanation of the use of the proportional compass which Galileo Galilei (1564–1642) had illustrated in his work, "Operations of the geometric and military compass" dated 1606. This instrument, having already been proposed in diverse earlier versions—for example by Fabrizio Mordente (1532–1608, "Modo di trovare l'astrolabio", Venice, 1567) and by Muzio Oddi (1569–1639, "Fabrica et uso del compasso polimetro", Milan, 1633)—was intended to provide a quick means of calculation for all operations pertaining to the art of war, such as ballistics or the measuring of distances and of places.

Ozanam's text ("L'Usage du Compas de Proportion") is divided into two parts (Fig. 5). The first regards the construction and the use of the sector, that versatile Galilean instrument of 1596. The second consists in a treatise on the division of surfaces. In the same year he also wrote a text on this universal instrument which would then be reprinted together with his volume on the proportional compass in 1769.

Ozanam also concluded a very ambitious undertaking in 1691. This was the compilation of a dictionary on mathematical terms. It contained definitions from algebra and geometry in addition to general notions of the disciplines which are founded on mathematics, such as naval architecture and music. It ends with an index which allows for the dictionary to be used for consultation, inserting it into the traditional genre of scientific dictionaries.

One whole work by Ozanam is dedicated to cartographic surveying. Dating to 1693, it illustrates in detail the instruments and techniques that are commonly used for land surveys. The text is divided into eleven chapters and describes the equipment, the methods, the instruments, and the types of charts used (Fig. 6).

From the point of view of diffusion, one of Ozanam's important texts is "Cours de mathématiques". Printed in 5 volumes in 1693, it contains an "Introduction to Mathematics and treatises on Arithmetic, Geometry and Fortifications, Mechanics and Perspectives, Geography, and Gnomonics."

Ozanam also studied the treatment of ballistics problems. This is seen in some of the figures in "Cours" (1693, vol. IV, plates IX and X, pages 74–76) which show a

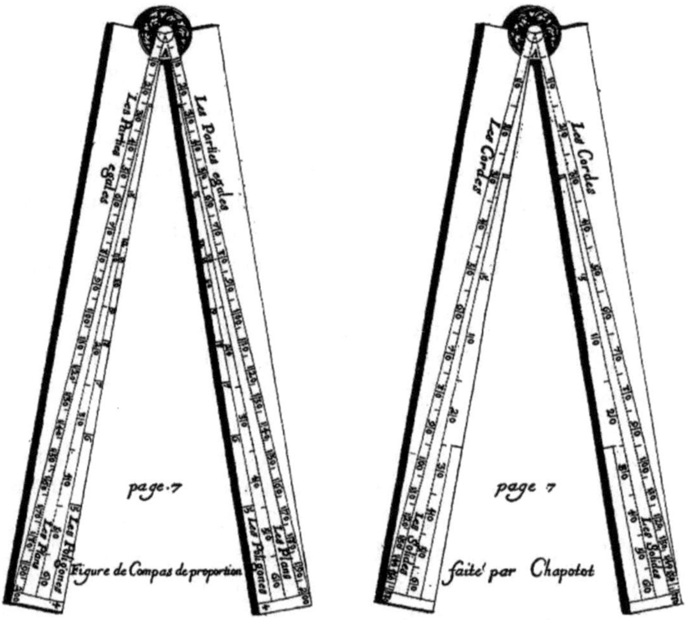

Fig. 5 Jacques Ozanam, "L'usage du compas de proportion", 1688

cannon whose parabolic trajectory is seen in a text by Bonaventura Cavalieri which defines the many different positions of the projectile according to Cartesian coordinates (1598–1647; "Lo specchio ustorio overo Trattato delle Settioni Coniche", Bologna 1632, 165 and Fig. XXV; see Ashworth 1987), even if Ozanam shows the cannon and the different positions of the projectile as well (Figs. 7 and 8).

"Récréations" (1694) is a sort of scientific encyclopedia that deals with various mathematical subjects presented in the form of problems (Fig. 9). The first edition is divided into two volumes. The first, split into seven parts, expounds on problems in Arithmetic, Geometry, Optics, Gnomonics, and the Cosmos. The second volume contains problems on Mechanics, Pyrotechnics, and Physics, followed by a treatise on elementary clocks.

This text written and translated from the manuscript by Domenico Martinelli of Spoleto (1632–?) had already been published in Italy under the title "Horologi elementari divisi in quattro parti. Nella prima parte fatti con l'acqua. Nella seconda con la terra. Nella terza con l'aria. Nella quarta con il fuoco. Alcuni muti, et alcuni col suono. Tutti facili, e molto commodi", Venice, 1669.

The 1778 edition was revised by Jean-Étienne Montucla (1725–1799) and the disciplines were introduced by expositions of a theoretical nature. The text was divided into four volumes. The first dealt with Arithmetic and Geometry with tables on the measurement of lengths and several appendixes on the subjects dealt with. The second volume was concerned with Mechanics, Optics, Acoustics, and Music. The third talks about Astronomy and Geography, Gnomonics, Navigation, Architecture, and Pyrotechnics. The fourth is concerned with Physics, Magnets,

Fig. 6 Jacques Ozanam, "Dictionnaire mathématique", 1691: pictured frontispiece

Electricity, and Chemicals with appendixes on phosphorous, on perpetual lamps, and on magic tricks.

In "Récréations" Ozanam does not waste time on long treatises about theory, but goes straight to the point, basing the text on practical examples. He even proposes some simplified systems for calculation, to encourage those who are not particularly

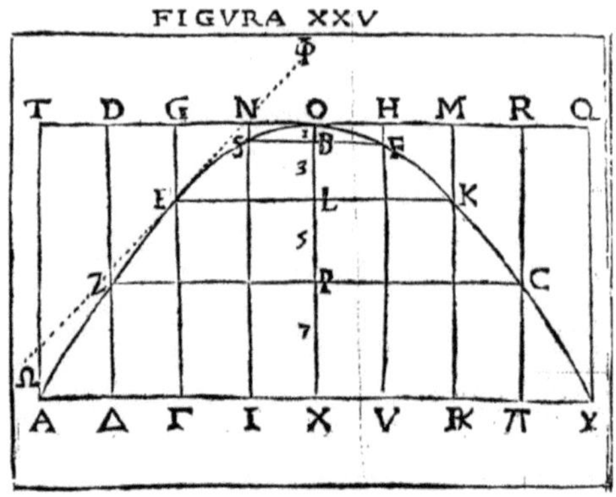

Fig. 7 Bonaventura Cavalieri, "Lo specchio ustorio overo Trattato delle Settioni Coniche", Bologna 1632, Fig. XXV

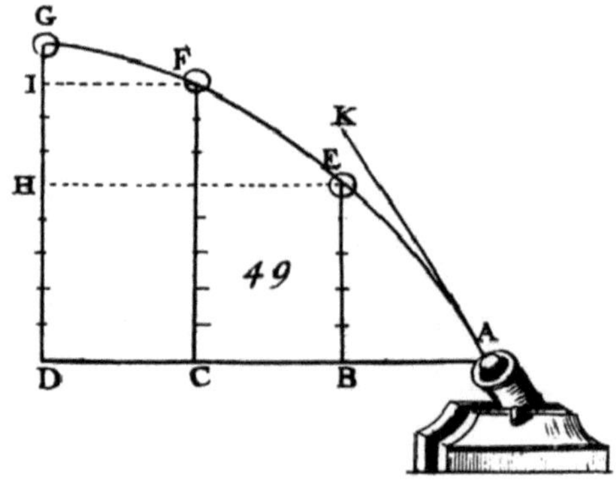

Fig. 8 Jacques Ozanam, "Cours de mathématiques", 1693, IV, tav. 9: cannon

gifted in mathematics, as for example his method of "multiplication par les doigts" ("Récréations", 1778 edition, first volume, Arithmétique).

The layout which he used for "Récréations" was inspired by works of previous authors among whom was Claude Gaspard Bachet De Méziriac (1581–1638, "Problèmes plaisants et délectables, qui se font par les nombres", Lyon, 1612). Other studies on the subject were published by the Jesuit, Jean Leurechon (1591–1670, "Récréation mathématique", Pont-à-Mousson, 1624), Claude Mydorge (1585–1647, "Récréations mathématiques", Paris, 1630), and by Daniel Schwenter (1585–1636, "Deliciae physico-mathematicae", Nürnberg, 1636).

In "Récréations" there are many treatises on other curious elements in science, such as optical illusions. Illustrated are diverse instruments designed for entertainment purposes, such as the anamorphosis (1694, Volume I, Problèmes d'Optique, pages 239–240), the dark room (page 235), and the magic lantern (page 252). He describes their functions in terms of show but also from an informative point of view (Brunetta 1997).

One part of "Récréations" (1778, third volume, part IX) is dedicated to architecture, but the chapter actually gathers diverse problems relating to mechanics, the geometry of vaults, and observations on practical applications. Here one also finds a suggestion for the creation and setting of wood beams (Problem I: Tirer d'un arbre

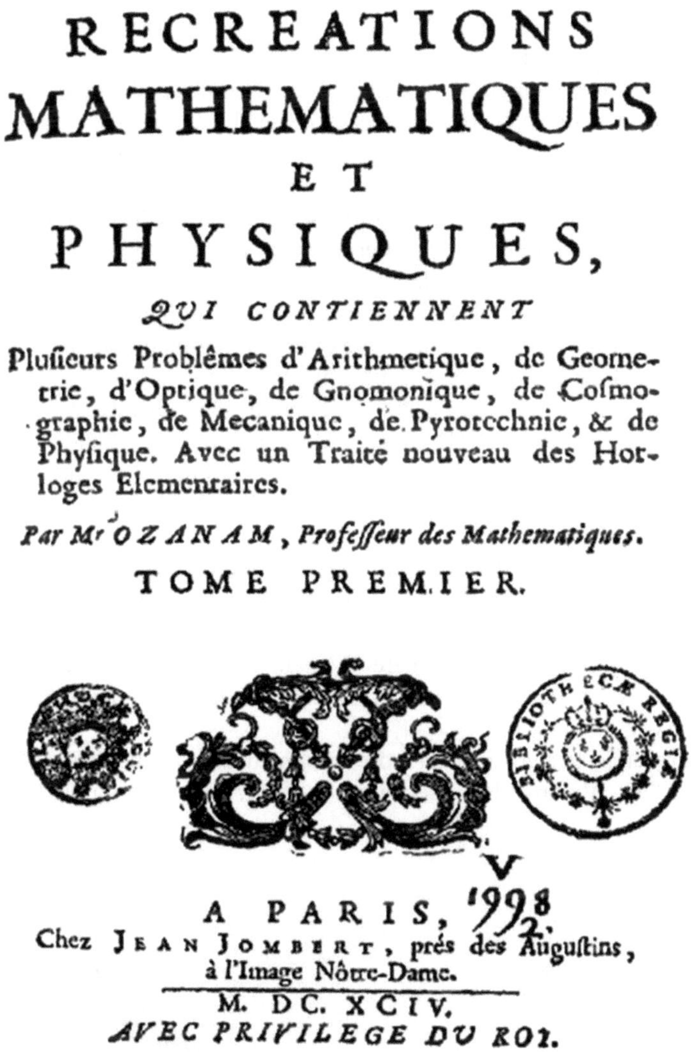

RECREATIONS
MATHEMATIQUES
ET
PHYSIQUES,
QUI CONTIENNENT
Plusieurs Problêmes d'Arithmetique, de Geometrie, d'Optique, de Gnomonique, de Cosmographie, de Mecanique, de Pyrotechnie, & de Physique. Avec un Traité nouveau des Horloges Elementaires.
Par Mr OZANAM, Professeur des Mathematiques.
TOME PREMIER.

A PARIS,
Chez JEAN JOMBERT, prés des Augustins, à l'Image Nôtre-Dame.
M. DC. XCIV.
AVEC PRIVILEGE DU ROI.

Fig. 9 Jacques Ozanam, "Récréations mathématiques et physiques", 1694: frontispiece of Volume I

la poutre de la plus grande résistance, 388). The subject—which is also relevant to Mechanics—is dealt with by citing Galileo's principles on the resistance of solids and adding practical notations designed to obtain the strongest beam carved out of a tree trunk. Successive treatises on the subject itself were inspired by this because of the clarity with which it was written (Vijver 2007), not to mention the originality of the contents.

Ozanam returns to the subject of gnomonics, which he had already written about in 1673, in "Récréations" (1694) whose publication in 1778 (third volume,

"Gnomonique") contains 34 problems accompanied by drawings which illustrate different types of sundials.

Ozanam was not inspired by recently acquired knowledge on the subject, which identify more general methods, such as those in the studies by Girard Desargues (Desargues 1640; Bosse 1643). Though he expounds on the astronomical and mathematical foundations which allow us to understand the phenomenon in its complexity (Bouchard 2006), Ozanam focuses to a larger extent on useful practical instruction for the building of the differing types of sundials.

In the fifth volume of "Cours de mathématiques" (1693), which is still dedicated to gnomonics, Ozanam illustrates the construction of an inclined equinox quadrant which can be used everywhere due to the option of regulating its inclination on the basis of the latitude in which it is used (Fig. 10).

Ozanam was also interested in the description of mechanical clocks (Fig. 11), such as hourglasses, for which Domenico Martinelli, the author cited in the body of "Récréations" (1694, Appendix to Volume I) recommends the use of powdered marble. Ozanam on the other hand felt that powder egg shells were better, as they were not affected by humidity.

"Récréations" also contained an explicit treatise dedicated to Mechanics (Volume II). In the 1778 edition, it offered 54 problems in which machines could be applied to solve practical problems. As seen in other disciplines as well, "Récréations" is not the place where Ozanam carries out blanket coverage. He instead focuses on the exposition of specific problems of recognized practical implications.

The subject is treated instead more systematically in the relevant chapter of "Cours" (1693, Volume IV, Mécanique), which was also printed as a single

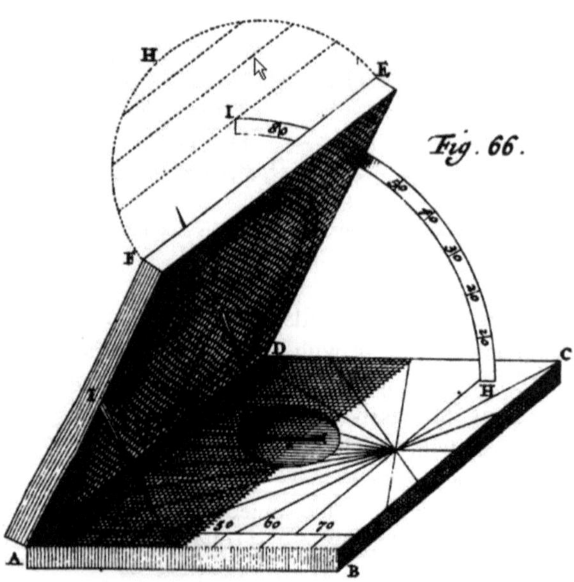

Fig. 10 Jacques Ozanam, "Cours de mathématiques", 1693, tomo V, Gnomonique, tav. 24: the inclined equinox quadrant

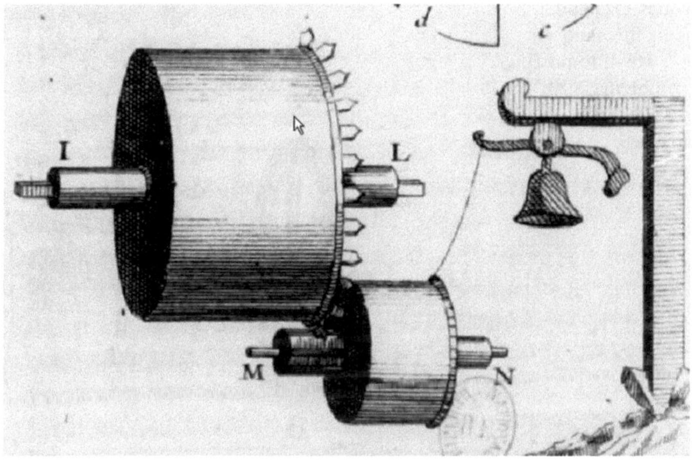

Fig. 11 Jacques Ozanam, "Récréations mathématiques et physiques", 1694, tomo II, horloges élémentaires, tav. 8: perpetual water clock

publication (1720). Here he described various machines—such as the scale, the lever, the pulley, and the inclined plane, in addition to a treatise on statics and on hydrostatics.

In "Récréations", however, several original tracts can be seen which, even though of limited size, show the innovative potential of the author which is expressed even in the mechanical sector. For example he describes a carriage that can be moved without the need of horses (1694, Volume II, Mécanique, problem XXI, plate 48, Fig. 142), with a system which was attributed to Dr. Élie Richard, a doctor in La Rochelle (Fig. 12).

Such a prodigious means of locomotion is re-elaborated and eloquently illustrated in the 1696 edition (Volume II, Mécanique, problem XXI, plate 61). Here the carriage is illustrated in its entirety with a driver holding the reins to maneuver the directional axle of the vehicle. Behind him, another man uses his legs to push two gears mounted on the axle which transmits rotation to the wheels (Fig. 13). In the 1778 edition (Volume II, Mécanique, problem XLV, 80–83, plate 10), the mechanism is critically studied, even if it can be identified to all extents and purposes as one of the precursors of the bicycle (Herlihy 2004).

Some subjects that have already been cited regarding theory and applications, can be found in the "Traité de l'arpentage" in 1699 which in the 1763 edition now contains diverse treatises ranging from Arithmetic to Elementary Geometry, Practical Geometry (with 12 problems regarding indirect measurement), Rectilinear Trigonometry, and a true surveying treatise, the "Traité de l'arpentage" with problems that provide applicative examples, the "Traité du nivelement", "Traité du toisé", and the "Traité de la géodesie". The text is accompanied by 12 figures. In the text, 12 pages are dedicated to the different units of measurement adopted in the

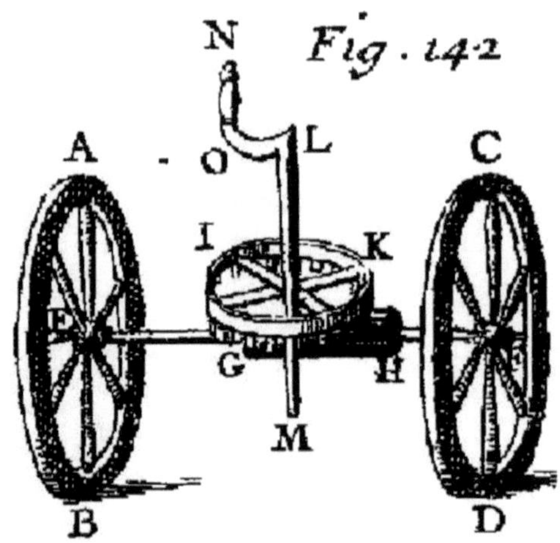

Fig. 12 Jacques Ozanam, "Récréations mathématiques et physiques", 1694, tomo II, Mécanique, problema XXI, tav. 48, Fig. 142: the horseless carriage

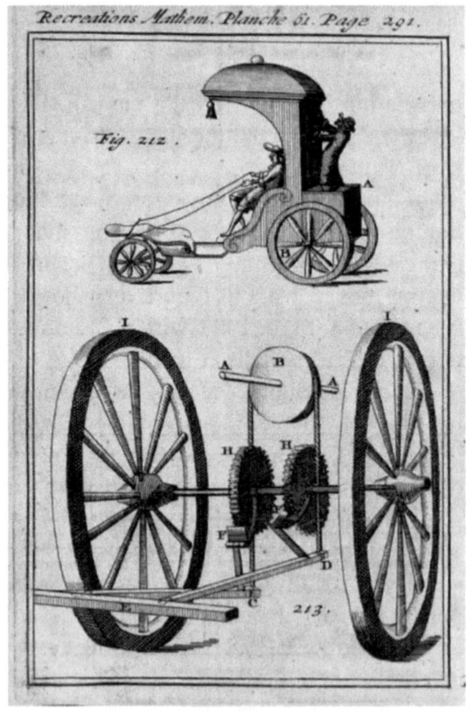

Fig. 13 Jacques Ozanam, "Récréations mathématiques et physiques", 1696, volume II, part III, Mécanique, problema XXI, tav. 61: detailed version of the horseless carriage

Kingdom of France, and it also shows the price of wood at the time for construction use, a demonstration of the practical character of the book.

In "Méthode de Lever les Plans et les Cartes" (1693) Ozanam deals, in a more systematic way, with the subject of surveying. He describes in great detail many different instruments such as, for example, the protractor and the Universal

Instrument. But his interest also extends to surveys done without instruments and to notation related to mapping.

For Ozanam's contribution to descriptive geometry, his "Perspective" was highly applauded by the critics as one of the first texts to enunciate the system of perspective construction, today called the measuring point, already described by Jacques Alleaume (1562–1627) in one of his posthumous works (Migon 1643; Andersen 2007).

Ozanam managed to set up the construction using general auxiliary vanishing points which allowed for the division of a segment into equal parts, without however having to check the true measurement of the segment and its parts. Following on this he indicated the distance points as true measuring points for the orthogonal direction of the drawing. And lastly, he added the identification of the measuring point which allows for the calculation of the extension of the segment, thanks to its configuration as a vanishing point for the direction corresponding to the reversal of the given line ("Cours de mathématiques", 1693, Volume IV, Perspective, plate 8) (Fig. 14).

A similar construction can be found in the works of Johan Heinrich Lambert (1728–1771) in 1759 (Andersen 2007, 654), and in those of Brook Taylor (1685–1731) in 1715. These show the interest generated on the theme of measurement which had already been dealt with by Ozanam in practical and survey geometry texts.

Another original notation is that constituted by the presence of a revolved viewpoint, and his efforts made in constructions based on the projection relationship which is generated by the real object and its perspective projection (plate 12) (Fig. 15). He was further inspired by the writings of Philippe de la Hire (1673; see Andersen 2007), in which he found such a construction, which can be identified by a logical correspondence which has for its center the revolved viewpoint, and for its axis, the ground line.

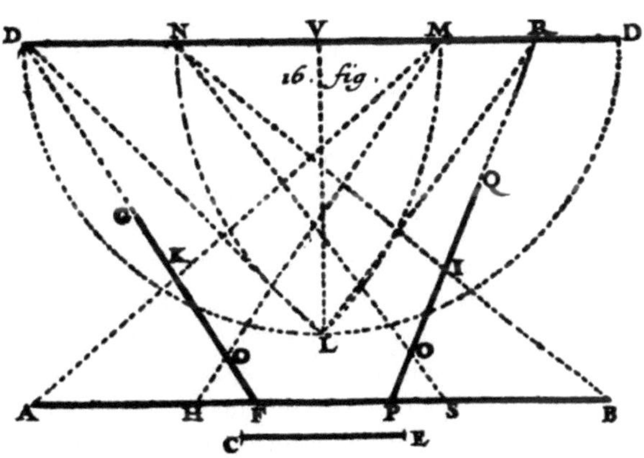

Fig. 14 Jacques Ozanam, "Cours de mathématiques", 1693, tomo IV, Perspective, tav. 8: the measuring point

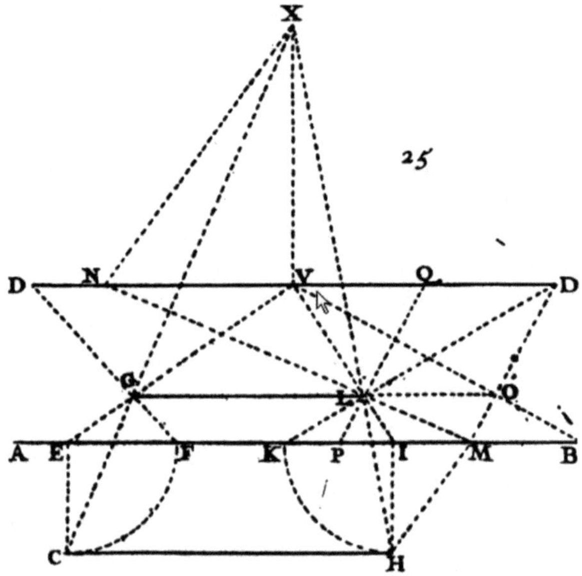

Fig. 15 Jacques Ozanam, "Cours de mathématiques", 1693, tomo IV, Perspective, tav. 12

On the Circulation and Influence of Works

His first biographies refer to Ozanam's ability to write about mathematics with ease and without corrections. His exercise of calculations and his continuous discussions of theoretical subjects can perhaps be attributed to his teaching jobs, which kept him busy until the start of his independent life, around 1661.

All the same, Ozanam was forced to carry out some compromises dictated by need. For example he had to ask for payment for lessons and for the compilation of educational books, which probably distracted him, taking away time from more significant editing of original works. Perhaps it is for this same reason that Ozanam was admitted only in 1707 to the Royal Academy of Sciences, in spite of the vast amount of work he had done. In any case these choices are also probably the basis of certain approximations which characterize studies on the life and the works of Ozanam.

In fact, repeated imprecisions in the analysis of some of Ozanam's texts can be identified. For example the attribution of the editing of the original trigonometric tables of 1670, when instead "Tabulae sinuum, tangentium et secantium et logarithmorum", as we have seen, constitute a posthumous publication of the original tables by Adriaan Vlacq del 1628, under the guidance of Ozanam himself. The original trigonometric tables however only emerged later in a text dated 1685 ("Tables des Sinus") which contains an explanation of the construction of the logarithmic tables based on the transformation of a product into a sum, and the use of decimal bases, in addition to an interesting correspondence of the algebraic interpretation with the geometric one.

Ozanam often discussed arithmetical problems in letters to other scholars as seen in the correspondence he exchanged with Jacques de Billy (Tannery, Henry 1912). Dialogue was also cultivated by discussions published on various occasions in the "Journal des Sçavans" which also has several reviews of texts by Ozanam (Table Générale des matière, 1757). This demonstrates the interest that his studies inspired among his contemporaries.

In particular, there is a eulogy by Gottfried Wilhelm von Leibniz (1646–1716) to the "Nouveaux Éléments d'Algèbre" published in 1702, which cites the usefulness of his algebraic methods for the reduction of irrational quantities (Niceron 1728; Zeitlinger, Sotheran 1921) and which is defined as one of the few independent studies done by René Descartes (1596–1650) and by his commentators ("Journal des Sçavans", 1702, 626).

Leibniz, who was in Paris from 1672 to 1676, says that he met Ozanam, who has surely been for him an important interlocutor. Their debates on diverse theoretical problems have been documented (G. W. Leibniz, "Sämtliche Schriften und Briefe", series III. vol. 1: 1672–1676, Berlin, Akademie-Verlag, 1988; see Probst 2006) even if, in 1686, Leibniz expresses bitterness for Ozanam's having published Leibniz's own method for squaring the circle (letter from Leibniz to Simon Foucher, see Strickland 2007). In spite of this, he praised him later, as has been seen, regarding his work.

As for "Récréations", it must be pointed out that the success of these books is due to the playful treatment given to the mathematics which brings out, in this case as well, the didactic setting of the work. The publication, expanded upon by Jean-Étienne Montucla (1778) was translated into English by Charles Hutton (1803). And in 1844, Edward Riddle (1788–1854) brought out a new edition in which he eliminated some of the parts that he felt outdated, inserting new elements to make it an updated manual for scientific recreation.

Many works have been inspired by Ozanam's "Récréations". This can be seen in the famous "Ladies' Diary" of 1775 by Charles Hutton who took care of the above-mentioned translation of Ozanam's "Récréations". In the XIX Century, several texts were published, such as that of Édouard Lucas (1842–1891, "Récréations mathématiques", 1882–1894) and Charles Lutwidge Dodgson (Lewis Carroll 1832–1898), which contains games of logic ("The Game of Logic", New York 1887).

Another interesting aspect of the circulation of his work is seen in the monographs dedicated to the instruments of measurement dealt with previously ("L'usage du compas de proportion", 1688; "L'Usage de l'instrument universel", 1688). While a more general overview on surveying and the instrumental measuring of places can be found in the "Traité de l'arpentage, et du Toisé" (1699) which is dedicated specifically to methods of measurement. Its contents are not particularly original, but were very practical and became widespread. The Barnabite monk, Jean Pierre Niceron (1685–1738) testifies to the existence of an edition of 1725 (Niceron 1728, 54), and there is also the 1779 amplified edition by Jacques Audierne (1701–1785) with several additions (Comolli 1791, 107).

Among the most popular text, as seen above, is also the "Dictionnaire mathématique". This contains, among other things, the description of an air bubble level (1691, 132) which is considered to be the invention of Mr. Chapotot, maker of scientific instruments (Boncompagni 1869), though it was also attributed to Melchisédech Thévenot (circa 1620–1692), who invented it in 1661 (Govi 1880).

Another text which became very widespread is "Cours de mathématiques" (1693). It was republished in various editions (1697 and 1699) and even translated into English (1712). Some of the subjects that are dealt with in "Cours" are also to be found in a separate edition, as is the case for "La Géographie et Cosmographie" (1711), "Méchanique" (1720), and "Perspective théorique et pratique" (1711), which was also published several times. The latter illustrates perspective constructions which are developed in the problems and in the practical exercises on the perspectives of elements belonging to the geometrical plane, in which, as we have said, inverse constructions frequently appear. This is followed by exercises dedicated to scenery, or, in other words, to perspectives of tridimensional elements. The text concludes with the treatment of shadows. Here general laws are given, along with practical exercises on sun shadows and on shadows generated by small sources of light.

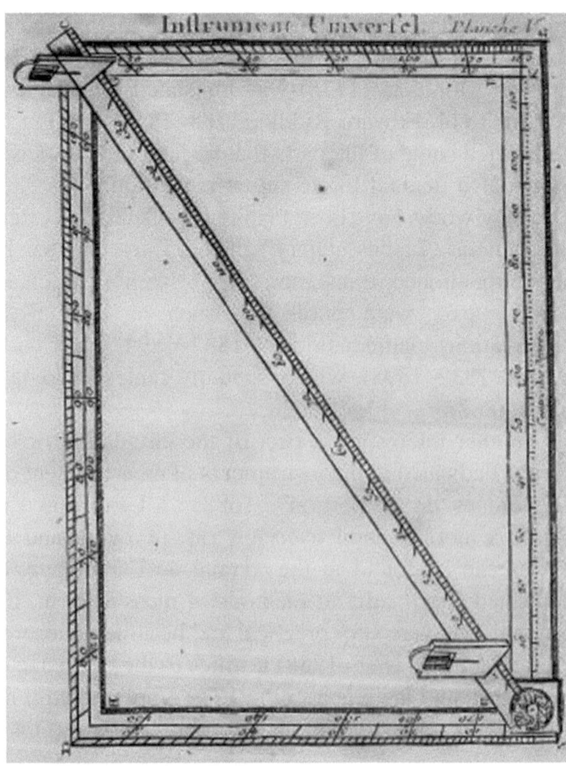

Fig. 16 Jacques Ozanam, "L'usage de l'instrument universel", 1688, tav. V: the universal instrument

Fig. 17 Jacques Ozanam, "L'usage de l'instrument universel", 1688, tav. IX: the applications for the universal instrument

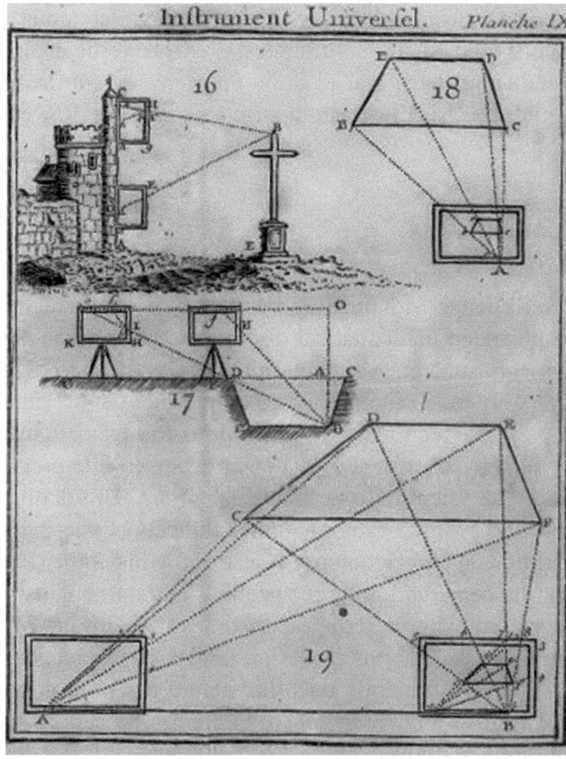

It is necessary, however, to remember that, even though he extended his writings to all branches of mathematics—both theoretical and practical—Ozanam often only redid ancient knowledge, ignoring the contemporary. This is seen in his cited "Perspective", in which new acquisitions, which inserted perspective into the vast setting of projective geometry, do not appear. This orientation on his part was, as is well known, pursued by some of his predecessors, such as the Jesuit, François D'Aguillon (1567–1617), who in his treatise on optics (1613) lumps together orthographic, stereographic, and scenographic projections. A more explicit intent is provided by Girard Desargues, who, during the course of his studies (1639 and 1640) defines the conic sections as central projections off the circle and demonstrates the identity of the base for the two types of projections, orthogonal and perspective.

Among the subjects dealt with in "Récréations" are many notations on the catoptrical applications for the measuring of heights (pages 198–199). Interest is shown for the proper instrumentation for this scope as seen by a declaration by Ozanam himself saying that he was the first to extend the use of the Universal Instrument, already used for land surveying, to various applications in practical geometry ("L'Usage de l'instrument universel", Preface, 1688).

The instrument consists in a rectangular tablet and in mobile and stationary rulers used to measure angles even on vertical planes, to trace parallel and perpendicular lines, and to draw up plans, as can be seen in plate IX of the text, in which he also illustrates its use in enlarging drawings (Figs. 16 and 17).

Conclusions

Jean-Étienne Montucla, in his history of mathematics (1758), affirms that Ozanam popularized mathematical disciplines, but he was not exclusively a communicator. On this subject, Montucla cites his text on lines of the second order, referring probably to "Traité des lignes du premier genre" published in 1687. In this work he uses conic sections to find solutions to higher than second-order equations.

His manuscript shows a clear desire to interpret the theory of mathematics as a common scientific base for applications. This setting can be identified in many of his works which gather together subjects of varying nature preceded by treatises on Arithmetic and Geometry. It also show his ability to elaborate on personal methods for the construction of perspectives and on new instruments for practical geometry, along with his description of innovative means of locomotion derived from the original applications of the principles of mechanics.

And so Ozanam's scientific nature has been delineated. He, who in spite of not always being ready to adhere to the most recent revolutions in the diverse scientific disciplines, shows, all the same, that he is informed about new aspects and the most updated scientific debates, in addition to his competence in classic knowledge.

List of Main Works

Les six livres de l'Arithmétique de Diophante d'Alexandrie, augmentés et réduits à la spécieuse par Ozanam, manoscritto (già appartenuto a Henry François D'Aguesseau, poi a Eugène Prouhet e a Baldassarre Boncompagni Ludovisi; attuale collocazione ignota), prima del 1687.

Adrian Vlacq, Jacques Ozanam, *Tabulae sinuum, tangentium et secantium et logarithmorum,* Lyon, Jean Thioly, 1670.

Méthode générale pour tracer des cadrans sur toute sorte de plans, Paris, Estienne Michallet, 1673; seconda edizione ampliata: Paris, Estienne Michallet, 1685; 1697; altra edizione: *Traité de gnomonique ou de la construction des cadran sur toute sorte de plans*, Paris, Claude Cramoisy, 1673.

La Géométrie pratique, contenant la trigonométrie théorique et pratique, la longimétrie, la planimétrie et la stéréométrie, par de nouvelles démonstrations très courtes et très faciles et de nouveaux abrégez pour mesurer exactement les plans et les solides, Paris, l'Auteur et Estienne Michallet, 1684.

Tables des Sinus, Tangentes et Sécantes et des Logarithmes des sinus et des tangentes; et des nombres depuis l'unité jusques à 10,000. Avec un traité de Trigonométrie par de nouvelles Démonstrations et des Pratiques très-faciles, tant pour la constructions des tables, que pour la supputation des triangles, Paris, Estienne Michallet, 1685; 1697; Paris, Antoine Jombert, 1720; 1741.

Traité des lignes du premier genre, expliquées par une méthode nouvelle et facile, Paris, Estienne Michallet, 1687.

Traité des lieux géométriques expliqués par une méthode courte et facile, Paris, Estienne Michallet, 1687.

Traité de la construction des équations, pour la solution des problèmes indéterminés, Paris, Estienne Michallet, 1687.

Traité de la sphère du monde. Par le sr. Boulenger, Nouvelle édition corrigée et augmentée par M. Ozanam, Paris, Jean Jombert, 1688.

L'Usage du Compas de Proportion expliqué et démontré d'une manière courte et facile et augmenté d'un Traité de la division des Champs, Paris, Estienne Michallet, 1688; La Haye, Henry van Bulderen, 1691; *nouvelle édition corrigée et augmentée*, Paris, Jean Jombert, 1700.

L'Usage de l'instrument universel, pour résoudre promptement et très exactement tous les problèmes de Géométrie Pratique sans aucun calcul, Paris, Estienne Michallet, 1688; La Haye, Henry van Bulderen, 1691.

L'Usage du Compas de Proportion et de l'Instrument universel Paris, Claude-Antoine Jombert, 1769.

Dictionnaire mathématique ou Idée générale des mathématiques: dans lequel l'on trouve outre les termes de cette science plusieurs termes des Arts et des autres sciences avec des raisonnements qui conduisent peu à peu l'esprit à une connaissance universelle des mathématiques, Paris, Estienne Michallet, Antoine Lambin, 1691; edizione inglese: *A Mathematical Dictionary or a Compendious Explication of All Mathematical Terms, translated and abridged by Joseph Raphson*, London, J. Nicholson, 1702.

La Géométrie pratique du Sr Boulenger ou nouvelle méthode pour toiser et arpenter promptement et facilement toutes sortes de grandeurs, sans se servir de Fractions, de Réductions, ny mêmes d'aucune Division, quoy qu'on se serve de la mesure ordinaire et de ses parties nouvelle édition augmentée de plusieurs notes et d'un Traité de l'Arithmétique par Géométrie par Mr Ozanam, Paris, Michel David, 1691.

Méthode de Lever les Plans et les Cartes de Terre et de Mer, avec toutes sortes d'Instruments, et sans Instruments. La description et l'usage de ces Instruments, qui sont le Demi-cercle, la Planchette de diverses façons, la Boussole, l'Instrument universel, et le Recipiangle. Et la manière de faire les remarques des marées, courants, écueils, etc. *et de lever les Plans des Villes ennemies*, Paris, Estienne Michallet, 1693; 1700.

Cours de mathématiques, qui comprend toutes les parties de cette science les plus utiles et les plus nécessaires a un homme de guerre, et a tous ceux qui se veulent perfectionner dans les mathématiques, Paris, Jean Jombert, 1693; Paris, 1697;

Amsterdam, 1699; edizione inglese: *Cursus mathematicus or a Compleat Course of the Mathematics*, 5 vols., London, J. Nicholson, 1712.

Traité de fortification contenant les méthodes anciennes et modernes pour la construction et la défense des places, et la manière de les attaquer, expliquée au plus long qu'elle n'a été jusques à présent, Paris, Jean Jombert, 1694; traduzione inglese di J. T. Desaguliers, *Treatise of Fortification,* Oxford, 1711, 1727.

Récréations mathématiques et physiques, qui contiennent plusieurs problèmes d'arithmétique, de géométrie, de musique, d'optique, de gnomonique, de cosmographie, de mécanique, de pyrotechnique, et de physique. Avec un traité des horloges élémentaires, Paris, Jean Jombert, 1694 (2 voll.); Paris, Jean Jombert, 1696; Paris, Jean Jombert, 1698; Amsterdam, George Gallet, 1698 (2 voll.); Paris, Claude Jombert, 1723; nuova edizione a cura di Jean-Étienne Montucla: *Récréations mathématiques et physiques,* Paris, Claude-Antoine Jombert, 1778, 4 vols.; traduzione inglese dell'edizione 1778 con aggiunte di Charles Hutton: London, 1803, 1814; edizione revisionata da Edward Riddle*: Recreations in science and natural philosophy,* London, Thomas Tegg, 1844.

Nouvelle Trigonométrie ou l'on trouve la manière de calculer toutes sortes de triangles rectilignes, sans les Tables de Sinus, et aussi par le Table de Sinus. Avec une application de la Trigonométrie à la mesure des Lignes droites accessibles et inaccessibles sur la terre, Paris, Jean Jombert, 1697

Traité de l'arpentage et du toisé, ou Méthode facile pour arpenter ou mesurer toutes sortes de superficies, et pour toiser exactement la maçonnerie, les vidanges des terres, et tous les autres corps, dont on peut avoir besoin dans la pratique; avec le toise du bois de charpente selon la Coutume de Paris, et un traité de la séparation des terres, Paris, Jean Jombert, 1699; *nouvelle édition mise dans une nouvel ordre par M. Audierne*, Paris, 1779.

Nouveaux Éléments d'Algèbre ou Principes généraux pour résoudre toutes sortes de problèmes de mathématique, Amsterdam, G. Gallet, 1702.

La Fortification Régulière Et Irrégulière Qui Comprend la Construction L'attaque et la Défense de Toutes Sortes de Places, Paris, Claude Jombert, 1711.

La Géographie et Cosmographie ... tirée du Cours de mathématique, Paris, Claude Jombert, 1711

La Perspective Théorique et Pratique, où l'on enseigne la manière de mettre toutes sortes d'objets en perspective, e d'en représenter les ombres causées par le Soleil ou par une petite lumière. Tirée du Cours de Mathématique, Paris, Claude Jombert, 1711.

Les Eléments d'Euclide du R.P. Dechales. Démontrée d'une manière nouvelle et facile, par M. Ozanam de l'Académie des Sciences. Nouvelle édition revue, corrigée et augmentée d'un grand nombre de propositions et d'usages, par M. Audierne, Paris, Charles-Antoine Jombert, 1735 e 1753; trad. it.: *Gli elementi di Euclide spiegati d'una maniera nuova e facile con l'uso di ciascuna proposizione per tutte le parti di matematica dal P. Dechales [...] riveduti, corretti, ed accresciuti dall'Ozanam... tradotti dal francese, terza edizione corretta e illustrata*, Bergamo, Francesco Locatelli, 1785.

La Méchanique, où il est traité des Machines simples et composée. Tirée du Cours de Mathématique, Paris, Claude Jombert, 1720

La gnomonique, ou l'on donne par un principe général la manière de faire des cadrans sur toutes sortes de surfaces, et d'y tracer les heures astronomiques, babyloniennes et italiques, les arcs des signes, les cercles des hauteurs, les verticaux et les autres cercles de la sphère. Tirée du Cours de Mathématique, Paris, Charles-Antoine Jombert, 1746

Acknowledgments The author would like to thank:

Antonio Becchi, Max-Planck-Institut für Wissenschaftsgeschichte.

Ron B. Thomson, Fellow Emeritus of Pontifical Institute of Mediaeval Studies (PIMS), University of Toronto.

Menso Folkerts, Deutsches Museum (Bibliotheksbau) München.

Patrizia Trucco, Andrea Bruzzo, Biblioteca Politecnica, Università degli Studi di Genova.

Irene Friedl, Ludwig-Maximilians-Universität, Universitätsbibliothek, München.

Bibliography

Académie des Sciences (s.d.) Liste des membres, correspondants et associés étrangers de l'Académie des sciences depuis sa création en 1666. http://www.academiesciences.fr/academie/membre/memO.pdf

Andersen K (2007) The geometry of an art. The History of the Mathematical Theory of Perspective from Alberti to Monge. Sources and Studies in the History of Mathematics and Physical Sciences. Copenhagen, Springer. ISBN 9780387489469

Ashworth WB Jr (1987) Iconography of a new physics. Hist Technol 4:267–297

Becchi A, Rousteau-Chambon H, Sakarovitch J (eds) (2013) Philippe de La Hire 1640–1718. Entre architecture et sciences, Picard, Paris. ISBN 9782708409422

Boncompagni B (1869) Bullettino di bibliografia e storia delle scienze matematiche e fisiche, Tomo III, 1869, p 347

Bosse A (1643) Manière universelle de Mr. Desargues, Lyonnois, pour poser l'essieu, et placer les heures et autres choses aux cadrans au soleil. Philippe Des Hayes, Paris

Bouchard AE (2006) La grande tradition française en Gnomonique ou la présence en Amérique d'une pensée scientifique dans l'art des cadrans solaire. Le Gnomoniste XIII(1):15–22

Bremner A (2009) A problem of Ozanam. Proc Edinburgh Math Soc (Series 2) 52:37–44. doi:10.1017/S0013091507000065

Brook T (1715) Linear perspective or a new method of representing. Knaplock, London

Brunetta GP (1997) Il viaggio dell'iconauta. Dalla camera oscura di Leonardo alla luce di Lumière. Marsilio, Venezia

Camerota F (2004) Il compasso geometrico e militare di Galileo Galilei. Firenze, Istituto e Museo di Storia della Scienza

Cantor MB (1894–1908) Vorlesungen über die Geschichte der Mathematik. B. G. Teubner, Leipzig. II, 770; III, 102–103, 270, 364

Cappelletti V (1969) Boncompagni Ludovisi, Baldassarre, in Dizionario Biografico degli Italiani, vol 11, pp 704–709

http://www.treccani.it/enciclopedia/baldassarre-boncompagni-ludovisi_%28Dizionario-Biografico%29/

Ceccarelli Marco (1998) Mechanism schemes in teaching: a historical overview. ASME J Mech Des 120:533–541

Ceccarelli M, Cigola M (2001) Trends in the drawing of mechanisms since the early middle ages. J Mech Eng Sci 215:269–289. Professional Engineering Publishing Limited, Suffolk UK. ISSN: 0954-4062

Cigola M (2012) The influence of the society of jesus on the spread of european mechanical knowledge in china in the XVIth and XVIIth centuries. In: Koetsier T, Ceccarelli M (eds) Explorations in the history of machines and mechanisms proceedings of HMM2012, Part 1. History of mechanisms and machine science, vol 15. Springer, London, New York, pp 69–79

Cigola M (2012) In praise of parallel theories: descriptive geometry and applied mechanics. In: Carlevaris L, Filippa M (eds) In praise of theory. The fundamentals of the disciplines of representation and survey. Gangemi editore, Roma, pp 39–46. ISBN: 978-88492-2519-8

Cigola M, Ceccarelli M (1995) On the evolution of mechanisms drawing. In: Proceedings of IXth IFToMM World Congress, Politecnico di Milano 1995, vol 4, pp 3191–3195

Comolli A (1791) Bibliografia storico-critica dell'architettura civile ed arti subalterne, Roma, Stamperia Vaticana, 1788-1792, vol III; anastatica: Milano: Labor, 1964

D'Aguesseau JBP (1785) Catalogue des livres imprimés et manuscrits de la bibliothèque de feu monsieur d'Aguesseau, doyen du Conseil, commandeur des ordres du roi etc., disposé par ordre des matières, avec une table des auteurs. Gogué et Née de la Rochelle, Paris

D'Aguillon F (1613) Opticorum Libri VI. Jo. Moreti, Anversa

Del Monte G (1577) Mechanicorum liber, Pesaro

Desargues G (1639) Brouillon project d'une atteinte aux événements des rencontres du cône avec un plan, Paris

Desargues G (1640) Brouillon project [...] pour la coupe des pierres [...] & de tracer tous Quadrants plat d'heures égales du soleil, Paris

Dhombres J, Dahan-Dalmedico A, Bkouche R, Houzel C, Guillemot M (1987) Mathématiques au fil des ages. Gauthier-Villars, Paris

Fontaine JP (2013) La bibliothèque du chancelier Henry-François d'Aguesseau. http://histoire-bibliophilie.blogspot.it/2013/07/la-bibliotheque-du-chancelier-henri.html

Fontenelle Bernard le Bovier de (1719) Eloge de M. Ozanam, Histoire de l'Académie royale des sciences. Année 1717. Avec les mémoires de Mathématique et de Physique, pour la même année. Tires des registres de cette Académie. Imprimerie Royale, Paris, pp 86–92; poi in: Oeuvres complètes de Monsieur de Fontenelle. 1818. Tome I, in *Histoire de l'Académie royale des sciences*, pp 252–256

Galilei, Galileo, Le meccaniche, ms. del 1649, in Salani A. 1935. a cura di. Opere, Torino: Utet

Giraud J (2010) Discussion: Jacques Ozanam. http://fr.wikipedia.org/wiki/Discussion:Jacques_Ozanam, http://fr.wikipedia.org/wiki/Jacques_Ozanam

Gorni V (s.d.) La formula di Ozanam. http://xoomer.virgilio.it/vannigor/ozanam.htm

Govi G (1870) Recherches historiques sur l'invention du niveau à bulle d'air, in Bullettino di bibliografia e storia delle scienze matematiche e fisiche. Tomo III, pp 1–15

Henry C (1880) Recherches sur le manuscrits de Pierre de Fermat suivies de fragments inédits de Bachet et de Malebranche, in Bullettino di bibliografia e storia delle scienze matematiche e fisiche, pp 447–532

Herlihy DV (2004) Bicycle: the history. Yale University Press

de la Hire P (1673) Nouvelle Méthode en Géométrie pour les sections des superficies coniques et cylindriques, Paris

Hoefer JCF, sous la direction de (1852) Aguesseau (Henry François D'). In *Nouvelle biographie universelle, depuis les temps les plus reculés jusqu'à nos jours*, vol I, 426–434. MM, Paris. Firmin-Didot Frères

Hofmann JE (1969) Leibniz und Ozanams Problem. Studia Leibnitiana 1:103–126

Hutton C (1745) Philosophical and Mathematical Dictionar. 2, London, 1745, pp 184–185

Institut de France (1939) Index Biographique des Membres et Correspondants de l'Académie des Sciences de 1666 à 1939. Gauthier-Villars, Paris, p 346

Lambert JH (1759) Die Freye Perspektive, Zurich; La perspective affranchie de l'embaras du plan geometral, Zurich

List of alumni of Jesuit educational institutions (s.d.) http://en.wikipedia.org/wiki/List_of_alumni_of_Jesuit_educational_institutions

Loria G (1902) Bollettino di bibliografia e storia delle scienze matematiche. Carlo Clausen, Torino, anno V, pp 19–20

Migon E (1643) La Perspective speculative et pratique [...] De l'invention du sieur Aleaume, Paris

Monfalcon J-B (1862) Ozanam Jacques. In Nouvelle biographie générale: depuis les temps les plus reculés jusqu'à nos jours, vol XXXVIII, pp 1017–1018. MM, Paris. Firmin-Didot Frères

Montucla J-É (1758) Historie des mathematiques, Paris, II, p 168

Nastasi P, Scimone A (1994) Pietro Mengoli and the six-square problem. Historia Mathematica 21:10–27

Narducci E (1892) Catalogo di manoscritti ora posseduti da d. Baldassarre Boncompagni, seconda edizione notabilmente accresciuta, contenente una descrizione di 249 manoscritti non indicati niella prima, e corredata di un copioso indice. Tipografia delle scienze matematiche e fisiche, Roma

Niceron JP (1728) Mémoires pour servir à l'histoire des hommes illustres, dans la république des lettres, avec un catalogue raisonné de leurs ouvrages. Briasson, Paris, pp 1727–1745. tome VI, 45–55

Niceron JP (1739) Mémoires pour servir à l'histoire des hommes illustres, dans la république des lettres, avec un catalogue raisonné de leurs ouvrages. Briasson, Paris, pp 1727–1745. tome XL, 185

O'Connor JJ, Robertson EF (2006) Jacques de Billy, MacTutor history of mathematics archive, University of St Andrews. http://www-history.mcs.st-andrews.ac.uk/Biographies/Billy.html

O'Connor JJ, Robertson EF (2002) Jacques Ozanam, MacTutor history of mathematics archive, University of St Andrews. http://www-history.mcs.st-andrews.ac.uk/Biographies/Ozanam.html

Ozanam C (2011) http://gw.geneanet.org/charlesozanam?lang=it&v=OZANAM&m=N

Pares J (1988) La gnomonique de Desargues à Pardies . Essai sur l'évolution d'un art scientifique 1640–1673. Société française d'histoire des sciences et des techniques. Cahiers d'histoire et de philosophie des sciences, Paris

Probst S (2006) Ozanam-Leibniz. http://mathforum.org/kb/message.jspa?messageID=4602183#reply-tree

Mayer U, Probst S (2012) Gottfried Wilhelm Leibniz. http://www.gwlb.de/Leibniz/Leibnizarchiv/Veroeffentlichungen/VII6.pdf

Rashed R (ed) (1975) L'Art de l'Algèbre de Diophante (trans Qustā ibn Lūqā). Bibliothèque Nationale, Le Caire

Reboux O (s.d.) Mesurer l'inaccessible. Arpentage et géométrie pratique. http://assprouen.free.fr/fichiers/classesP/7-inaccessible.pdf

Riddle E (1844) Preface. In: Recreations in science and natural philosophy. Dr. Hutton's translation of Montucla's edition of Ozanam, London, V–VI

Rohr RRJ (1986) Les cadrans solaires universels de Jacques Ozanam. Centaurus 29:165–177

Schaaf WL (s.d.) Number games and other mathematical recreations. In: Enciclopaedia britannica (s.d.). http://www.britannica.com/EBchecked/topic/422300/number-game

Schaaf WL (2008) Ozanam, Jacques. Complete dictionary of scientific biography. Encyclopedia.com: http://www.encyclopedia.com/doc/1G2–2830903260.html

Singmaster D (1996) Chronology of recreational mathematics. http://www.eldar.org/~problemi/singmast/recchron.html

Strickland L (2007) Leibniz to Simon Foucher. http://www.leibniz-translations.com/foucher1686.htm

Sturdy DJ (1995) Science and social status: the members of the academie des sciences. 1666–1750, Woodbridge, Boydell et Brewer, pp 379–381

Styan GPH, Boyer Christian, Chu Ka Lok (2009) Some comments on Latin squares and on Graeco-Latin squares, illustrated with postage stamps and old playing cards. Stat Pap 50:917–941. doi:10.1007/s00362-009-0261-5

Tannery P, Henry C (1912) Oeuvres de Fermat. tome IV, Gauthier-Villars, Paris

Table Générale des matière contenues dans le journal des savans de l'édition de Paris, depuis l'année 1665, qu'il a commencé, jusqu'en 1750. 1757, VII. Briasson, Paris

Thomson RB, Folkerts M (2013) Boncompagni Manuscripts: Present Shelfmarks, Beta Version 1.8, May 2013. http://warburg.sas.ac.uk/fileadmin/images/library/3orientation/Boncompagni.pdf

Van De Vijver D (2007) Tentare licet. The Theresian academy's question on the theory of beams of 1783. Nexus Netw J 9(2):263–275. doi:10.1007/S0000400700421

Westfall RS (s.d.) Jacques Ozanam. http://galileo.rice.edu/Catalog/NewFiles/ozanam.html

Zeitlinger H, Sotheran HC (eds) (1921) Bibliotheca chemico-mathematica: catalogue of works in many tongues on exact and applied science. Henry Sotheran & Co, London, vol 2, p 643

MIX
Papier aus verantwortungsvollen Quellen
Paper from responsible sources
FSC® C105338
FSC
www.fsc.org

If you have any concerns about our products,
you can contact us on
ProductSafety@springernature.com

In case Publisher is established outside the EU,
the EU authorized representative is:
Springer Nature Customer Service Center GmbH
Europaplatz 3, 69115 Heidelberg, Germany

Printed by Libri Plureos GmbH
in Hamburg, Germany